Epidemiology and Control of Plant Diseases

Epidemiology and Control of Plant Diseases

Editor

Alessandro Vitale

MDPI • Basel • Beijing • Wuhan • Barcelona • Belgrade • Manchester • Tokyo • Cluj • Tianjin

Editor
Alessandro Vitale
Department of Agriculture,
Food and Environment
University of Catania
Catania
Italy

Editorial Office
MDPI
St. Alban-Anlage 66
4052 Basel, Switzerland

This is a reprint of articles from the Special Issue published online in the open access journal *Plants* (ISSN 2223-7747) (available at: www.mdpi.com/journal/plants/special_issues/Epidemiology_Control_Plant_Diseases).

For citation purposes, cite each article independently as indicated on the article page online and as indicated below:

LastName, A.A.; LastName, B.B.; LastName, C.C. Article Title. *Journal Name* **Year**, *Volume Number*, Page Range.

ISBN 978-3-0365-6953-6 (Hbk)
ISBN 978-3-0365-6952-9 (PDF)

Contents

About the Editor vii

Preface to "Epidemiology and Control of Plant Diseases" ix

Alessandro Vitale
Epidemiology and Control of Plant Diseases
Reprinted from: *Plants* **2023**, *12*, 793, doi:10.3390/plants12040793 1

Penny Makhumbila, Molemi Rauwane, Hangwani Muedi and Sandiswa Figlan
Metabolome Profiling: A Breeding Prediction Tool for Legume Performance under Biotic Stress Conditions
Reprinted from: *Plants* **2022**, *11*, 1756, doi:10.3390/plants11131756 5

Sandiswa Figlan and Learnmore Mwadzingeni
Breeding Tools for Assessing and Improving Resistance and Limiting Mycotoxin Production by *Fusarium graminearum* in Wheat
Reprinted from: *Plants* **2022**, *11*, 1933, doi:10.3390/plants11151933 21

Thomas Thomidis, Konstantinos Michos, Fotis Chatzipapadopoulos and Amalia Tampaki
Evaluation of Two Predictive Models for Forecasting Olive Leaf Spot in Northern Greece
Reprinted from: *Plants* **2021**, *10*, 1200, doi:10.3390/plants10061200 35

Tao Ji, Luca Languasco, Ming Li and Vittorio Rossi
Effects of Temperature and Wetness Duration on Infection by *Coniella diplodiella*, the Fungus Causing White Rot of Grape Berries
Reprinted from: *Plants* **2021**, *10*, 1696, doi:10.3390/plants10081696 51

Jacobus T. J. Verhoeven, Marleen Botermans, Ruben Schoen, Harrie Koenraadt and Johanna W. Roenhorst
Possible Overestimation of Seed Transmission in the Spread of Pospiviroids in Commercial Pepper and Tomato Crops Based on Large-Scale Grow-Out Trials and Systematic Literature Review
Reprinted from: *Plants* **2021**, *10*, 1707, doi:10.3390/plants10081707 65

Sofia Bertacca, Andrea Giovanni Caruso, Daniela Trippa, Annalisa Marchese, Antonio Giovino and Slavica Matic et al.
Development of a Real-Time Loop-Mediated Isothermal Amplification Assay for the Rapid Detection of Olea Europaea Geminivirus
Reprinted from: *Plants* **2022**, *11*, 660, doi:10.3390/plants11050660 79

Anna A. Lukianova, Peter V. Evseev, Alexander A. Stakheev, Irina B. Kotova, Sergey K. Zavriev and Alexander N. Ignatov et al.
Quantitative Real-Time PCR Assay for the Detection of *Pectobacterium parmentieri*, a Causal Agent of Potato Soft Rot
Reprinted from: *Plants* **2021**, *10*, 1880, doi:10.3390/plants10091880 93

Tarek A. Shalaby, Naglaa A. Taha, Dalia I. Taher, Metwaly M. Metwaly, Hossam S. El-Beltagi and Adel A. Rezk et al.
Paclobutrazol Improves the Quality of Tomato Seedlings to Be Resistant to *Alternaria solani* Blight Disease: Biochemical and Histological Perspectives
Reprinted from: *Plants* **2022**, *11*, 425, doi:10.3390/plants11030425 105

Patchareeya Withee, Sukanya Haituk, Chanokned Senwanna, Anuruddha Karunarathna, Nisachon Tamakaew and Parichad Pakdeeniti et al.
Identification and Pathogenicity of *Paramyrothecium* Species Associated with Leaf Spot Disease in Northern Thailand
Reprinted from: *Plants* **2022**, *11*, 1445, doi:10.3390/plants11111445 . . . . **121**

Khalid A. Alhudaib, Sherif M. El-Ganainy, Mustafa I. Almaghasla and Muhammad N. Sattar
Characterization and Control of *Thielaviopsis punctulata* on Date Palm in Saudi Arabia
Reprinted from: *Plants* **2022**, *11*, 250, doi:10.3390/plants11030250 . . . . **141**

Dawei Li, Foysal Ahmed, Nailong Wu and Arlin I. Sethi
YOLO-JD: A Deep Learning Network for Jute Diseases and Pests Detection from Images
Reprinted from: *Plants* **2022**, *11*, 937, doi:10.3390/plants11070937 . . . . **157**

Dalia Aiello, Alberto Fiorenza, Giuseppa Rosaria Leonardi, Alessandro Vitale and Giancarlo Polizzi
Fusarium nirenbergiae (*Fusarium oxysporum* Species Complex) Causing the Wilting of Passion Fruit in Italy
Reprinted from: *Plants* **2021**, *10*, 2011, doi:10.3390/plants10102011 . . . . **171**

Francesca Calderone, Alessandro Vitale, Salvina Panebianco, Monia Federica Lombardo and Gabriella Cirvilleri
COS-OGA Applications in Organic Vineyard Manage Major Airborne Diseases and Maintain Postharvest Quality of Wine Grapes
Reprinted from: *Plants* **2022**, *11*, 1763, doi:10.3390/plants11131763 . . . . **179**

About the Editor

Alessandro Vitale

Alessandro Vitale is Associate Professor in Plant Pathology at the Department of Agriculture, Food and Environment of the University of Catania. He received his PhD in Plant Health Technologies and Protection in 2004 at the University of Catania, where he worked on fungal diseases of ornamental crops and the occurrence of fungicide resistance within Calonectria populations. His research is focused on fungal diseases of vegetables, ornamental, fruit, and tropical crops. He is actually working on the diagnosis and characterization, epidemiology, fungicide resistance, and, above all, environmentally friendly and sustainable management of airborne and soilborne plant diseases.

Preface to "Epidemiology and Control of Plant Diseases"

Global climatic changes (GCCs) have a significant impact on the ecology and epidemiology of plant pathogens, which, affect crop productivity and food security. GCC-induced alterations in temperature, rainfall patterns, and extreme weather events can create favorable conditions for the growth and spread of plant pathogens, leading to increased levels of disease infections.

In addition, high CO_2 rates due to GCCs can directly affect plant–pathogen interactions, altering plant physiology and defense mechanisms, as well as pathogen virulence and reproduction rates. This can lead to deep changes in the composition and diversity of plant pathogen communities, with potential negative impacts on disease management strategies. To mitigate these adverse effects, plant protection and disease management strategies need to be adapted to the changing environmental conditions.

All of this implies that the timing of treatments should be considered, as well as the development of innovative and integrated disease management strategies that are resilient to GCC-induced environmental stresses. It is also crucial to promote sustainable agricultural practices that reduce greenhouse gas emissions and enhance climate change resilience in agro-ecosystems.

Alessandro Vitale
Editor

plants

Editorial

Epidemiology and Control of Plant Diseases

Alessandro Vitale

Department of Agriculture, Food and Environment, University of Catania, Via Santa Sofia, 100-95123 Catania, Italy; alevital@unict.it

I am pleased to present this edition of the Special Issue of Plants, dedicated to multi-faceted topic of epidemiology and control of plant diseases in agricultural systems.

Regarding this point, it is noteworthy that some plant disease epidemics are currently having shocking effects on agricultural and forestry heritage and food crop production all over the world. Moreover, global climatic changes caused by human activities have progressively induced a significant increase of global mean temperature with direct repercussions on the epidemiology of plant pathogens. As a consequence, life cycles of plant pathogens could be deeply altered or an increase of emerging plant diseases could be observed on many commercial crops in different agroecosystems [1,2]. Thus, a better understanding of plant pathogen epidemiology, as well as individuation of new bio-based products, forecasting models, deep learning and reducing chemicals use approaches exploited for plant protection could better satisfy the development of ecofriendly, safer, and sustainable management strategies [3–5]. In light of above considerations, researchers and technicians are invited to investigate on epidemiological cycles of airborne and soilborne plant pathogens with the aim of ensuring good agricultural productivity, maintaining effective management of plant diseases and, simultaneously, good economic and environmental sustainability. This Special Issue underlines the performances of new eco-sustainable means of control to increase knowledge for plant pathologists, international scientific communities, and industries, and will provide a better understanding of the mode of action and timing application procedure of control means in different soil–plant systems. At the same time, an early and deep understanding of plant disease epidemiology is needed to tackle future challenges ahead and to relate directly with the disease control strategies.

Comprehensively, the Special Issue collected 13 original contributions (1 review, 1 perspective, and 11 research papers).

The review of Makhumbila et al. [6] provides a progress update on metabolomic studies of legumes in response to different biotic stresses. In their contribution, metabolome annotation and data analysis platforms are discussed together with future perspectives. The integration of metabolomics with other "omics" tools in breeding programs can aid greatly in ensuring food security through the production of stress tolerant cultivars.

In a perspective article, Figlan et al. [7] highlights the importance of breeding tools for improving resistance to Fusarium head blight disease caused by *Fusarium graminearum* in wheat as well as for limiting mycotoxin contamination to reflect the current state of affairs. According to the outlined scenario, only by combining these aspects in wheat research and development will it be possible to promote sustainable quality grain production and safeguard human and livestock health from mycotoxicoses.

In their research article, Thomidis et al. [8] compare two models, a generic and a polynomial model, respectively, for the forecasting in field appearance of olive leaf spot caused by *Venturia oleaginea*. The results obtained with both models correctly predicted infection periods, although differences regarding the severity parameter were reported according to the goodness-of-fit for the data collected on olive leaves in 2016, 2017, and 2018. Specifically, the generic model predicted lower severity values, which fits well with the incidence of the disease symptoms on unsprayed trees. Otherwise, the polynomial model

Citation: Vitale, A. Epidemiology and Control of Plant Diseases. *Plants* **2023**, *12*, 793. https://doi.org/10.3390/plants12040793

Received: 28 January 2023
Accepted: 7 February 2023
Published: 10 February 2023

predicted high severity levels of infection, but these did not fit well with the incidence of disease symptoms. This paper also establishes that a temperature within 5–25 °C range was appropriate for conidial germination, being 20 °C the optimum. At the latter temperature, it is also found that at least 12 h duration of leaf wetness is needed to start the germination of *V. oleaginea* conidia.

Similarly, Ji et al. [9] also investigate the effects of temperature and wetness duration on the infection severity of *Coniella diplodiella,* causal agent of grapevine white rot, by using artificial inoculation of grape berries through via infection pathways (uninjured and injured berries, and through pedicels). Results show that injured berries were affected sooner than uninjured ones, and disease increased as the inoculum dose increased. Irrespective of the infection pathway, 1 h of wetness is sufficient to cause infection at any temperature tested (10–35 °C), with the optimal temperature being 23.8 °C. The length of incubation is shorter for injured berries than for uninjured ones, and it is shorter at 25–35 °C than at lower temperatures. Finally, authors develop mathematical equations that fit well with the data, infection through any infection pathway, and incubation on injured berries, which could be used to predict infection period and, thus, to schedule fungicide applications.

The research paper of Verhoeven et al. [10] presents data on seed transmission of pospiviroids from large-scale grow-out trials of infested pepper and tomato seed lots produced under standard seed-industry conditions. Moreover, the paper shows the results of a systematic review of published data on seed transmission and outbreaks in commercial pepper and tomato crops. Based on the data of the grow-out trials and review of the literature, it is concluded that the role of seed transmission in the spread of pospiviroids in practice is possibly overestimated.

In another research paper, Bertacca et al. [11] develop a real-time loop-mediated isothermal amplification (LAMP) assay for simple, rapid, and efficient detection of the Olea europaea geminivirus (OEGV), a virus recently reported in different olive cultivation areas worldwide. This real-time LAMP assay results are also suitable for phytopathological laboratories with limited facilities and resources, as well as for direct OEGV detection in the field, representing a reliable method for rapid screening of olive plant material.

The research paper of Lukianova et al. [12] reports on the development of a specific and sensitive detection protocol based on a real-time PCR with a TaqMan probe for *Pectobacterium parmentieri*, a plant-pathogenic bacterium, recently attributed as a separate species and causing soft rot of potato tubers. Since no cross-reaction with the non-target bacterial species, or loss of sensitivity, is observed, this specific and sensitive diagnostic tool may reveal a wider distribution and host range for this bacterium, and will expand knowledge of the life cycle and environmental preferences of this pathogen.

In the research contribution of Shalaby et al. [13], the foliar application of paclobutrazol (PBZ) at different rates of 25, 50, and 100 mg L^{-1} enhances the quality of tomato seedlings and induces resistance to early blight (*Alternaria solani*) disease post inoculation at 7, 14, and 21 days under greenhouse conditions. Higher values in chlorophyll content, enzyme activities, cuticle thickness of stems and numbers, and thickness of stomata are recorded on PBZ-treated tomato plants.

In the research paper of Withee et al. [14], 16 isolates of *Paramyrothecium* spp. retrieved from 14 host species across nine plant families are collected and identified. In detail, a new species *Paramyrothecium vignicola* sp. nov. is identified according to morphological features and concatenated (ITS, cmdA, rpb2, and tub2) phylogeny. Further, *P. breviseta* and *P. foliicola* represent novel geographic records for Thailand, while *P. eichhorniae* represents a novel host record (*Psophocarpus* sp., *Centrosema* sp., *Aristolochia* sp.). These *Paramyrothecium* species fulfill Koch's postulates and, moreover, cross pathogenicity assays on *Coffea arabica* L., *Commelina benghalensis* L., *Glycine max* (L.) Merr., and *Dieffenbachia seguine* (Jacq.) Schott reveal a potential multiple host range for these pathogens.

In another research paper, Alhudaib et al. [15] focus their attention on a devastating disease, i.e., the black scorch, for date palm plantations in Saudi Arabian Peninsula. The authors report variable symptoms as well as neck bending, leaf drying, tissue necrosis, wilting, and mortality of the entire tree in the Saudi Arabia peninsula. Based on morphological, molecular and pathogenicity assays, the causal agent of above mentioned symptoms is identified in *Thielaviopsis punctulata*. Further indications about the disease control are provided by authors who suggest using fosetyl Al- and difenoconazole-based formulates. These fungicides can be used in integrated disease management strategies to control black scorch disease.

Li et al. [16] develop in an interesting study a deep learning network called YOLO (You Only Look Once) for detecting jute diseases (-JD) from images. In the main architecture of YOLO-JD, the authors integrate three new modules, the Sand Clock Feature Extraction Module (SCFEM), the Deep Sand Clock Feature Extraction Module (DSCFEM), and the Spatial Pyramid Pooling Module (SPPM) to extract image features effectively. The authors also build a new large-scale image dataset for jute diseases and pests with ten classes. Compared with other state-of-the-art experiments, YOLO-JD achieves the best detection accuracy, with an average mAP (mean average precision) of 96.63%.

Aiello et al. [17] describe for the first time the new wilt symptoms on passion fruit caused by *Fusarium nirenbergiae*. This report also focuses on the phytopathological implications of this fungal pathogen, which may represent a future significant threat for the expanding passion fruit production in Italy and Europe.

In the last research paper, Calderone et al. [18] demonstrate that a chitosan oligosaccharide (COS–) oligogalacturonides (OGA) complex, applied alone or combined with arbuscular mycorrhizal fungi (AMF), is able to reduce significantly powdery mildew (*Erysiphe necator*), gray mold (*Botrytis cinerea*), and sour rot infections on red berried Nero d'Avola and white berried Inzolia wine grape cultivars. Overall, this strategy can be proposed as a valid and safer option for the sustainable management of the main grapevine pathogens in organic agroecosystems.

Conflicts of Interest: The author declares no conflict of interest.

References

1. Ristaino, J.B.; Anderson, P.K.; Bebber, D.P.; Brauman, K.A.; Cunniffe, N.J.; Fedoroff, N.V.; Finegold, C.; Garrett, K.A.; Gilligan, C.A.; Jones, C.M.; et al. The persistent threat of emerging plant disease pandemics to global food security. *Proc. Natl. Acad. Sci. USA* **2021**, *118*, e2022239118. [CrossRef] [PubMed]
2. Castello, I.; D'Emilio, A.; Raviv, M.; Vitale, A. Soil solarization as a sustainable solution to control tomato Pseudomonads infections in greenhouses. *Agron. Sustain. Dev.* **2017**, *37*, 59. [CrossRef]
3. Saleem, M.H.; Potgieter, J.; Arif, K.M. Plant disease detection and classification by deep learning. *Plants* **2019**, *8*, 468. [CrossRef] [PubMed]
4. Fragalà, F.; Castello, I.; Puglisi, I.; Padoan, E.; Baglieri, A.; Montoneri, E.; Vitale, A. New insights into municipal biowaste derived products as promoters of seed germination and potential antifungal compounds for sustainable agriculture. *Chem. Biol. Technol. Agric.* **2022**, *9*, 69. [CrossRef]
5. Castello, I.; D'Emilio, A.; Baglieri, A.; Polizzi, G.; Vitale, A. Management of Chrysanthemum Verticillium wilt through VIF soil mulching combined with fumigation at label and reduced rates. *Agriculture* **2022**, *12*, 141. [CrossRef]
6. Makhumbila, P.; Rauwane, M.; Muedi, H.; Figlan, S. Metabolome profiling: A breeding prediction tool for legume performance under biotic stress conditions. *Plants* **2022**, *11*, 1756. [CrossRef] [PubMed]
7. Figlan, S.; Mwadzingeni, L. Breeding tools for assessing and improving resistance and limiting mycotoxin production by *Fusarium graminearum* in wheat. *Plants* **2022**, *11*, 1933. [CrossRef] [PubMed]
8. Thomidis, T.; Michos, K.; Chatzipapadopoulos, F.; Tampaki, A. Evaluation of Two Predictive Models for Forecasting Olive Leaf Spot in Northern Greece. *Plants* **2021**, *10*, 1200. [CrossRef] [PubMed]
9. Ji, T.; Languasco, L.; Li, M.; Rossi, V. Effects of Temperature and Wetness Duration on Infection by *Coniella diplodiella*, the Fungus Causing White Rot of Grape Berries. *Plants* **2021**, *10*, 1696. [CrossRef] [PubMed]
10. Verhoeven, J.T.J.; Botermans, M.; Schoen, R.; Koenraadt, H.; Roenhorst, J.W. Possible Overestimation of Seed Transmission in the Spread of Pospiviroids in Commercial Pepper and Tomato Crops Based on Large-Scale Grow-Out Trials and Systematic Literature Review. *Plants* **2021**, *10*, 1707. [CrossRef] [PubMed]
11. Bertacca, S.; Caruso, A.G.; Trippa, D.; Marchese, A.; Giovino, A.; Matic, S.; Noris, E.; Ambrosio, M.I.F.S.; Alfaro, A.; Panno, S.; et al. Development of a Real-Time Loop-Mediated Isothermal Amplification Assay for the Rapid Detection of Olea Europaea Geminivirus. *Plants* **2022**, *11*, 660. [CrossRef] [PubMed]

12. Lukianova, A.A.; Evseev, P.V.; Stakheev, A.A.; Kotova, I.B.; Zavriev, S.K.; Ignatov, A.N.; Miroshnikov, K.A. Quantitative Real-Time PCR Assay for the Detection of *Pectobacterium parmentieri*, a Causal Agent of Potato Soft Rot. *Plants* **2021**, *10*, 1880. [CrossRef] [PubMed]
13. Shalaby, T.A.; Taha, N.A.; Taher, D.I.; Metwaly, M.M.; El-Beltagi, H.S.; Rezk, A.A.; El-Ganainy, S.M.; Shehata, W.F.; El-Ramady, H.R.; Bayoumi, Y.A. Paclobutrazol Improves the Quality of Tomato Seedlings to Be Resistant to *Alternaria solani* Blight Disease: Biochemical and Histological Perspectives. *Plants* **2022**, *11*, 425. [CrossRef]
14. Withee, P.; Haituk, S.; Senwanna, C.; Karunarathna, A.; Tamakaew, N.; Pakdeeniti, P.; Suwannarach, N.; Kumla, J.; Suttiprapan, P.; Taylor, P.W.J.; et al. Identification and Pathogenicity of *Paramyrothecium* Species Associated with Leaf Spot Disease in Northern Thailand. *Plants* **2022**, *11*, 1445. [CrossRef] [PubMed]
15. Alhudaib, K.A.; El-Ganainy, S.M.; Almaghasla, M.I.; Sattar, M.N. Characterization and Control of *Thielaviopsis punctulata* on Date Palm in Saudi Arabia. *Plants* **2022**, *11*, 250. [CrossRef]
16. Li, D.; Ahmed, F.; Wu, N.; Sethi, A.I. YOLO-JD: A Deep Learning Network for Jute Diseases and Pests Detection from Images. *Plants* **2022**, *11*, 937. [CrossRef] [PubMed]
17. Aiello, D.; Fiorenza, A.; Leonardi, G.R.; Vitale, A.; Polizzi, G. *Fusarium nirenbergiae* (*Fusarium oxysporum* Species Complex) Causing the Wilting of Passion Fruit in Italy. *Plants* **2021**, *10*, 2011. [CrossRef] [PubMed]
18. Calderone, F.; Vitale, A.; Panebianco, S.; Lombardo, M.F.; Cirvilleri, G. COS-OGA Applications in Organic Vineyard Manage Major Airborne Diseases and Maintain Postharvest Quality of Wine Grapes. *Plants* **2022**, *11*, 1763. [CrossRef] [PubMed]

plants

Review

Metabolome Profiling: A Breeding Prediction Tool for Legume Performance under Biotic Stress Conditions

Penny Makhumbila [1,*], Molemi Rauwane [1], Hangwani Muedi [2] and Sandiswa Figlan [1]

[1] Department of Agriculture and Animal Health, School of Agriculture and Life Sciences, College of Agriculture and Environmental Sciences, University of South Africa, 28 Pioneer Ave, Florida Park, Roodeport 1709, South Africa; rauwaneme@gmail.com (M.R.); figlas@unisa.ac.za (S.F.)
[2] Research Support Services, North West Provincial Department of Agriculture and Rural Development, 114 Chris Hani Street, Potchefstroom 2531, South Africa; hmuedi@nwpg.gov.za
* Correspondence: 57994463@mylife.unisa.ac.za

Abstract: Legume crops such as common bean, pea, alfalfa, cowpea, peanut, soybean and others contribute significantly to the diet of both humans and animals. They are also important in the improvement of cropping systems that employ rotation and fix atmospheric nitrogen. Biotic stresses hinder the production of leguminous crops, significantly limiting their yield potential. There is a need to understand the molecular and biochemical mechanisms involved in the response of these crops to biotic stressors. Simultaneous expressions of a number of genes responsible for specific traits of interest in legumes under biotic stress conditions have been reported, often with the functions of the identified genes unknown. Metabolomics can, therefore, be a complementary tool to understand the pathways involved in biotic stress response in legumes. Reports on legume metabolomic studies in response to biotic stress have paved the way in understanding stress-signalling pathways. This review provides a progress update on metabolomic studies of legumes in response to different biotic stresses. Metabolome annotation and data analysis platforms are discussed together with future prospects. The integration of metabolomics with other "omics" tools in breeding programmes can aid greatly in ensuring food security through the production of stress tolerant cultivars.

Keywords: legumes; metabolomics; biotic stress; stress tolerance; metabolome annotation

Citation: Makhumbila, P.; Rauwane, M.; Muedi, H.; Figlan, S. Metabolome Profiling: A Breeding Prediction Tool for Legume Performance under Biotic Stress Conditions. *Plants* **2022**, *11*, 1756. https://doi.org/10.3390/plants11131756

Academic Editor: Alessandro Vitale

Received: 25 May 2022
Accepted: 22 June 2022
Published: 1 July 2022

1. Introduction

Leguminous crops such as *Arachis hypogaea* (groundnut), *Glycine max* (soybean), *Phaseolus vulgaris* (common bean), *Pisum sativum* (common pea), *Cicier arietinum* (chickpea), *Vigna anguiculata* (cowpea), *Vicia faba* (faba bean), *Lens culinaris* (lentil), *Cajanus cajan* (pigeon pea), *Lupinus* spp. (lupin), and *Vigna subterranean* (bambara bean) contribute to the improvement of ecosystems [1–3], nutrition and food security [4–7]. Although legumes contribute greatly to food security, their production globally is hindered by biotic stresses that include nematodes, viruses, insect pests, and bacterial and fungal pathogens [8–10]. The occurrence of biotic stresses in legume production systems has impacted negatively on production and has resulted in significant yield losses globally [11–13]. In many breeding programmes, the key objective is to develop crop varieties that are adaptable to an array of stressors in order to meet global food demands [14–16], thus addressing sustainable development goals 1 and 2 of the United Nations [17]. Legume programmes have been improving gradually over the years and have advanced from traditional methods of breeding to using genomic tools [18]. Traditional breeding techniques rely mostly on manual selection and the crossing of genotypes with desirable traits, and although these methods have contributed greatly to legume breeding, the genetic gain was often not statistically significant [19].

Contemporary biotechnology tools including next generation sequencing (NGS) platforms have aided many breeding programmes with provision of genetic data that traditional breeding techniques cannot fully reveal [20]. Biotechnological "omics" approaches have contributed greatly to breeding aimed at the improvement of plant stress tolerance by providing insight into genetic diversity, genotype variations, genetic maps and other useful information pertaining to the genetics of plant populations [21,22]. Despite the importance of genomic data generated by the other omics platforms (transcriptomics, transgenomics, epigenomics), plants produce molecular compounds with molecular weights expressed in abundance and are responsible for biochemical functions under different environments [23]. Metabolomics highlights metabolite expressions and changes, together with their interactions and phenotypic characters of plants under stress conditions. When plants are exposed to stress, metabolic homeostasis alterations occur, requiring the plant to adjust its metabolomic pathways, and this phenomenon is referred to as acclimation [24–27]. When this process occurs, the plant activates signal transduction pathways that set off the assembly of proteins and metabolomic compounds that aid in reaching a new homeostasis [28,29]. Furthermore, metabolome analysis provides information on the metabolomic pathways that are responsible for complex processes that occur when a plant is exposed to stress conditions [20].

A detailed review of metabolomic studies focused on specific biotic stressors of legumes can aid in identifying gaps and create an interactive platform for researchers to conduct, and possibly collaborate on, more studies aimed at improving legume production in the world. This is because the dimensionality of large data sets generated through metabolomics can be interpreted holistically utilising multivariate data analysis [30]. This will further highlight the importance of metabolite detection in breeding programmes and techniques that can be employed for different objectives since metabolites relate to phenotypic and genomic data [9]. This review reports on metabolomics as a breeding prediction tool in legume breeding under biotic stress. We also briefly discuss the impact of metabolomics in legume breeding programmes aimed at improving biotic stress tolerance.

2. Biotic Stressors of Legumes

2.1. Insect Pests

Insect pests attack legume crops by boring, webbing and damaging plant parts such as the leaves, pods, stems and roots [31,32]. In addition to attacking plants, insect pests may also act as vectors for pathogens that negatively impact crop production systems [33]. Insect pests such as aphids [33,34], pod borers [31,35], thrips [36,37] and whiteflies [38,39] have been reported to feed on legume crops, among others. The use of biological enemies of pests, cultural control (crop rotation, mulching, intercropping, etc.), mechanical control (water hosing at high pressure), chemical application and integrated pest management strategies have been recommended for the control of insect pests in legumes [39–42]. These efforts have been found to be effective in reducing insect severity in legumes [39,43]. However, the insects are constantly adapting to control measures used in production systems [44]. Breeding for tolerance to insect pests is the most sustainable approach and this requires an understanding of the plant's signal pathways that respond to insect attack [45].

Pathways expressed in rice infested with caterpillars included flavonoids, phenolic acids, amino acids and derivatives. These improved the production of cytosolic calcium ions that signal herbivore attack to the plant [46]. Maize infested with *Monolepta hieroglyphica* revealed significant up-/down-regulation of metabolites derived from sugar and amino acid pathways that might be responsible for resistance. Similar results were reported in cabbage infested with aphids [47]. Insect–plant metabolomic response of leguminous crops has been conducted for red clover, pea and alfalfa in a composite study with aphid infestation. Triterpene, flavonoid and saponin enriched pathways were found to be responsive to aphid attack [34]. Flavonoids and amino acids have also been found to be significantly enriched in alfalfa infested with thrips [48]. However, limited studies have been conducted on the host-plant metabolomic response of leguminous crops to insects, as well as to other biotic stressors. These studies could have far-reaching impacts on stress biomarker identification with potential benefits in legume improvement programmes.

2.2. Diseases of Legumes

2.2.1. Bacterial Diseases

Bacterial diseases of legumes can be categorised into leaf blights, leaf spots/bacterial wilts and other multiple symptoms of sprout rot and dwarfism [49]. Their symptoms are based on the tissues that they infiltrate (leaves, stems and roots) [50]. Legume bacterial diseases are known to cause yield losses of up to 50%, which negatively impacts economic gains and food security [51]. The two plant bacterial pathogens *Xanthomonas axonopodis* and *Pseudomonas syringae* are known worldwide for causing bacterial blight [49,52]. Symptoms of infection usually occur on all aerial parts of the plant, and in severe incidences, defoliation and wilting occur [52,53]. Like bacterial blight, another disease that threatens legume production is bacterial wilt, caused by *Curtobacterium flaccumfaciens pv. Flaccumfaciens* [54]. The pathogen has created new variants that cause damage to legume crops worldwide by causing leaf chlorosis in plants. In fields where the disease occurs, upon plant maturation and shattering of seeds, the infected seed replants itself and allows the pathogen to thrive from generation to generation [54,55]. The control of bacterial diseases has relied on integrated approaches that limit the survival of pathogens. This includes crop rotation and the use of pathogen free certified seed [52]. These measures are only effective to a limited extent, and detecting pathogens in seed is not an easy task for farmers. A promising and more long-term method for the control of bacterial diseases would be the utilisation/breeding of tolerant varieties [56,57].

The evaluation of metabolite profiles in citrus infected with huanlongbing caused by the bacterium '*Candidatus Liberibacter asiaticus*' reported distinct sugars as well as amino and organic acids expressed in the roots, thus giving insight on resistance [58]. Metabolomic compounds synthesized from flavonoids, amino and phenolic acids act as protective agents in the xylem of oat plants when infected with halo blights caused by *P. syringae pv.* by repairing the cell wall [59]. Similar metabolomic pathways including phenols and acetates have been reported in tomato infected with bacterial wilt caused by *Ralstonia solanacearum* [60]. To date, there is little to no information from metabolomic studies on the response of leguminous crops to bacterial disease infection to aid breeders with biomarker discovery.

2.2.2. Fungal Diseases

The occurrence of fungal diseases in legume production areas is known to cause substantial yield losses of up to 100% [59]. Fungal pathogens can cause infection at any plant growth stage (emergence, seedling, vegetative and reproductive stage) by attacking organs and tissues that are involved in the transportation of water and nutrients [61,62]. Upon infection, these pathogens degrade the plant cell wall, which consequently results in the death of the plant, especially if the variety grown does not have any resistant genes [63]. Root rot caused by *Rhizoctonia solani*, *Fusarium solani*, *Fusarium oxysporum* and *Aphanomyces euteiches* and fungal wilt caused by *Formae speciales* are some of the most destructive fungal diseases that limit the productivity of legume crops worldwide [64]. The pathogen *R. solani* is considered one of the most destructive fungal pathogens that usually infects the roots and hypocotyl of the plant through penetration of the appressoria [63]. At pre-emergence and post-emergence plant growth stages, *R. solani* causes symptoms of damping-off, root rot and stem canker [65]. Under greenhouse conditions, the seedling survival of some leguminous crops may be less than 5% [66]. The pathogen may further infect the plant's fruits in highly humid conditions, thus reducing crop quality and yield [67]. *Fusarium* spp. are also predominant pathogens that interfere with plant growth by causing damping off and root rot [68]. In African small-scale farms, yield losses of up to 100% caused by the *F. solani* pathogen in common bean have been reported [69]. In addition, *A. euteiches* is a soil-borne fungal pathogen that poses a threat to legume production by causing wilting, root rot and consequently yield losses of up to 80% [70,71].

The management of fungal diseases is problematic due to the complexity of these pathogens [72]. Over the years, management has been implemented by integrating conventional methods such as crop rotations, increased greenhouse temperatures, biological enemies and chemical use [73]. The use of fungicides has been a promising avenue for the control of fungal pathogens. However, chemicals used to control pathogens have an immense economic and environmental impact [74]. This has led to the exploration of using biological control measures such as bacterium and fungal strains as environmentally friendly alternatives to control pathogens that attack plants [75]. *Trichoderma* spp. are widely used strains for the biological control of fungal diseases. Beneficial strains of *T. velutinum* have been found to be an effective biological control measure that promotes the accumulation of metabolites that are responsible for defence in common bean infected with *F. solani*. Even though numerous strains have been found to be effective in controlling fungal diseases, legislation in many countries regarding the use of biopesticides and their shelf life is still a challenge [76,77]. The development of disease-resistant cultivars using genomic technologies can aid in improving legume productivity worldwide [54]. Legume metabolomics focussed on breeding for disease resistance can be beneficial to breeding programmes by increasing the availability of resistant genotypes that are released to farmers [78].

The metabolomic profiling of leguminous crops has been conducted in common bean and provided major findings in relation to metabolomic pathways including amino acids, flavonoids, isoflavanoids, purines and proline metabolism, which were shown to promote plants' potential for defence against *Fusarium* pathogens [79]. In addition, Mayo-Prieto et al. [80] also reported amino acids, peptides, carbohydrates, flavonoids, lipids, phenols, terpenes and glycosides that were up-/down-regulated as a defence mechanism by the common bean plant against the pathogen *R. solani*. Similar results have been reported in other leguminous crops including chickpea infected with *F. oxysporum*, soybean infected with *Aspergillus oryzae/Rhizopus oligosporus*, pea infected with *Dydymella pinodes* and *R. solani* (Table 1) [81,82]. Intensifying the fungal–legume metabolomic research worldwide will aid in understanding the biochemical properties of these leguminous crops in response to disease stress.

2.2.3. Viral Diseases

Viral pathogens attack many crops, including legumes, by causing the yellowing of leaves, stunting and poor pod setting, which result in poor yields [65]. Major viral diseases causing production losses in legumes belong to the *Nanoviridae*, *Luteovridae* and *Poltyvridae* families. These diseases cause the necrosis of plants, and their identification requires molecular techniques. Over the years, the accurate identification of viruses has improved because of an increasing number of available genomic platforms. [49,66]. Viruses attach themselves to specific sites of vectors such as insects (aphids, beetles, etc.) and remain there until transmission to their host occurs [67]. The control of viral diseases is difficult and thus requires adherence to quarantine prescripts, removal of inoculum sources, adjustments of planting dates, intercropping, crop rotation, chemical application aimed at controlling pests (elimination of vectors) and the use of tolerant/resistant genotypes [68].

Utilising metabolomic techniques on the *Citrus tristeza* virus of Mexican lime *Citrus aurantifolia* revealed up-/down-regulation of amino acids, alkaloids and phenols during infection, thus signalling pathogen defence when different strains of the virus were utilised [83]. In stems of *Amarathus hypochondriacus* L. infected with *Ageratum enation* virus, alkaloids, amino acids, dicarboxylic acids, glutamine and sugars may increase or decrease in concentration as a mechanism to improve overall respiratory metabolism [84]. Studies on the response of leguminous crops to viral disease infection are limited, thus requiring more research in order to fully understand the underlying information relating to metabolites expressed under virus pressure.

2.3. *Parasitic Weeds*

Unlike "normal" weeds that disadvantage the plant greatly, parasitic weeds on the other hand extensively extract moisture, nutrients, photosynthates and other resources from the host plant [69]. When parasitic weeds are not controlled, the extraction of resources continues, consequently extinguishing the crop [70]. Roomrape species, *Striga gesnerioides* and *Alectra vogelii* are problematic parasitic weeds that cause yield losses in many legume production areas in Sub-Saharan Africa [71]. Biological control [69], intercropping [72], chemical application and cultural practices (timely planting) are recommended for the control of parasitic weeds [73]. However, these are often not successful, and the fight against parasitic weeds lies within breeding for resistance [71,73]. Although breeding for resistance will aid in controlling parasitic weeds, the complexity and low heritability is a challenge that breeders face when breeding for parasitic weed resistance [71,73,74]. Initiatives to use breeding prediction tools such as metabolomic techniques for parasitic weed resistance have been explored in rice to study and dissect *S. hermonthica* resistance [85]. This study reported the phenylpropanoid pathway, which contributes to the formation of lignin in rice, to be an important pathway that can be utilised for resistance to *S. hermonthica*. There is a deficit on metabolomic experiments that evaluate the performance of legumes under parasitic weed conditions.

2.4. Parasitic Nematodes

Legumes are famous for their ability to fix nitrogen by using rhizobium, which is a mutualist bacterium [75]. However, the presence of parasitic nematodes reduces rhizobia activity, which leads to poor nodulation [76]. Parasitic nematodes invade the roots of plants and form an indefinite feeding area, which, in turn, can affect root development, thus leading to poor plant growth [77]. *Heterodera* and *Globodera* spp. are root knot and cyst nematodes that affect many crops including legumes, resulting in over 12% yield losses [78]. The presence of parasitic nematodes often leads to infection by other pathogens including *fusarium* spp.; therefore, the utilisation of sustainable control strategies for other pathogens is essential for legumes [74]. Soybean evaluated under *Melodegyne pinodes* and *Heterodera glycines* pressure exhibited phenylpropanoids, cysteine, methionine, alkaloid and tropane pathways that can be attributed to resistance properties of the crop to nematodes [86]. The in-depth exploration of metabolites of other crops including legumes would be beneficial to understanding nematode–crop biological interactions.

Table 1. Summary of metabolomic studies conducted in response to biotic stress in leguminous crops using different platforms such as GC-MS, LC-QqQ-MS, LC-MS, LC-obitrap-MS, UHPLC-MS, ^{1}H NMR and GC-MS/TOF.

Legume	Biotic Stress	Classification	Method	Total Metabolites	Reference
C. arietinum	*Fusarium oxysporum*	Fungal	GC-MS	72	[87]
G. max	*Aspergillus oryzae/ Rhizopus oligosporus*	Fungal	LC-QqQ-MS	489	[88]
	Heterodera glycines	Nematode	GC-MS	20	[86]
M. sativa	*Thysanoptera* spp.	Insect	LC-MS	772	[48]
	Acyrthosiphon pisum Harris	Insect	LC-Obitrap-MS/UHPLC-MS	107	[34]
P. sativum	*Acyrthosiphon pisum Harris*	Insect	LC-Obitrap-MS/UHPLC-MS	57	[34]
	Didymella pinodes	Fungal	LC-MS/MS	31	[89]
	Rhizoctonia solani	Fungal	^{1}H NMR	126	[81]
	Didymella pinodes	Fungal	GC-MS/TOF	39	[82]
P. vulgaris	*Fusarium solani*	Fungal	UPLC	743	[79]
	Trichoderma velutinum/ Rhizoctotonia solani	Fungal	LC-MS	216	[80]
T. pratense	*Acyrthosiphon pisum Harris*	Insect	LC-Obitrap-MS/UHPLC-MS	103	[34]
V. faba	*Acyrthosiphon pisum Harris*	Insect	LC-Obitrap-MS/UHPLC-MS	13	[34]

3. Legume Metabolomics

3.1. Metabolome Profiling Techniques

The use of metabolomics has been applauded for its ability to provide detailed prospects by in-depth study of crop biology. Information that is derived from metabolomic tools can be translated to assess phenotypic changes/biomarkers, gene changes and, also, to distinctively support other genomic experiments [79,80]. Furthermore, metabolomic studies can be applied for polygenic traits and prediction of epistatic effects [79,88]. The overall success of detecting metabolites and their changes depends on utilising analytical techniques that can detect compound concentrations, proportions and molecular weights [81,82,89]. The concept of metabolome profiling was introduced with the use of mass spectrometry and at a later stage, gas chromatography was also introduced [87]. Since the inception of the latter, metabolome profiling using both spectrometric and chromatographic techniques have been improving [30,90]. Different strategies are utilised for compound profiling in metabolomics, including metabolite profiling, fingerprinting and target analysis [91,92]. Metabolite fingerprinting compares "fingerprints" of metabolites [93]. The profiling analyses broader groups of metabolites that are related to specific pathways or compound classes, while target analysis is utilised for targeting specific metabolic pathways and observes the occurrences of modifications [94]. Protocols for both metabolite profiling and fingerprinting in stress experiments involve the sample acquisition from a stressed plant (leaves, stems or roots; Figure 1A) that are cut and placed in a labelled tube (Figure 1B). Dewar with liquid nitrogen is ideal for snap freezing samples in the field and a laboratory ultra-freezer with a temperature above −60 °C is recommended for sample preservation to avoid dehydration (Figure 1C). The stored samples are then crushed, and extraction is conducted in preparation for metabolite analysis, using the appropriate technology that generates spectral data (Figure 1D–F).

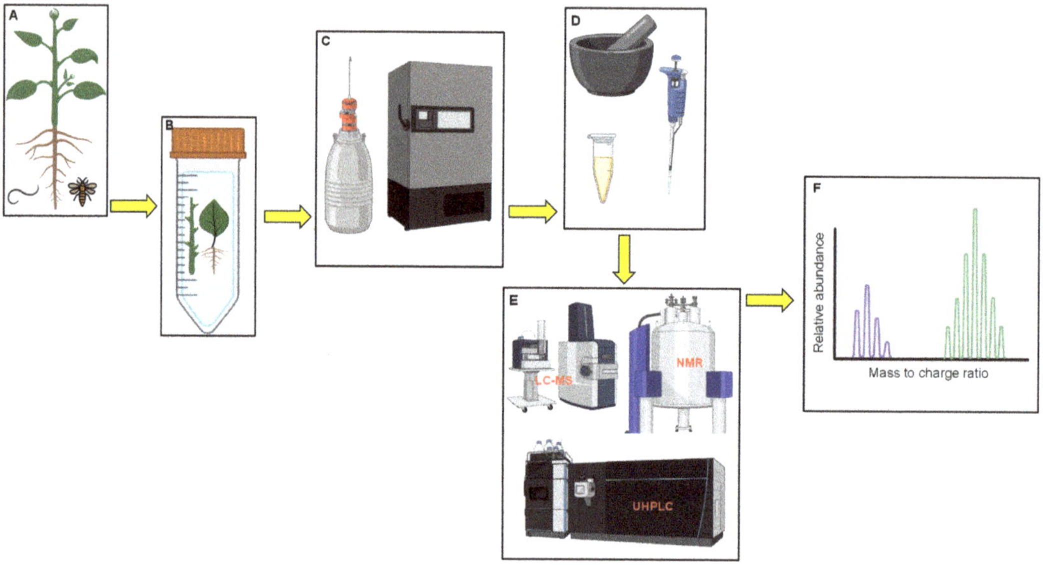

Figure 1. Flow diagram summarizing steps taken for metabolomic sample analysis in biotic stress experiments. Plant under biotic stress (**A**), samples from selected plant parts in a tube (**B**), snap freezing samples in liquid nitrogen and later stored in an ultra-freezer (**C**), extraction of metabolites in accordance with recommended protocols (**D**), metabolome analysis technologies (E), generation of raw spectral data (**F**).

3.2. *Metabolite Profiling*

Metabolite profiling is important in studying organisms' biochemical pathways [88]. Numerous technologies such as gas chromatography-mass spectrometry (GC-MS), liquid chromatography-mass spectrometry (LC-MS), nuclear magnetic resonance (NMR), capillary electrophoresis-MS (CE-MS) and Fourier transform-infrared (FT-IR) spectroscopy are commonly used analytical platforms for metabolite profiling [49,95]. The unique properties of these profiling techniques together with their applications, limitations and successes in plant metabolomics have been discussed by numerous researchers [30,96–99]. There are limited studies on the metabolome profiling of legume crops evaluated under insect stress. Although not a model for legume crops, metabolomic profiling has been performed on *Medicago sativa* (a close relative of the model legume crop *M. truncatula*) under insect stress (Table 1) [34,48]. In plant–insect interactions, a metabolome profiling study on alfalfa cultivars reported the production of numerous up-regulated metabolites in response to infestation by thrips using LC-MS (Table 1). Among the metabolite classes were amino acids together with derivatives that produced toxic amino acids released by the plant in response to insect attack [48]. Similar metabolites analysed using UHPLC-MS were also reported for pea (*P. sativum*), red clover (*Trifolium pratense*) and other alfalfa genotypes in response to biotic stress [34]. In addition, Narula et al. [87] reported a large number of metabolites that were up-regulated and down-regulated when chickpea was infected with *F. oxysporum* using GC-MS as a metabolome profiling tool. Similar results were also reported for common bean infected with *F. solani* [79], *T. velutinum* and *R. solani* [80] (Table 1). Among the primary metabolites reported, amino acids, alcohols and alkaloids were upregulated. Precursor molecules of these metabolites were found to be responsible for defence and energy provision for the plant [91]. More studies have been reported on *P. sativum* focusing on metabolite profiling under biotic stress (Table 1), particularly fungal pathogens [92,100,101]. For example, using ^{1}H NMR, young pea plants showed a heightened production of amino acids that signal the production of the metabolite proline during fungal infection [81]. However, as the plant grows older, its energy requirements change, and proline production reduces. Overall, the down-regulation of metabolites can be used as a guideline for selecting resistant/tolerant varieties. Varieties resistant to pathogens also produce sulphur as a defence strategy. Resistant cultivars tend to have increased sulphur assimilation with high energy accumulation from sugar metabolites (nitrogen mobilization) for restoration of damaged plant cells [92].

4. Metabolome Data Processing and Annotation Tools Used in Legume Stress Tolerance

Metabolome usage has grown rapidly because of its provision of the cellular function data of small molecules (<1500 Da) linked to more than 40,000 metabolites that are registered on numerous databases [102]. Data generated by metabolomic technologies such as GC-MS, LC-MS and NMR, amongst others, are enormous and require software tools that are able to visualise, detect peaks, normalize/transform the sample data, annotate, identify, quantify and statistically analyse targeted/untargeted metabolite variations, in accordance with applied algorithms for univariate/multivariate analysis (Figure 2) [103,104]. There is no single tool that can unravel information from a metabolome profile; thus, analysis integrates numerous databases and requires algorithms that are provided by an array of tools [105]. Studies of metabolites in crops use an array of statistical platforms to evaluate variations of metabolites in different stress environment [106]. In legumes, metabolome data processing platforms (Table 2) used in studies of biotic stress for legumes include R and SIMCA [48,81]. Software such as SIMCA, Analyst software, STAT GRAPHICS Centurion, Labsolutions, ChromaTOF and agilent software MassHunter require licensing for metabolome data processing. However, there are numerous web-based accessible platforms that can be used for data processing, metabolome annotation and visualisation such as R, XCMS, MetaboAnalyst, METLIN, KEGG, HMBD, MeV, MetLAB and others (Tables 2 and 3) [103].

The representation of biological networks is important in metabolomics, as it gives representation of relationships or patterns that occur in metabolomic pathways. There are numerous metabolomic pathway databases that aid in grouping metabolites with similar functions. Metabolomic pathway databases including KEGG, cytoscape, MapMan and iPath, among others, are applicable to plants [103,107].

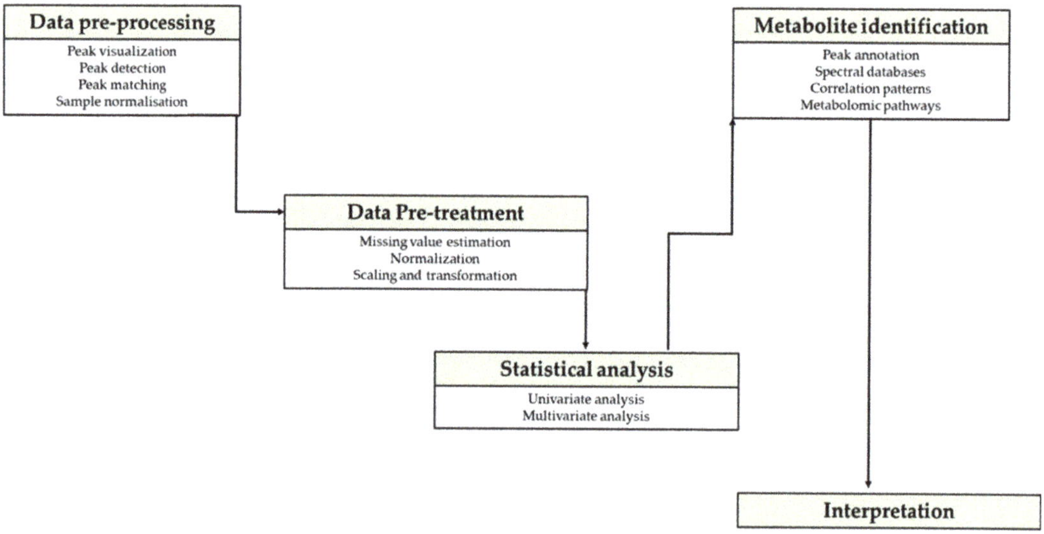

Figure 2. Flow diagram illustrating data handling steps for metabolomic experiments. After acquiring raw data, pre-processing, pre-treatment and statistical analysis are required prior to interpretation of results.

Table 2. Statistical tools and databases used for metabolome data processing and annotation in legume biotic stress studies.

Legume	Statistical Tool/Database Name	Access Domain (URL, Accessed on 28 April 2022)	Function	Reference
P. vulgaris	Analyst software	https://sciex.com/products/software/analyst-software	Data processing Metabolite annotation	[79]
	R	https://www.r-project.org/	Data processing	
	KEGG	https://www.genome.jp/kegg/kegg2.html	Metabolomic pathways	
	Agilent MassHunter	https://www.agilent.com/en/promotions/masshunter-mass-spec	Data processing	[80]
	Pubchem	https://pubchem.ncbi.nlm.nih.gov/	Metabolite annotation	
	HMBD	https://hmdb.ca/	Metabolite annotation	
	CAS	https://www.cas.org/	Metabolite annotation	
	ChemSpider	http://www.chemspider.com/	Metabolite annotation	
	METLIN	https://metlin.scripps.edu/landing_page.php?pgcontent=mainPage	Metabolite annotation	

Table 2. *Cont.*

Legume	Statistical Tool/Database Name	Access Domain (URL, Accessed on 28 April 2022)	Function	Reference
M. sativa	Analyst software	https://sciex.com/products/software/analyst-software	Data processing Metabolite annotation	[48]
	R	https://www.r-project.org/	Data processing	
	KEGG	https://www.genome.jp/kegg/kegg2.html	Metabolomic pathway analysis	
	XCMS	https://xcmsonline.scripps.edu/landing_page.php?pgcontent=institute	Data processing	[34]
	MetaboAnalyst	https://www.metaboanalyst.ca/	Data processing	
	R	https://www.r-project.org/	Data processing	
	METLIN	https://metlin.scripps.edu/landing_page.php?pgcontent=mainPage	Metabolite annotation	
	MassBank	https://massbank.eu/MassBank/	Metabolite annotation	
	HMBD	https://hmdb.ca/	Metabolite annotation	
	LipidMaps	https://www.lipidmaps.org/	Metabolite annotation	
	KEGG	https://www.genome.jp/kegg/kegg2.html	Metabolomic pathways	
	Labsolutions	https://www.shimadzu.com/an/products/software-informatics/software-option/labsolutions-cs/index.html	Data Processing Metabolite annotation	
P. sativum	COVAIN toolbox	https://bio.tools/covain	Data processing Metabolite annotation	[89]
	STATGRAPHICS Centurion	https://www.statgraphics.com/	Data processing	
	R Studio	https://www.rstudio.com/	Data processing	
	ChromaTOF	https://www.leco.com/product/chromatof-software	Data processing and Metabolite annotation	[82]
	SIMCA	https://www.sartorius.com/en/products/process-analytical-technology/data-analytics-software/mvda-software/simca	Data processing and Metabolite annotation	
	JMP software	https://www.jmp.com/support/downloads/JMPG101_documentation/Content/JMPGUserGuide/IN_G_0018.htm	Data processing and Metabolite annotation	[81]
	SIMCA	https://www.sartorius.com/en/products/process-analytical-technology/data-analytics-software/mvda-software/simca	Data processing and Metabolite annotation	
	R	https://www.r-project.org/	Data processing	
	KEGG	https://www.genome.jp/kegg/kegg2.html	Metabolomic pathway analysis	
C. ariethium	MeV	https://mev.tm4.org/#/about	Data processing and Metabolite annotation	[87]
	XLSAT software	https://www.xlstat.com/en/	Data processing	

Table 3. Statistical tools and databases used for metabolome data processing and annotation in legume biotic stress studies.

Legume	Statistical Tool/Database Name	Access Domain (URL, Accessed on 28 April 2022)	Function	Reference
L. japonicus	MapMan/PageMan	https://mapman.gabipd.org/mapman	Data processing Metabolite annotation	[108,109]
	MeV	https://mev.tm4.org/#/about	Data processing Metabolite annotation	
	Microsoft Excel	https://www.microsoft.com/en-za/	Data processing	
	MetaGeneAlyse	https://metagenealyse.mpimp-golm.mpg.de/	Data processing Metabolite annotation	
L. corniculatus *L. creticus* *L. tenius* *L. burttii* *L. uligino* *L. filicaulis*	GRaphPad (Prism)	https://www.graphpad.com/	Data processing	[110]
	MeV	https://mev.tm4.org/#/about	Data processing Metabolite annotation	
	MetaGeneAlyse	https://metagenealyse.mpimp-golm.mpg.de/	Data processing Metabolite annotation	
	Microsoft Excel	https://www.microsoft.com/en-za/	Data processing	
Stylosanthes	Microsoft Excel	https://www.microsoft.com/en-za/	Data processing	[26]
	SPSS	https://www.ibm.com/products/spss-statistics	Data processing	
	R	https://www.r-project.org/	Data processing	
	KEGG	https://www.genome.jp/kegg/kegg2.html	Metabolomic pathways	
P. vulgaris	MapMan	https://mapman.gabipd.org/mapman	Data processing and Metabolite annotation	[80]
	KEGG	https://www.genome.jp/kegg/kegg2.html	Metabolomic pathways	

5. Conclusions

Legume crops are grown in most regions of the world because they provide food security for many households. With the current climate crisis, the production of crops that are adaptable to biotic and abiotic stress is paramount. Legumes are produced in semi-arid environments and in these production areas, multiple stressors are prevalent. Plant stress response is a very complex phenomenon that researchers are constantly striving to understand by making use of high-throughput techniques. The integration and application of omics tools in agriculture has evolved and broadened the understanding of the underlying biochemical and molecular mechanisms of crops grown in diverse environments. Metabolomic studies are already becoming one of the omics tools used for breeding strategies. However, strong bioinformatics skills are needed for the processing and manipulation of the data. Furthermore, metabolomic database availability should be improved in order to accelerate information availability for legume crops. Additionally, studies that integrate metabolomics with other omics tools should aim to elaborate on the metabolomic aspects. For example, in many studies integrating transcriptomics and metabolomics, the information tends to be denser for gene expression than for metabolomics. In such cases, metabolome specific papers should be published separately to avoid complexity of integrating all the data and suppressing metabolomic information. Overall, the integration of metabolomics with other omics tools provides a powerful strategy to unravel plant–pest/pathogen interaction in biotic stress environments.

Author Contributions: Conceptualization, S.F. and M.R.; writing—original draft preparation, P.M., S.F., M.R. and H.M.; writing—review and editing, P.M., S.F., M.R. and H.M.; funding acquisition, S.F. All authors have read and agreed to the published version of the manuscript.

Funding: This research received no external funding.

Institutional Review Board Statement: Not applicable.

Informed Consent Statement: Not applicable.

Data Availability Statement: All the data are included in the main text.

Conflicts of Interest: The authors declare no conflict of interest.

References

1. Gao, D.; Wang, X.; Fu, S.; Zhao, J. Legume plants enhance the resistance of soil to ecosystem disturbance. *Front. Plant Sci.* **2017**, *8*, 1295. [CrossRef] [PubMed]
2. Makarov, M.; Onipchenko, V.; Malysheva, T.; Zuev, A.; Tiunov, A. Symbiotic nitrogen fixation by legumes in alpine ecosystems: A vegetation experiment. *Russ. J. Ecol.* **2021**, *52*, 9–17. [CrossRef]
3. Zhao, J.; Wang, X.; Wang, X.; Fu, S. Legume-soil interactions: Legume addition enhances the complexity of the soil food web. *Plant Soil* **2014**, *385*, 273–286. [CrossRef]
4. Jimenez-Diaz, R.; Castillo, P.; Jimenez-Gasco, M.; Landa, B.; Navas-Cortes, J. Fusarium wilt of chickpeas: Biology, ecology and management. *Crop Prot.* **2015**, *73*, 16–27. [CrossRef]
5. Kulkarni, K.; Tayade, R.; Asekova, S.; Song, J.; Shannon, J.; Lee, J. Harnessing the potential of forage legumes, alfalfa, soybean, and cowpea for sustainable agriculture and global food security. *Front. Plant Sci.* **2018**, *9*, 1314. [CrossRef] [PubMed]
6. Olanrewaju, O.; Oyatomi, O.; Babalola, O.; Abberton, M. Breeding potentials of bambara groundnut for food and nutrition security in the face of climate change. *Front. Plant Sci.* **2022**, *12*, 798993. [CrossRef] [PubMed]
7. Sauer, C.; Mason, N.; Maredia, M.; Mofya-Mukuka, R. Does adopting legume-based cropping practices improve the food security of small-scale farm households? panel survey evidence from Zambia. *Food Secur.* **2018**, *10*, 1463–1478. [CrossRef]
8. Kudapa, H.; Ramalingam, A.; Nayakoti, S.; Chen, X.; Zhuang, W.; Liang, X.; Kahl, G.; Edwards, D.; Varshney, R. Functional genomics to study stress responses in crop legumes: Progress and prospects. *Funct. Plant Biol.* **2013**, *40*, 1221–1233. [CrossRef]
9. Ramalingam, A.; Kudapa, H.; Pazhamala, L.; Weckwerth, W.; Varshney, R. Proteomics and metabolomics: Two emerging areas for legume improvement. *Front. Plant Sci.* **2015**, *6*, 1116. [CrossRef]
10. Vrignon-Brenas, S.; Celette, F.; Piquet-Pissaloux, A.; David, C. Biotic and abiotic factors impacting establishment and growth of relay intercropped forage legumes. *Eur. J. Agron.* **2016**, *81*, 169–177. [CrossRef]
11. Dita, M.; Rispail, N.; Prats, E.; Rubiales, D.; Singh, K. Biotechnology approaches to overcome biotic and abiotic stress constraints in legumes. *Euphytica* **2006**, *147*, 1–24. [CrossRef]
12. Sathya, A.; Vijayabharathi, R.; Gopalakrishnan, S. Plant growth-promoting actinobacteria: A new strategy for enhancing sustainable production and protection of grain legumes. *3 Biotech* **2017**, *7*, 102. [CrossRef] [PubMed]
13. Caldas, D.; Konzen, E.; Recchia, G.; Pereira, A.; Tsai, S. Functional genomics of biotic and abiotic stresses in *Phaseolus vulgaris*. In *Abiotic and Biotic Stress in Plants—Recent Advances and Future Perspectives*; Shanker, A., Shanker, C., Eds.; IntechOpen: London, UK, 2016; p. 151. ISBN 978-953-51-2250-0.
14. Hussain, B. Modernization in plant breeding approaches for improving biotic stress resistance in crop plants. *Turk. J. Agric. For.* **2015**, *39*, 515–530. [CrossRef]
15. Zivy, M.; Wienkoop, S.; Renaut, J.; Pinheiro, C.; Goulas, E.; Carpentier, S. The quest for tolerant varieties: The importance of integrating "Omics" techniques to phenotyping. *Front. Plant Sci.* **2015**, *6*, 448. [CrossRef]
16. Saed-Moucheshi, A.; Pessarakli, M.; Mozafari, A.; Sohrabi, F.; Moradi, M.; Marvasti, F. Screening barley varieties tolerant to drought stress based on tolerant indices. *J. Plant Nutr.* **2022**, *45*, 739–750. [CrossRef]
17. Khanal, U.; Wilson, C.; Rahman, S.; Lee, B.; Hoang, V. Smallholder farmers' adaptation to climate change and its potential contribution to UN's sustainable development goals of zero hunger and no poverty. *J. Clean. Prod.* **2021**, *281*, 124999. [CrossRef]
18. Jacob, C.; Carrasco, B.; Schwember, A. Advances in breeding and biotechnology of legume crops. *Plant Cell Tissue Organ Cult.* **2016**, *127*, 561–584. [CrossRef]
19. Pratap, A.; Das, A.; Kumar, S.; Gupta, S. Current perspectives on introgression breeding in food legumes. *Front. Plant Sci.* **2021**, *11*, 589189. [CrossRef]
20. Singh, D.; Chaudhary, P.; Taunk, J.; Singh, C.; Singh, D.; Tomar, R.; Aski, M.; Konjengbam, N.; Raje, R.; Singh, S. Fab advances in fabaceae for abiotic stress resilience: From "Omics" to artificial intelligence. *Int. J. Mol. Sci.* **2021**, *22*, 535. [CrossRef]
21. Afzal, M.; Alghamdi, S.; Migdadi, H.; Khan, M.; Nurmansyah; Mirza, S.; El-Harty, E. Legume genomics and transcriptomics: From classic breeding to modern technologies. *Saudi J. Biol. Sci.* **2020**, *27*, 543–555. [CrossRef]
22. Varshney, R.; Dubey, A. Novel genomic tools and modern genetic and breeding approaches for crop improvement. *J. Plant Biochem. Biotechnol.* **2009**, *18*, 127–138. [CrossRef]

23. Hong, J.; Yang, L.; Zhang, D.; Shi, J. Plant metabolomics: An indispensable system biology tool for plant science. *Int. J. Mol. Sci.* **2016**, *17*, 767. [CrossRef] [PubMed]
24. Lugan, R.; Niogret, M.; Leport, L.; Guegan, J.; Larher, F.; Savoure, A.; Kopka, J.; Bouchereau, A. Metabolome and water homeostasis analysis of *Thellungiella salsuginea* suggests that dehydration tolerance is a key response to osmotic stress in this halophyte. *Plant J.* **2010**, *64*, 215–229. [CrossRef] [PubMed]
25. Heyneke, E.; Watanabe, M.; Erban, A.; Duan, G.; Buchner, P.; Walther, D.; Kopka, J.; Hawkesford, M.; Hoefgen, R. Characterization of the wheat leaf metabolome during grain filling and under varied N-supply. *Front. Plant Sci.* **2017**, *8*, 2048. [CrossRef]
26. Luo, J.; Liu, Y.; Zhang, H.; Wang, J.; Chen, Z.; Luo, L.; Liu, G.; Liu, P. Metabolic alterations provide insights into stylosanthes roots responding to phosphorus deficiency. *BMC Plant Biol.* **2020**, *20*, 85. [CrossRef]
27. Joshi, J.; Hasnain, G.; Logue, T.; Lynch, M.; Wu, S.; Guan, J.; Alseekh, S.; Fernie, A.; Hanson, A.; McCarty, D. A core metabolome response of maize leaves subjected to long-duration abiotic stresses. *Metabolites* **2021**, *11*, 797. [CrossRef]
28. Xing, T.; Jordan, M. Genetic engineering of plant signal transduction mechanisms. *Plant Mol. Biol. Report.* **2000**, *18*, 309–318. [CrossRef]
29. Shulaev, V.; Cortes, D.; Miller, G.; Mittler, R. Metabolomics for plant stress response. *Physiol. Plant.* **2008**, *132*, 199–208. [CrossRef]
30. Wolfender, J.-L.; Marti, G.; Thomas, A.; Bertrand, S. Current approaches and challenges for the metabolite profiling of complex natural extracts. *J. Chromatogr. A* **2015**, *1382*, 136–164. [CrossRef]
31. Kumar, A.; Jaiwal, R.; Sreevathsa, R.; Chaudhary, D.; Jaiwal, P. Transgenic cowpea plants expressing *Bacillus thuringiensis* Cry2Aa insecticidal protein imparts resistance to *Maruca vitrata* legume pod borer. *Plant Cell Rep.* **2021**, *40*, 583–594. [CrossRef]
32. Singh, S.; Singh, P. Biochemical factors associated with resistance to spotted pod borer, *Maruca vitrata* (Fabricius) in green gram. *Legume Res.* **2021**, *44*, 1398–1401. [CrossRef]
33. Kamphuis, L.; Zulak, K.; Gao, L.; Anderson, J.; Singh, K. Plant-aphid interactions with a focus on legumes. *Funct. Plant Biol.* **2013**, *40*, 1271–1284. [CrossRef] [PubMed]
34. Sanchez-Arcos, C.; Kai, M.; Svatos, A.; Gershenzon, J.; Kunert, G. Untargeted metabolomics approach reveals differences in host plant chemistry before and after infestation with different pea aphid host races. *Front. Plant Sci.* **2019**, *10*, 188. [CrossRef] [PubMed]
35. Sharma, H. Bionomics, host plant resistance, and management of the legume pod borer, *Maruca Vitrata*—A review. *Crop Prot.* **1998**, *17*, 373–386. [CrossRef]
36. Pobozniak, M. The occurrence of thrips (Thysanoptera) on food legumes (Fabaceae). *J. Plant Dis. Prot.* **2011**, *118*, 185–193. [CrossRef]
37. Singh, C.; Singh, N. Occurrence of insect-pests infesting cowpea (*Vigna unguiculata* walpers) and their natural enemy complex in associations with weather variables. *Legume Res.* **2014**, *37*, 658–664. [CrossRef]
38. Kalyan, R.; Ameta, O. Impact of abiotic factors on seasonal incidence of insect pests of soybean. *Legume Res.* **2017**, *40*, 762–767. [CrossRef]
39. Rajawat, I.; Kumar, A.; Alam, M.; Tiwari, R.; Pandey, A. Insect pests of black gram (*Vigna mungo* L.) and their management in Vindhya region. *Legume Res.* **2021**, *44*, 225–232. [CrossRef]
40. Ahmed, K.; Awan, M. Integrated management of insect pests of chickpea *Cicer arietinum* (L. Walp) in South Asian countries: Present status and future strategies—A review. *Pak. J. Zool.* **2013**, *45*, 1125–1145.
41. Abudulai, M.; Kusi, F.; Seini, S.; Seidu, A.; Nboyine, J.; Larbi, A. Effects of planting date, cultivar and insecticide spray application for the management of insect pests of cowpea in Northern Ghana. *Crop Prot.* **2017**, *100*, 168–176. [CrossRef]
42. Mofokeng, M.; Gerrano, A. Efforts in breeding cowpea for aphid resistance: A review. *Acta Agric. Scand. Sect. B—Soil Plant Sci.* **2021**, *71*, 489–497. [CrossRef]
43. Kooner, B.; Malhi, B.; Cheema, H. *Insect Pest Management of Mungbean*; Shanmugasundaram, S., Ed.; Springer: Berlin/Heidelberg, Germany, 2004; pp. 214–235.
44. Edwards, O.; Singh, K. Resistance to insect pests: What do legumes have to offer? *Euphytica* **2006**, *147*, 273–285. [CrossRef]
45. Van Dam, N.; van der Meijden, E. A role for metabolomics in plant ecology. In *Biology of Plant Metabolomics*; Hall, R., Ed.; Wiley: Hoboken, NJ, USA, 2011; Volume 43, pp. 87–107. ISBN 1460-1494.
46. Wang, Y.; Liu, Q.; Du, L.; Hallerman, E.; Li, Y. Transcriptomic and metabolomic responses of rice plants to *Cnaphalocrocis medinalis* caterpillar infestation. *Insects* **2020**, *11*, 705. [CrossRef] [PubMed]
47. Kusnierczyk, A.; Winge, P.; Jorstad, T.; Troczynska, J.; Rossiter, J.; Bones, A. Towards global understanding of plant defence against aphids—Timing and dynamics of early arabidopsis defence responses to cabbage aphid (brevicoryne brassicae) attack. *Plant Cell Environ.* **2008**, *31*, 1097–1115. [CrossRef]
48. Zhang, Z.; Chen, Q.; Tan, Y.; Shuang, S.; Dai, R.; Jiang, X.; Temuer, B. Combined transcriptome and metabolome analysis of alfalfa response to thrips infection. *Genes* **2021**, *12*, 1967. [CrossRef]
49. Rubiales, D.; Fondevilla, S.; Chen, W.; Gentzbittel, L.; Higgins, T.J.V.; Castillejo, M.A.; Singh, K.B.; Rispail, N. Achievements and challenges in legume breeding for pest and disease resistance. *Crit. Rev. Plant Sci.* **2015**, *34*, 195–236. [CrossRef]
50. Mgbechi-Ezeri, J.; Porter, L.; Johnson, K.; Oraguzie, N. Assessment of sweet cherry (*Prunus avium L.*) genotypes for response to bacterial canker disease. *Euphytica* **2017**, *213*, 145. [CrossRef]
51. Obasa, K.; Haynes, L. Two new bacterial pathogens of peanut, causing early seedling decline disease, identified in the Texas Panhandle. *Plant Dis.* **2022**, *106*, 648–653. [CrossRef]

52. Chen, N.; Ruh, M.; Darrasse, A.; Foucher, J.; Briand, M.; Costa, J.; Studholme, D.; Jacques, M. Common bacterial blight of bean: A model of seed transmission and pathological convergence. *Mol. Plant Pathol.* **2021**, *22*, 1464–1480. [CrossRef]
53. Tugume, J.; Tusiime, G.; Sekamate, A.; Buruchara, R.; Mukankusi, C. Diversity and interaction of common bacterial blight disease-causing bacteria (*Xanthomonas* spp.) with *Phaseolus vulgaris* L. *Crop J.* **2019**, *7*, 1–7. [CrossRef]
54. Osdaghi, E.; Young, A.; Harveson, R. Bacterial wilt of dry beans caused by *Curtobacterium flaccumfaciens* pv. *flaccumfaciens*: A new threat from an old enemy. *Mol. Plant Pathol.* **2020**, *21*, 605–621. [CrossRef] [PubMed]
55. Vaghefi, N.; Adorada, D.; Huth, L.; Kelly, L.; Poudel, B.; Young, A.; Sparks, A. Whole-genome data from *Curtobacterium flaccumfaciens* pv. *flaccumfaciens* strains associated with tan spot of mungbean and soybean reveal diverse plasmid profiles. *Mol. Plant. Microbe Interact.* **2021**, *34*, 1216–1222. [CrossRef] [PubMed]
56. Nair, R.; Pandey, A.; War, A.; Hanumantharao, B.; Shwe, T.; Alam, A.; Pratap, A.; Malik, S.; Karimi, R.; Mbeyagala, E. Biotic and abiotic constraints in mungbean production-progress in genetic improvement. *Front. Plant Sci.* **2019**, *10*, 1340. [CrossRef] [PubMed]
57. Gonzalez, A.; Yuste-Lisbona, F.; Fernandez-Lozano, A.; Lozano, R.; Santalla, M. Genetic mapping and QTL analysis in common bean. In *Common Bean Genome*; DeLaVega, M., Santalla, M., Marsolais, F., Eds.; Springer: Cham, Switzerland, 2017; pp. 69–107. ISBN 2199-4781.
58. Padhi, E.; Maharaj, N.; Lin, S.; Mishchuk, D.; Chin, E.; Godfrey, K.; Foster, E.; Polek, M.; Leveau, J.; Slupsky, C. Metabolome and microbiome signatures in the roots of citrus affected by huanglongbing. *Phytopathology* **2019**, *109*, 2022–2032. [CrossRef]
59. Andam, A.; Azizi, A.; Majdi, M.; Abdolahzadeh, J. Comparative expression profile of some putative resistance genes of chickpea genotypes in response to ascomycete fungus, *Ascochyta rabiei* (Pass.) Labr. *Braz. J. Bot.* **2020**, *43*, 123–130. [CrossRef]
60. Gerlin, L.; Escourrou, A.; Cassan, C.; Macia, F.; Peeters, N.; Genin, S.; Baroukh, C. Unravelling physiological signatures of tomato bacterial wilt and xylem metabolites exploited by *Ralstonia solanacearum*. *Environ. Microbiol.* **2021**, *23*, 5962–5978. [CrossRef]
61. Pandey, A.; Burlakoti, R.; Kenyon, L.; Nair, R. Perspectives and challenges for sustainable management of fungal diseases of mungbean [*Vigna radiata* (L.) R. wilczek var. radiata]: A review. *Front. Environ. Sci.* **2018**, *6*, 53. [CrossRef]
62. Kowalska, B. Management of the soil-borne fungal pathogen—*Verticillium dahliae* Kleb. causing vascular wilt diseases. *J. Plant Pathol.* **2021**, *103*, 1185–1194. [CrossRef]
63. Anderson, J.; Lichtenzveig, J.; Oliver, R.; Singh, K. *Medicago truncatula* as a model host for studying legume infecting *Rhizoctonia solani* and identification of a locus affecting resistance to root canker. *Plant Pathol.* **2013**, *62*, 908–921. [CrossRef]
64. Infantino, A.; Kharrat, M.; Riccioni, L.; Coyne, C.; McPhee, K.; Grunwald, N. Screening techniques and sources of resistance to root diseases in cool season food legumes. *Euphytica* **2006**, *147*, 201–221. [CrossRef]
65. Marzouk, T.; Chaouachi, M.; Sharma, A.; Jallouli, S.; Mhamdi, R.; Kaushik, N.; Djebali, N. Biocontrol of *Rhizoctonia solani* using volatile organic compounds of solanaceae seed-borne endophytic bacteria. *Postharvest Biol. Technol.* **2021**, *181*, 111655. [CrossRef]
66. Wang, H.; Chang, K.; Hwang, S.; Gossen, B.; Turnbull, G.; Howard, R.; Strelkov, S. Response of lentil cultivars to *Rhizoctonia* seedling diseases in Canada. *J. Plant Dis. Prot.* **2006**, *113*, 219–223. [CrossRef]
67. Wolfgang, A.; Taffner, J.; Guimarães, R.A.; Coyne, D.; Berg, G. Novel strategies for soil-borne diseases: Exploiting the microbiome and volatile-based mechanisms toward controlling *Meloidogyne*-based disease complexes. *Front. Microbiol.* **2019**, *10*, 1296. [CrossRef] [PubMed]
68. Willsey, T.; Chatterton, S.; Heynen, M.; Erickson, A. Detection of interactions between the pea root rot pathogens *Aphanomyces euteiches* and *Fusarium* spp. using a multiplex QPCR assay. *Plant Pathol.* **2018**, *67*, 1912–1923. [CrossRef]
69. Diaz, L.; Arredondo, V.; Ariza-Suarez, D.; Aparicio, J.; Buendia, H.; Cajiao, C.; Mosquera, G.; Beebe, S.; Mukankusi, C.; Raatz, B. Genetic analyses and genomic predictions of root rot resistance in common bean across trials and populations. *Front. Plant Sci.* **2021**, *12*, 629221. [CrossRef] [PubMed]
70. Gaulin, E.; Jacquet, C.; Bottin, A.; Dumas, B. Root rot disease of legumes caused by *Aphanomyces euteiches*. *Mol. Plant Pathol.* **2007**, *8*, 539–548. [CrossRef]
71. Wu, L.; Chang, K.; Conner, R.; Strelkov, S.; Fredua-Agyeman, R.; Hwang, S.; Feindel, D. *Aphanomyces euteiches*: A threat to Canadian field pea production. *Engineering* **2018**, *4*, 542–551. [CrossRef]
72. Smolinska, U.; Kowalska, B. Biological control of the soil-borne fungal pathogen *Sclerotinia sclerotiorum*—A review. *J. Plant Pathol.* **2018**, *100*, 1–12. [CrossRef]
73. You, M.; Lamichhane, J.; Aubertot, J.; Barbetti, M. Understanding why effective fungicides against individual soilborne pathogens are ineffective with soilborne pathogen complexes. *Plant Dis.* **2020**, *104*, 904–920. [CrossRef]
74. Bilkiss, M.; Shiddiky, M.J.A.; Ford, R. Advanced diagnostic approaches for necrotrophic fungal pathogens of temperate legumes with a focus on *Botrytis* spp. *Front. Microbiol.* **2019**, *10*, 1889. [CrossRef]
75. Kisiel, A. Biological control as an alternative method of protecting crops against fungal pathogens. *Rocz. Ochr. Srodowiska* **2019**, *21*, 1366–1377.
76. Chandler, D.; Bailey, A.; Tatchell, G.; Davidson, G.; Greaves, J.; Grant, W. The development, regulation and use of biopesticides for integrated pest management. *Philos. Trans. R. Soc. B—Biol. Sci.* **2011**, *366*, 1987–1998. [CrossRef] [PubMed]
77. Quiroz, R.; Maldonado, J.; Alanis, M.; Torres, J.; Saldivar, R. Fungi-based biopesticides: Shelf-life preservation technologies used in commercial products. *J. Pest Sci.* **2019**, *92*, 1003–1015. [CrossRef]
78. Harveson, R. Improving yields and managing dry bean bacterial diseases in Nebraska with new copper-alternative chemicals. *Plant Health Prog.* **2019**, *20*, 14–19. [CrossRef]

79. Chen, L.; Wu, Q.; He, T.; Lan, J.; Ding, L.; Liu, T.; Wu, Q.; Pan, Y.; Chen, T. Transcriptomic and metabolomic changes triggered by *Fusarium solani* in common bean (*Phaseolus vulgaris* L.). *Genes* **2020**, *11*, 177. [CrossRef]
80. Mayo-Prieto, S.; Marra, R.; Vinale, F.; Rodríguez-González, Á.; Woo, S.L.; Lorito, M.; Gutiérrez, S.; Casquero, P.A. Effect of *Trichoderma velutinum* and *Rhizoctonia solani* on the metabolome of bean plants (*Phaseolus vulgaris* L.). *Int. J. Mol. Sci.* **2019**, *20*, 549. [CrossRef]
81. Copley, T.; Aliferis, K.; Kliebenstein, D.; Jabaji, S. An integrated RNAseq-H-1 NMR metabolomics approach to understand soybean primary metabolism regulation in response to *Rhizoctonia* foliar blight disease. *BMC Plant Biol.* **2017**, *17*, 84. [CrossRef]
82. Turetschek, R.; Desalegn, G.; Epple, T.; Kaul, H.; Wienkoop, S. Key metabolic traits of *Pisum sativum* maintain cell vitality during *Didymella pinodes* infection: Cultivar resistance and the microsymbionts' influence. *J. Proteom.* **2017**, *169*, 189–201. [CrossRef]
83. Pérez-Clemente, R.M.; Montoliu, A.; Vives-Peris, V.; Arbona, V.; Gómez-Cadenas, A. Hormonal and metabolic responses of mexican lime plants to CTV infection. *J. Plant Physiol.* **2019**, *238*, 40–52. [CrossRef]
84. Srivastava, S.; Bisht, H.; Sidhu, O.; Srivastava, A.; Singh, P.; Pandey, R.; Raj, S.; Roy, R.; Nautiyal, C. Changes in the metabolome and histopathology of *Amaranthus hypochondriacus L.* in response to *Ageratum enation* Virus Infection. *Phytochemistry* **2012**, *80*, 8–16. [CrossRef]
85. Mutuku, J.; Cui, S.; Hori, C.; Takeda, Y.; Tobimatsu, Y.; Nakabayashi, R.; Mori, T.; Saito, K.; Demura, T.; Umezawa, T. The structural integrity of lignin is crucial for resistance against *Striga hermonthica* parasitism in rice. *Plant Physiol.* **2019**, *179*, 1796–1809. [CrossRef] [PubMed]
86. Kang, W.; Zhu, X.; Wang, Y.; Chen, L.; Duan, Y. Transcriptomic and metabolomic analyses reveal that bacteria promote plant defense during infection of soybean cyst nematode in soybean. *BMC Plant Biol.* **2018**, *18*, 98. [CrossRef]
87. Narula, K.; Elagamey, E.; Abdellatef, M.; Sinha, A.; Ghosh, S.; Chakraborty, N.; Chakraborty, S. Chitosan-triggered immunity to *Fusarium* in chickpea is associated with changes in the plant extracellular matrix architecture, stomatal closure and remodeling of the plant metabolome and proteome. *Plant J.* **2020**, *103*, 561–583. [CrossRef] [PubMed]
88. Uchida, K.; Sawada, Y.; Ochiai, K.; Sato, M.; Inaba, J.; Hirai, M. Identification of a unique type of isoflavone O-methyltransferase, GmIOMT1, based on multi-Omics analysis of soybean under biotic stress. *Plant Cell Physiol.* **2020**, *61*, 1974–1985. [CrossRef] [PubMed]
89. Sistani, N.; Kaul, H.; Desalegn, G.; Wienkoop, S. Rhizobium impacts on seed productivity, quality, and protection of *Pisum sativum* upon disease stress caused by *Didymella pinodes*: Phenotypic, proteomic, and metabolomic traits. *Front. Plant Sci.* **2017**, *8*, 1961. [CrossRef] [PubMed]
90. Makkouk, K.M. Plant pathogens which threaten food security: Viruses of chickpea and other cool season legumes in West Asia and North Africa. *Food Secur.* **2020**, *12*, 495–502. [CrossRef]
91. Wood, C.; Pilkington, B.; Vaidya, P.; Biel, C.; Stinchcombel, J. Genetic conflict with a parasitic nematode disrupts the legume-rhizobia mutualism. *Evol. Lett.* **2018**, *2*, 233–245. [CrossRef] [PubMed]
92. Davis, E.L.; Mitchum, M.G. Nematodes. sophisticated parasites of legumes. *Plant Physiol.* **2005**, *137*, 1182–1188. [CrossRef] [PubMed]
93. Abo, M.; Alegbejo, M. Strategies for sustainable control of viral diseases of some crops in Nigeria. *J. Sustain. Agric.* **1997**, *10*, 57–80. [CrossRef]
94. Sauerborn, J.; Muller-Stover, D.; Hershenhorn, J. The role of biological control in managing parasitic weeds. *Crop Prot.* **2007**, *26*, 246–254. [CrossRef]
95. Rodenburg, J.; Demont, M.; Zwart, S.; Bastiaans, L. Parasitic weed incidence and related economic losses in rice in Africa. *Agric. Ecosyst. Environ.* **2016**, *235*, 306–317. [CrossRef]
96. Hooper, A.; Hassanali, A.; Chamberlain, K.; Khan, Z.; Pickett, J. New genetic opportunities from legume intercrops for controlling *Striga spp.* parasitic weeds. *Pest Manag. Sci.* **2009**, *65*, 546–552. [CrossRef] [PubMed]
97. Perez-de-Luque, A.; Flores, F.; Rubiales, D. Differences in crenate broomrape parasitism dynamics on three legume crops using a thermal time model. *Front. Plant Sci.* **2016**, *7*, 1910. [CrossRef] [PubMed]
98. Rubiales, D.; Flores, F.; Emeran, A.; Kharrat, M.; Amri, M.; Rojas-Molina, M.; Sillero, J. Identification and multi-environment validation of resistance against broomrapes (*Orobanche crenata* and *Orobanche foetida*) in faba bean (*Vicia faba*). *Field Crops Res.* **2014**, *166*, 58–65. [CrossRef]
99. Arora, P.; Yadav, R.; Dilbaghi, N.; Chaudhury, A. Biological nitrogen fixation: Host-rhizobium interaction. In *Frontiers on Recent Developments in Plant Science*; Goyal, A., Maheshwari, P., Eds.; Bentham Books: Sharjah, United Arab Emirates, 2012; Volume 1, pp. 39–59. ISBN 2589-1464.
100. Hassan, S.; Behm, C.; Mathesius, U. Effectors of plant parasitic nematodes that re-program root cell development. *Funct. Plant Biol.* **2010**, *37*, 933–942. [CrossRef]
101. Steinfath, M.; Strehmel, N.; Peters, R.; Schauer, N.; Groth, D.; Hummel, J.; Steup, M.; Selbig, J.; Kopka, J.; Geigenberger, P. Discovering plant metabolic biomarkers for phenotype prediction using an untargeted approach. *Plant Biotechnol. J.* **2010**, *8*, 900–911. [CrossRef]
102. Hirayama, A.; Wakayama, M.; Soga, T. Metabolome analysis based on capillary electrophoresis-mass spectrometry. *TrAC—Trends Anal. Chem.* **2014**, *61*, 215–222. [CrossRef]
103. Patel, M.; Pandey, S.; Kumar, M.; Haque, M.; Pal, S.; Yadav, N. Plants metabolome study: Emerging tools and techniques. *Plants* **2021**, *10*, 2409. [CrossRef]

104. Junot, C.; Fenaille, F.; Colsch, B.; Becher, F. High resolution mass spectrometry based techniques at the crossroads of metabolic pathways. *Mass Spectrom. Rev.* **2014**, *33*, 471–500. [CrossRef]
105. Sun, X.; Weckwerth, W. COVAIN: A toolbox for uni- and multivariate statistics, time-series and correlation network analysis and inverse estimation of the differential Jacobian from metabolomics covariance data. *Metabolomics* **2012**, *8*, 81–93. [CrossRef]
106. Ghatak, A.; Chaturvedi, P.; Weckwerth, W. Metabolomics in plant stress physiology. In *Plant Genetics and Molecular Biology*; Varshney, R., Pandey, M., Chitikineni, A., Eds.; Springer: Berlin/Heidelberg, Germany, 2018; Volume 164, pp. 187–236. ISBN 0724-6145.
107. Fukushima, A.; Kusano, M. Recent progress in the development of metabolome databases for plant systems biology. *Front. Plant Sci.* **2013**, *4*, 73. [CrossRef] [PubMed]
108. Sanchez, D.; Szymanski, J.; Erban, A.; Udvardi, M.; Kopka, J. Mining for robust transcriptional and metabolic responses to long-term salt stress: A case study on the model legume *Lotus japonicus*. *Plant Cell Environ.* **2010**, *33*, 468–480. [CrossRef] [PubMed]
109. Sanchez, D.; Schwabe, F.; Erban, A.; Udvardi, M.; Kopka, J. Comparative metabolomics of drought acclimation in model and forage legumes. *Plant Cell Environ.* **2012**, *35*, 136–149. [CrossRef]
110. Sanchez, D.; Pieckenstain, F.; Escaray, F.; Erban, A.; Kraemer, U.; Udvardi, M.; Kopka, J. Comparative ionomics and metabolomics in extremophile and glycophytic *Lotus* species under salt stress challenge the metabolic pre-adaptation hypothesis. *Plant Cell Environ.* **2011**, *34*, 605–617. [CrossRef] [PubMed]

plants

Perspective

Breeding Tools for Assessing and Improving Resistance and Limiting Mycotoxin Production by *Fusarium graminearum* in Wheat

Sandiswa Figlan [1,*] and Learnmore Mwadzingeni [2]

1 Department of Agriculture and Animal Health, Science Campus, University of South Africa, Corner Christiaan De Wet and Pioneer Avenue, Private Bag X6, Florida 1709, South Africa
2 Seed Co Limited, Rattray Arnold Research Station, Chisipite, Harare P.O. Box CH142, Zimbabwe; mwadzingenil@yahoo.com
* Correspondence: figlas@unisa.ac.za

Abstract: The recently adopted conservation and minimum tillage practices in wheat-production systems coupled with the concomitant warming of the Earth are believed to have caused the upsurges in Fusarium head blight (FHB) prevalence in major wheat-producing regions of the world. Measures to counter this effect include breeding for resistance to both initial infection of wheat and spread of the disease. Cases of mycotoxicosis caused by ingestion of wheat by-products contaminated with FHB mycotoxins have necessitated the need for resistant wheat cultivars that can limit mycotoxin production by the dominant causal pathogen, *Fusarium graminearum*. This manuscript reviews breeding tools for assessing and improving resistance as well as limiting mycotoxin contamination in wheat to reflect on the current state of affairs. Combining these aspects in wheat research and development promotes sustainable quality grain production and safeguards human and livestock health from mycotoxicosis.

Keywords: contamination; health; infection; molecular techniques; selection

Citation: Figlan, S.; Mwadzingeni, L. Breeding Tools for Assessing and Improving Resistance and Limiting Mycotoxin Production by *Fusarium graminearum* in Wheat. *Plants* **2022**, *11*, 1933. https://doi.org/10.3390/plants11151933

Academic Editor: Alessandro Vitale

Received: 30 May 2022
Accepted: 20 July 2022
Published: 26 July 2022

1. Introduction

Breeding wheat for Fusarium head blight (FHB) resistance involves systematic genetic manipulation of the crop to incorporate superior biochemical and morpho-physiological traits that safeguard it against the damaging effects of the dominant causal species, *Fusarium graminearum*. Infection of crops by *F. graminearum* does not only reduce yield, but also exposes the grain to contamination by mycotoxins. Mycotoxin contamination in grain crops intended for processing food, feed and beverages often results in the accumulation of these toxic fungal metabolites in foodstuffs, causing health hazards to both human beings and livestock. *F. graminearum* species complex infects grain crops including wheat, barley and maize. Breeding for resistance against FHB aims to reduce the impact of the pathogen on crop yield as well as mycotoxin contamination in infected grain. Various strategies for breeding against Fusarium head blight have been embarked on because resistance against the disease is multigenic and is further confounded by the large influence of genotype by environment interactions [1,2]. Resistance against FHB is conferred by more than 250 quantitative trait loci (QTL) distributed across the entire chromosome cascade of the wheat genome [3–5]. To effectively compart the negative effects of the disease, strong background knowledge is needed on various aspects including the importance of FHB as a grain disease, mycotoxin contamination of infected grain, breeding strategies to reduce mycotoxin contamination in grain as well as the tools used to assess and limit mycotoxin contamination during breeding, selection and the entire wheat value chain.

Fusarium head blight, also known as 'scab', is a wheat disease that is mainly caused by the fungal complex called *F. graminearum* Schwabe (teleomorph *Gibberella zeae* Schwein.

Petch). It is one of the most common diseases affecting bread wheat (*Triticum aestivum* L.) around the world. Epidemics of FHB occur in cycles of four or five years worldwide [6] and in shorter periods under favorable conditions, particularly where no-till or minimum tillage practices, high humidity and/or high temperature coincide with early flowering to the soft dough stages of susceptible wheat cultivars. In addition to the enormous grain yield losses, *F. graminearum* infection is associated with the accumulation of mycotoxins that put the health of human beings and livestock consuming infected grain at risk [7]. Ingestion of huge amounts of mycotoxin contaminated grain may lead to mycotoxicosis, which under severe circumstances may cause death. It is important to note that there is very high genetic variability of *F. graminearum* species, which results in high resilience and complicates efforts towards breeding for FHB resistance due to genotype by isolate and isolate by environment interactions [8].

F. graminearum produces two groups of toxins, namely zearalenone and trichothecenes. Zearalenone (previously referred to as F-2 toxin) is one of the most prevalent estrogenic mycotoxins produced through the polyketide pathway. This mycotoxin is denoted as 6-[10-hydroxy-6-oxo-*trans*-1-undecenyl]-B-resorcyclic acid lactone. Zearalenone derives its name from *Gibberella zea* and, resorcyclic acid lactone because of the C-1′ to C-2 'double bonds. The '-one' denotes its ketone group [9]. The toxicity of zearalenone is through binding to estrogen receptors ending up in estrogenicity, occasionally causing hyperestrogenism in livestock and human beings, especially women. Eventually, the toxicity of zearalenone may lead to myelofibrosis, reproductive system disorders, cancers, skeletal malformations and weakening [10], nervous disorders [11] and various other physiological malfunctions. Trichothecenes, on the other hand, are chemically tricyclic sesquiterpenes, which have double bonds at the C-9, 10 position and a C-12, 13 epoxy functional group. The most common contaminants of cereals are type-A and type-B trichothecenes [12,13]. Type-A trichothecenes are different from type-B by the absence of a carbonyl group at C-8 and hydroxylation at C-7. Type-A mycotoxins include diacetoxyscirpenol, T-2 and HT-2 toxins while type-B trichothecenes include fusarenone-X, nivalenol and deoxynivalenol. The effects of trichothecene ingestion through contaminated foodstuffs by animals and human beings include diarrhea, vomiting and death when the toxicosis is severe.

Various national and multi-national organisations have drafted guidelines on food safety to ensure that consumers are safe from the risks of eating contaminated food. The regulating bodies include the European Food Safety Authority, Codex Alimentarius and the USA Food and Drug Administration. Realizing that it is not possible to produce mycotoxin-free wheat grain, the regulating bodies have set threshold limits, which are practically attainable to reduce the incidence of mycotoxins in wheat products and other foodstuffs. The threshold regulations mainly protect the health of animals and human beings from the dangers caused by mycotoxins. Contamination of wheat with mycotoxins occurs during infection by mycotoxin, producing fungi such as *F. graminearum*, and further toxin accumulation may occur postharvest during grain storage [14–17]. Various interventions are necessary to limit mycotoxin contamination of the wheat grain by *F. graminearum* mycotoxins.

FHB can be managed using various strategies including cultural, biological and chemical control methods as well as breeding for resistance against the disease. Wheat production has thrived for ages through selection for superior traits and painstaking efforts to incorporate disease resistance. With increased efforts to incorporate FHB resistance into wheat, disease incidence and the spread of infection decrease, resulting in a subsequent reduction in mycotoxin contamination. Moreover, resistance may be specific to reduce mycotoxin production by the infecting *F. graminearum*. Various breeding strategies are being embarked upon to ensure minimal mycotoxin contamination of wheat grain. It is also important to develop laboratory tools to assist the selection of wheat varieties that suppress mycotoxin production as well as to ensure compliance with wheat grain safety standards. This review discusses these aspects beginning with various breeding strategies employed against FHB. Emphasis has been put on traditional breeding strategies, new techniques of resistance breeding and tools for monitoring mycotoxin levels in the harvested wheat grain.

2. Resistance against Fusarium Head Blight in Wheat

Resistance to FHB is categorised into various types of which the most prominent ones are type I and type II [18]. Type I refers to the resistance against initial infection and is exhibited by the ability of the cultivar to create a barrier to initial entry of the pathogen into the plant. On the other hand, Type II resistance is resistance to the spread of the pathogen after it has gained entry into the plant. The later type of resistance is more stable. Type I and II resistance can be tested under both field and artificial environments [19]. Usually, screening for resistance against FHB takes place in the advanced generations like F_4 onwards [20]. Select breeding lines are chosen and are artificially inoculated with the pathogen isolate(s)/races(s) to screen for resistance [21]. Assessment of resistance to FHB is done through generally visualizing discolouration of the spikes and by precisely assessing the intensity and number of affected grains. Affected grain may have a pinkish discolouration, sometimes with a chalky appearance. Assessment covers both the proportion of kernels that are diseased and the level of mycotoxins in the affected grain [22]. Resistance against mycotoxin accumulation is called type III resistance, which requires special tools for assessment, unlike type I and II which can be assessed visually. Both type I and type II resistance have indirect effects on toxin accumulation, but resistance to toxin accumulation, type III resistance, still has to be a targeted breeding objective on its own. Generally, genotypes to be used as donors of resistance in FHB breeding programmes and ultimate varieties must (1) resist initial infection (type I), (2) limit the pathogen spread in infected spikes (type II), (3) reduce mycotoxin accumulation in the grain (type III)), (4) resist kernel damage (type IV) and (5) tolerate the presence of the disease without much yield penalty (type V) [19]. Knowledge of the genetic basis underlying these observable types of resistance is slowly being demystified through advanced biotechnology and genetics.

3. Breeding Focus against Fusarium Head Blight

With the development of settlements for human beings and crop domestication, early farmers selected plants that had desirable traits and the resulting gene pool formed the basis of today's domesticated crops. Natural selection for superior agronomic traits was accelerated by the active mating and selection of offspring with desirable traits. Crops progressively improved, hence, huge monoculture practices were established to what has become modern agriculture. Wheat is one of the crops that has been extensively bred over the years leading, notably, to the Green Revolution of the 1960s. After a prolonged period of painstaking breeding efforts, Dr. Norman Borlaug, the Father of the Green Revolution, developed high yielding wheat varieties in India and Pakistan, a move that averted massive hunger. Despite this milestone, various diseases continue to threaten the crop, particularly wheat rusts and Fusarium head blight. Breeding for disease resistance continued to protect yields of high yielding varieties, among other control strategies. The wheat disease resistance breeding strategy at the International Centre for Maize and Wheat Improvement (CIMMYT) systematically grouped breeding needs of various regions in the world into mega-environments [23]. Breeding for resistance against FHB falls within the needs of mega-environment 2, which is characterized by high rainfall. China has been a significant source of resistance to FHB and hundreds of wheat lines carrying resistance have been shared with CIMMYT. Among the Chinese lines that carry FHB resistance are Sumai#3, Shanghai#5, Suzhoe#6, Yangmai#6, Wuhan#3 Ning 7840, and Chuanmai 18, which have been developed using traditional breeding methods. Genes for resistance against FHB are mostly additive, requiring a meticulous programme for resistance incorporation and selection [24].

Genetic variation for FHB resistance breeding is large. Therefore, there is a wide pool of sources of resistance. This makes it easy for resistance to be incorporated into wheat with options from exotic and native sources. However, Asian sources of resistance against FHB such as the Chinese spring wheat, Sumai#3, are prominently used worldwide. Resistance to FHB is mostly additive, being controlled by the effects of multiple genes. Quantitative trait loci controlling FHB across all 21 bread wheat chromosomes have been mapped and identified, with just a few validated and used in breeding [4,5,25]. These QTL are prevalent in Chinese genotypes derived from Sumai#3 and they contain *Fhb1*, *Fhb2*, as well as *Qfhs.ifa-5A* [26–32]. Nevertheless, other resistance QTL do exist outside of Sumai#3. The presence of *Fhb1* (Sumai#3) and *Qfhs.nau-2DL* (breeding line CJ9306), which confer resistance to both type II and type III resistance, are of particular interest. *Fhb1* improves the detoxification of deoxynivalenol (DON) to DON-3-glucoside [33]. *Qfhs.ifa-5A* confers type III resistance by suppressing mycotoxin accumulation. Although resistance to FHB acquired from sources such as Sumai#3 has been useful, its use has been moderate and therefore new sources of resistance are desperately needed, especially resistance to curb toxin accumulation in wheat infected with *F. graminearum*. The current shortfalls in breeding for resistance against FHB therefore require radical use of new technologies. These technologies will help to improve wheat productivity to meet the needs of the growing global population.

Wheat breeding programs against FHB also aim to reduce mycotoxin production by the infecting fungus *F. graminearum*. From a food safety concern, this is an important breeding objective to ensure that harvested grain is strictly below the mycotoxin threshold level. To breed for resistance against FHB, a reliable inoculation method is needed. This allows repeatable assessment of resistance to ensure selection of resistant lines under high and uniform disease pressure. It is also important to use a cocktail of isolates/races for inoculation to ensure selection for broad-spectrum or multi-race resistance, preferably using races prevalent in the area where the resistant cultivars will be released. Isolates that produce higher levels of DON, a type-B trichothecene, are found to be more aggressive and could be useful for effective selection for type III resistance [34–39]. Resistance of wheat to DON accumulation is acquired through the ability of the plant to degrade the mycotoxin, for example, the possession of a putative deoxynivalenol-glycosyl transferase that detoxifies DON [33,40]. Newer strategies for resistance breeding have been adopted over the years and progress has been made ever since the adoption of these technologies. Breeding programs that aim to limit DON production by *F. graminearum* in wheat have greatly benefited from these new technologies.

4. Traditional Crop Breeding against Fusarium Head Blight

Conventional breeding is a systematic hybridization and selection strategy aimed to release superior genotypes. In certain instances, the trait of interest is transferred from a wild relative of the crop to be improved and this is termed wide crossing. Breeding for disease resistance often takes a different strategy from conventional breeding for complex agronomic traits such as yield. There has to be a source of resistance, which donates the resistance gene/genes to the recipient genotype containing most of the desirable agronomic traits, except for the resistant gene(s) of interest. In such a scenario, backcross breeding, which is the most prominent classical breeding technique against plant diseases, is used to recover most of the recipient genotype's genome. In certain instances, the resistance incorporated into a cultivar against FHB may be race-specific, though in most cases it is race non-specific. It is always important to adopt a clear resistance breeding strategy so that broad-spectrum and durable resistance may be incorporated into the cultivar. When using traditional breeding techniques, it is critical to select effectively in the early generations for FHB resistance; otherwise the promising gene combinations are lost irretrievably [41]. Thus, the selection efficiency increases when the breeding method can be used to select successfully in the early generations of selection [41]. Following the vast research investments that were put towards FHB resistance, backcross breeding is no longer sorely classical but is now fused with various molecular marker techniques for effective and timely selection as well as gene and QTL introgression.

5. Molecular Breeding Techniques

The use of resistant cultivars remains a valuable tool for the control of FHB. It therefore remains imperative to intensify breeding efforts and optimize breeding and selection strategies for resistance against FHB and mycotoxin production. The development and improvement, in recent years, of molecular techniques like real-time polymerase chain reaction (PCR), marker-assisted selection, marker-assisted QTL backcrossing, next generation sequencing technologies and genetic engineering, are boosting research on FHB resistance and its associated mycotoxicosis. Screening for resistance against FHB usually takes place in advanced generations like F_4 onwards when select breeding lines are chosen and artificially inoculated with the pathogen to screen for resistance [42]. This task is very laborious and requires time for completion. In this case, advanced molecular techniques are required to monitor levels of inoculation, to select for resistance in genotypes to be used as parents in breeding for resistance to FHB and to introgress resistance genes into elite genotypes. These molecular tools are therefore useful in wheat pre-breeding and breeding against FHB.

5.1. RNA Interference to Reduce Mycotoxin Contamination in Fusarium graminearum Infected Wheat

The discovery of more sophisticated biotechnological approaches such as ribonucleic acid (RNA) interference (RNAi) offers new transformation opportunities to enhance resistance against *F. graminearum* and other invading wheat pathogens [43]. This is achieved through induced silencing of target virulent genes. RNA interference is an essential cellular system involved in gene regulation and protection of eukaryotes against infection by viruses [44]. It is an important systematic mechanism that can be employed to fight mycotoxigenic plant pathogenic fungi like *F. graminearum*. RNAi post-transcriptionally converts double stranded RNA molecules into short-stranded RNA duplexes of about 21 to 28 nucleotides often termed short interfering RNAs (siRNAs), which then cleaves to complimentary mRNA, effecting gene silencing or regulation [45–48]. RNA interference pathways are often triggered by the presence of viral RNAs providing gene regulated defense against specific RNA viruses. In this case, the mechanism will be termed virus-induced gene silencing (VIGS), whose success is highly dependent on designing effective vectors that will produce complementary siRNA species, efficient uptake of siRNAs by the fungus and amplification of the silencing effect within the target organism [43]. Silencing of target genes has recently been proved to be effective against plant pathogenic fungi [49] and has been demonstrated on *Puccinia* in wheat among other crop species and their respective fungal pathogens. Machado et al. [50] reviewed the recent advances in RNAi-mediated FHB control and suppression of mycotoxin contamination in a number of cereals. This involves the use of the barley stripe mosaic virus (BSMV) vector. *P. striiformis* genes were also observed to be silenced using the host-induced RNA interference mechanism [51]. In a more recent study, Cheng et al. [52] reported that wheat resistance against pathogenic fungi can be improved through RNAi sequences originating from chitin synthase (Chs) 3b gene originating from *F. graminearum*. These sequences are used for host-induced silencing of the chitin synthase gene in plant pathogenic fungi. This is one of the techniques that holds future promise for the incorporation of resistance against *F. graminearum* in wheat.

5.2. Gene Transfer in General and Specifically against Fusarium Head Blight

Gene transfer technologies that insert foreign genes in plants are another molecular breeding strategy with potential to enhance wheat resistance to FHB [53]. These technologies include particle bombardment or biolistic transformation and *Agrobacterium*-mediated genetic transformation [54]. The former bombards deoxyribonucleic acid (DNA)-coated gold or tungsten micro-projectiles into the target crop's genome using a particle gun, thereby inserting foreign genes. The later technique uses *A. tumefaciens* as a vector that copies and transfers the transfer DNA (T-DNA) molecules on a tumour-inducing (Ti) plasmid into the nucleus of target plant cells, thereby incorporating foreign DNA that is eventually inserted and becomes part of the plant genome. *Agrobacterium* transformation, however, works effectively with selected plant species, and inserts mostly three genes, including two T-DNA molecules and a selectable marker per transformation construct [55]. Biolistic transformation non-randomly targets AT-rich regions with matrix attachment region (MAR) motifs that are nuclear matrix prone eukaryotic DNA elements [56,57]. The

MARs create open chromatin, allowing the host plant genome to be accessible to transgenes. An advantage shared by both *Agrobacterium* transformation and biolistic transformation is that they can integrate two trans-genes into the target host genome [58].

The *Agrobacterium*-mediated transformation stages involve initiation, which includes identification, isolation and insertion of the gene of interest into a suitable functional construct consisting of the gene expression promoter, gene of interest, selectable marker and codon modification. This is followed by *Agrobacterium*-mediated transformation or bacterium-to-plant transfer and finally nucleus targeting [59–61]. During gene transfer within the plant cell, the transformed *Agrobacterium* facilitates the transfer of T-DNA molecules into the plant genome, then the transgene is randomly incorporated into the plant chromosome. Integration of T-DNA into the plant DNA sequence is then facilitated by non-homologous end-joinings.

Transfer of foreign genes that enhance FHB resistance into wheat is a viable alternative which has, in recent years, been used extensively to increase not only the crops' genomic variability, but also the fitness of wheat against *F. graminearum*. Among first genes to be transferred since 1992 was the *Bar* gene used as a selective marker and various others including the *TaPIMP1* gene [62], the *Yr10* gene [63] and the *TcLr19PR1* gene [64]. Various genes that encode pathogenicity related proteins (PR proteins) could be the new sources of wheat resistance against FHB. These PR proteins are defensins, which have a broad range of antifungal properties [65]. Defensin $RsAFP_2$ with growth inhibitory characteristics against *F. graminearum* was incorporated into variety Yangmai 12 using biolistic particle bombardment [66]. The success of the transformation was confirmed using PCR and Southern blot analysis. Expression of the $RsAFP_2$ genes in transformed wheat lines was confirmed using RT-PCR and Western blotting. Disease resistance was assessed, and the transformed lines showed resistance against *F. graminearum* compared to the untransformed control lines [66]. The low transformation efficiency using the biolistic particle bombardment, however, warrants the need for other gene transformation techniques alongside. *Agrobacterium*-mediated transformation is one such technique that has been used successfully to introduce foreign genes into the wheat plant with improved transformation efficiency.

In one effort, chitinase and #beta#-1,3-glucosanase genes were transformed into wheat to improve resistance against FHB. The transformation of chitinase and #beta#-1,3-glucosanase genes (constructed into binary vector pCAMBIA3301) was mediated by *Agrobacterium* and the resultant transgenic lines showed resistance against FHB in the field [67]. Transformation of plant cells with exotic genes mediated with *Agrobacterium* is the initial step in introducing genes into plant cells that generate into adult plants capable of producing normal seeds. However, this process is difficult with wheat because of its complex hexaploid genome. Therefore, a more efficient protocol for wheat transformation called, 'Pure Wheat', was introduced [68]. This technique has renewed hope in accelerating transgenic wheat plants with superior traits such as FHB resistance and its associated ability to limit mycotoxin production.

5.3. Genome Editing for FHB Resistance

Major improvements in wheat will likely be brought about by genome editing, which promises to supersede the traditional random mutagenesis and conventional breeding. Genome editing technologies include the clustered regularly interspaced short palindromic repeat-associated endonucleases (CRISPR/Cas) technique, which is gaining much popularity, and other sequence-specific nucleases (SSNs) such as the transcription activator-like effector nucleases (TALENs) and zinc-finger nucleases (ZFNs). These technologies offer the benefits of gene knock-out, knock-in, replacement, activation and DNA repair [69–72]. Among these genome editing technologies, the CRISPR/Cas technology seems to hold more promise with regards to FHB resistance. The Cas nuclease system has been used with success in understanding fungal biology, with various reports in *Neurospora crassa* [73], *Aspergillus* spp. [74,75], *Penicillium chrysogenum* [76], *Alternaria alternata* [77], *Pyricularia oryzae* [78] and *Ustilago maydis* [79]. Following on these milestones, a Cas9-based genome editing system was established in *F. graminearum* [80] and hopefully this study will generate leads to a breakthrough in *F. graminearum* control.

Several research groups have made concerted efforts to develop transgenic and mutagenic lines that confer resistance to FHB. Table 1 summarizes some of the genes that have been manipulated in wheat, barley, *Brachypodium* and *Arabidopsis* that were manipulated through advanced technologies and proved to confer reduced *F. graminearum* infection and DON accumulation. However, much effort is still needed to link the various research institutions with public and private seed companies to ensure that research and development are aimed at variety release to benefit farming communities in FHB prone areas. This effort should involve pre-commercial field-testing activities including multi-environmental trials and end-use quality analysis.

Table 1. Genetic transformation to enhance resistance against *Fusarium graminearum* causing Fusarium head blight in wheat and other cereals.

Crop	Technology	Gene Involved	Effect on Transformed Line	Reference
Wheat	Gene silencing	Chitin synthase *ChS3B*	Enhanced combined type I and II resistance against FHB by targeting chitin biosynthesis	[52]
Barley	Gene silencing	*FgCYP51A* and *FgCYP51B*	Reduced fungal growth by targeting Sterol biosynthesis	[81]
Brachypodium distachyon	Gene silencing	*Fg00677*, *Fg08731* and *CYP51*	Improved FHB resistance by silencing the genes through inhibiting *CYP51A*, *CYP51B*, and *CYP51C* genes and essential protein kinase biosynthesis	[82]
Wheat	Deletion mutation	*TaHRC*	Enhanced FHB resistance by silencing the gene that encodes a nuclear protein conferring FHB susceptibility	[83]
Wheat	Overexpression	*Barley HvUGT13248*	Decreased DON content in flour by increasing detoxification	[84]
Arabidopsis	Gene silencing	*CYP51*	Restricted fungal infection and reduced virulence by targeting Ergosterol biosynthesis	[85]
Barley	Gene silencing	*CYP51*	Restricted fungal infection and reduced virulence by targeting Ergosterol biosynthesis	[85]
Wheat	Overexpression	*AtLTP4.4*	Supressed DON induced reactive oxygen species and plant stress from infection	[86]
Wheat	Gene silencing	*FgSGE1*, *FgSTE12* and *FgPP1*	Reduced infection and DON accumulation by targeting DON biosynthesis, fungal penetration structure formation and essential phosphatase	[87]
Wheat	Epigenetic regulation of gene expression	Several	Reduced FHB severity and DON accumulation through methylation	[88]
Wheat	Trans-gene expression	*Bradi5g03300 UGT*	Conferred resistance both to initial infection and to spike colonization and reduce mycotoxin content	[89]

Table 1. *Cont.*

Crop	Technology	Gene Involved	Effect on Transformed Line	Reference
Wheat	Mutation/Deleation involving the 3′ exon	histidine-rich calcium-binding-protein gene	Resistance to FHB spread	[90]
Wheat	Trans-gene expresion	*HvUGT13248*	Increased resistance against *Fusarium graminearum*	[91]

5.4. Association Mapping to Find FHB Molecular Markers

Molecular breeding and selection for FHB resistance in wheat have largely benefited from association mapping of putative QTL through associating phenotypic reactions to genotypes. Currently, high-density wheat 90 K single nucleotide polymorphism (SNP) assays are being used in genome-wide association (GWAS) studies aimed to dissect the genetic basis of resistance to Fusarium head blight in wheat breeding populations [92]. Association mapping studies have enabled the discovery of several loci associated to the resistance to FHB spread and DON accumulation. Alternative to the GWAS approach, candidate-gene association mapping can be used by targeting associations of pre-specified FHB resistance genes and the observed phenotypic reaction [93]. A recent GWAS study identified 16 significant SNPs associated with Fusarium-damaged kernels and DON levels on wheat chromosomes and suggested that FHB severity can even be reduced by small-effect QTL [94]. Such studies form the basis of maker-assisted selection and marker-based gene and/or QTL introgression by identifying putative markers linked to genetic regions controlling particular traits. Quality phenotypic data, often with high heritability from multi-environmental trials, is required for effective association studies.

All these advanced technologies that can be employed to enhance FHB resistance have their own advantages and disadvantages when compared to traditional breeding methods. Table 2 highlights some of these pros and cons to guide future research. Generally, this indicates that the recent technologies can not completely be divorced from all aspects of traditional breeding, particularly phenotyping or field testing to account for the expression of introduced genes under real production conditions and assessing the ultimate impact on final yield.

Table 2. Pros and cons of using traditional breeding methods against using recent technologies.

Aspect	Traditional Methods	Recent Technologies
	Pros	**Cons**
Field expression of genes	Reliably confirmed each season	Gene may be present but not expressed as desired in the field [41]
Variety release	Often targeted towards FHB resistant variety release and commercialization across multiple environments	Mostly limited to research and laboratory experiments under controlled environments
Skills and reaserch facilities	Readily available	Still limited with most institutions outsourcing and licencing the technologies
Selection methods	Well established breeding and selection procedures	Procedures mostly still being developed and improved
Acceptability	Widely accepted	Some technologies like gene transformation are not widely accepted by policy makers and consumers

Table 2. *Cont.*

Aspect	Traditional Methods	Recent Technologies
	Cons	**Pros**
Time utilization	Takes long–up to 12 years to release a variety	Significantly reduced time depending on technology
Cost	Costly in terms of time and resources allocated to release a variety	Relatively cheap since the costs are concentrated over short space of time and less resources required
Environmental influence	FHB expression can be influenced by the environment during phenotyping [23]	Tracking of genes and transgenes at molecular level is more reliable
Space required	Several hectors of land are often required to handle breeding nurseries	Conversion and transformation often need lab and greenhouse space
Foreign genes	Restricted to the use of plants of the same genius or species (cross compatible)	FHB resistant genes can be transferred from different plant or micro species without fertilization barriers [53]
QTL conferring FHB resistance	Difficult to detect and transfer	Easy to detect and transfer [92–94]
Pyramiding and stacking multiple genes	Difficult	Easier with genetic engineering [87]

6. Tools to Assist Breeding for Resistance against FHB and Mycotoxin Contamination

Laboratory analytical tools are useful to assess toxin accumulation in wheat infected with *F. graminearum*. These tools can be used in breeding programmes to assess if resistance to mycotoxin accumulation by *F. graminearum* is incorporated and in monitoring the safety of food products made from wheat grain. To incorporate Fusarium head blight resistance in wheat, various assessment methods are employed for each breeding objective. Resistance against pathogen penetration and resistance against disease spread after initial infection can be monitored visually. Monitoring resistance against mycotoxin accumulation requires specialized equipment that is able to detect even trace amounts of the mycotoxins. For the purposes of the current review, real-time PCR, chromatography and mass spectrometry-based approaches are discussed as tools to assist selection.

6.1. Real-Time PCR

Inoculation with *F. graminearum* and then determining the quantity of the inoculum is done by real-time PCR. Real-time PCR is important for diagnoses using species-specific primers to detect a suspect pathogen and for quantifying pathogen titre in infected kernels [95–98]. The technique has the potential to unpack the gene expression in response to FHB infection through monitoring transcriptome expression patterns within specific plant tissue after inoculation. Newer genomic technologies, such as genome-wide single polymorphism mapping, genome sequencing, microarrays and RNA sequencing, have been instrumental in identifying genotypes with FHB resistance. These techniques have also been useful in identifying QTL, linking resistance with other phenotypic traits as well as detecting and validating diagnostic markers.

6.2. Chromatography and Mass Spectrometry-Based Approaches to Assist Selection

Regulatory standards with threshold prescriptions for wheat products such as the Codex Alimentarius Commission 2015 require that there are monitoring procedures to quantify the DON toxin in harvested wheat grain and grain products. Chromatography and mass spectrometry-based techniques become handy in such circumstances to ensure safety of wheat products in the market. Notably, high performance liquid chromatography (HPLC) is commonly used for separation, identification, and quantification of mycotoxin levels in flour, food and feed mixtures. Other techniques include gas chromatography–mass spectrometry (GC-MS) and thin-layer chromatography (TLC), which are also effective for early detection and quantification of DON in wheat. Equally important is the use of these quantitative techniques in screening breeding material and donor lines to be used in breeding against FHB, especially for type II resistance. Chromatography and mass spectrometry have been useful in identifying mycotoxin contaminants of wheat [96] and mycotoxin accumulation [99]. Because of their ability to detect and quantify contaminants and trace elements, chromatography and mass spectrometry-based techniques are useful in routine monitoring of grain safety to ensure compliance to prescribed standards. This could be the extension of the use of these techniques beyond research. With these state-of-the-art tools, breeding and selection of FHB resistant genotypes are becoming more efficient and reliable data are being produced on resistance to infection and mycotoxin contamination in the wheat grain.

7. Conclusions

The safety of wheat products is essential to ensure that human and animal lives are not endangered. Mycotoxins produced by the wheat-infecting *Fusarium graminearum* pathogen pose serious health risks to animals and human beings. It is therefore of the utmost importance to breed wheat varieties that are able to limit the accumulation of mycotoxins in wheat kernel that have been infected with *F. graminearum*. Traditional breeding techniques have been utilized to incorporate resistance against *F. graminearum* from resistance sources such as Sumai#3. However, the limitations of traditional plant breeding require integration of new and more sophisticated methods for cultivar improvement to fast-track *F. graminearum* resistance breeding. These techniques will also bolster resistance against mycotoxin accumulation. Clustered regularly interspaced short palindromic repeat-associated endonucleases (CRISPR/Cas) as well as RNA interference are some of the advanced tools that have revolutionized crop improvement efforts. Various molecular techniques like real-time PCR and biochemical analytical tools such as chromatography and mass spectrometry are also useful for detecting levels of infection by *F. graminearum*, and their use remains relevant for the future.

Author Contributions: Each author participated sufficiently in the completion of this work. Conceptualization, S.F.; writing—original draft preparation, S.F. and L.M.; writing—review and editing, S.F. and L.M. All authors have read and agreed to the published version of the manuscript.

Funding: This research has received no external funding.

Institutional Review Board Statement: Not applicable.

Informed Consent Statement: Not applicable.

Data Availability Statement: Not applicable.

Acknowledgments: The authors acknowledge the University of South Africa for the overall research support.

Conflicts of Interest: The authors declare no conflict of interest.

References

1. Kumar, S.; Saharan, M.S.; Panwar, V.; Chatrath, R.; Singh, G.P. Genetics of Fusarium head blight resistance in three wheat genotypes. *Indian J. Genet.* **2019**, *79*, 614–617. [CrossRef]
2. Li, H.; Zhang, F.; Zhao, J.; Bai, G.; Amand, P.S.; Bernardo, A.; Ni, Z.; Sun, Q.; Su, Z. Identification of a novel major QTL from Chinese wheat cultivar Ji5265 for Fusarium head blight resistance in greenhouse. *Theor. Appl. Genet.* **2022**, *31*, 1. [CrossRef] [PubMed]
3. Zhang, J.; Gill, H.S.; Brar, N.K.; Halder, J.; Ali, S.; Liu, X.; Bernardo, A.; St Amand, P.; Bai, G.; Gill, U.S.; et al. Genomic prediction of Fusarium head blight resistance in early stages using advanced breeding lines in hard winter wheat. *Crop J.* **2022**, 26. [CrossRef]
4. Liu, S.; Hall, M.D.; Griffey, C.A.; McKendry, A.L. Meta-analysis of QTL associated with Fusarium head blight resistance in wheat. *Crop Sci.* **2009**, *49*, 1955–1968. [CrossRef]
5. Buerstmayr, M.; Steiner, B.; Buerstmayr, H. Breeding for Fusarium head blight resistance in wheat—Progress and challenges. *Plant Breed.* **2020**, *139*, 429–454. [CrossRef]
6. Figueroa, M.; Hammond-Kosack, K.E.; Solomon, P.S. A review of wheat diseases—A field perspective. *Mol. Plant Pathol.* **2018**, *19*, 1523–1536. [CrossRef]

7. Mielniczuk, E.; Barbara, S.-B. Fusarium head blight, mycotoxins and strategies for their reduction. *Agronomy* **2020**, *10*, 509. [CrossRef]
8. de Arruda, M.H.M.; Zchosnki, F.L.; Silva, Y.K.; de Lima, D.L.; Tessmann, D.J.; Da-Silva, P.R. Genetic diversity of *Fusarium meridionale, F. austroamericanum*, and *F. graminearum* isolates associated with Fusarium head blight of wheat in Brazil. *Trop. Plant Pathol.* **2021**, *46*, 98–108. [CrossRef]
9. Sheini, A. Colorimetric aggregation assay based on array of gold and silver nanoparticles for simultaneous analysis of aflatoxins, ochratoxin and zearalenone by using chemometric analysis and paper based analytical devices. *Microchim. Acta* **2020**, *187*, 1–11. [CrossRef]
10. Zinedine, A.; Soriano, J.M.; Molto, J.C.; Manes, J. Review on the toxicity, occurrence, metabolism, detoxification, regulations and intake of zearalenone: An oestrogenic mycotoxin. *Food Chem. Toxicol.* **2007**, *45*, 1–18. [CrossRef]
11. Venkataramana, M.; Chandra Nayaka, S.; Anand, T.; Rajesh, R.; Aiyaz, M.; Divakara, S.T.; Murali, H.S.; Prakash, H.S.; Lakshmana Rao, P.V. Zearalenone induced toxicity in SHSY-5Y cells: The role of oxidative stress evidenced by N-acetyl cysteine. *Food Chem. Tox.* **2014**, *65*, 335–342. [CrossRef] [PubMed]
12. Schollenberger, M.; Müller, H.M.; Rüfle, M.; Suchy, S.; Plank, S.; Drochner, W. Natural occurrence of 16 Fusarium toxins in grains and feedstuffs of plant origin from Germany. *Mycopathologia* **2006**, *161*, 43–52. [CrossRef] [PubMed]
13. Vanhoutte, I.; Audenaert, K.; De Gelder, L. Biodegradation of mycotoxins: Tales from known and unexplored worlds. *Front. Microbiol.* **2016**, *7*, 561. [CrossRef] [PubMed]
14. Chammem, N.; Issaoui, M.; De Almeida AI, D.; Delgado, A.M. Food crises and food safety incidents in European Union, United States, and Maghreb Area: Current risk communication strategies and new approaches. *J. AOAC Int.* **2018**, *101*, 923–938. [CrossRef] [PubMed]
15. FAO; WHO. *Codex General Standard for Contaminants and Toxins in Food and Feed*; Codex Alimentarius: Rome, Italy, 1995.
16. EC. *Setting Maximum Levels for Certain Contaminants in Foodstuffs and Amendments*; European Commission: Brussels, Belgium, 2006.
17. FAO; WHO. *Codex Committee on Contaminants in Foods*; KoreaScience: Tokyo, Japan, 2018; p. 169.
18. Mesterhazy, A. Updating the breeding philosophy of wheat to Fusarium head blight (FHB): Resistance components, QTL identification, and phenotyping—A review. *Plants* **2020**, *9*, 1702. [CrossRef]
19. Kubo, K.; Kawada, N.; Fujita, M. Evaluation of Fusarium head blight resistance in wheat and the development of a new variety by integrating type I and II resistance. *Jpn. Agric. Res. Q.* **2013**, *47*, 9–19. [CrossRef]
20. Pumphrey, M.O.; Bernardo, R.; Anderson, J.A. Validating the *Fhb1* QTL for Fusarium head blight resistance in near-isogenic wheat lines developed from breeding populations. *Crop Sci.* **2007**, *47*, 200–206. [CrossRef]
21. Francesconi, S.; Angelo, M.; Giorgio, M.B. Different inoculation methods affect components of Fusarium head blight resistance in wheat. *Phytopathol. Mediterr.* **2019**, *58*, 679–691. [CrossRef]
22. Gaire, R.; Sneller, C.; Brown-Guedira, G.; Van Sanford, D.; Mohammadi, M.; Kolb, F.L.; Olson, E.; Sorrells, M.; Rutkoski, J. Genetic Trends in Fusarium Head Blight Resistance from 20 Years of Winter Wheat Breeding and Cooperative Testing in the Northern USA. *Plant Dis.* **2022**, *106*, 364–372. [CrossRef]
23. Crespo-Herrera, L.A.; José, C.; Mateo, V.; Hans-Joachim, B. Defining Target Wheat Breeding Environments. In *Wheat Improvement*; Springer: Cham, Switzerland, 2022; pp. 31–45.
24. Singh, R.P.; Rajaram, S. Breeding for disease resistance in wheat. In *Bread Wheat: Improvement and Production*; No. CIS-3621. CIMMYT; FAO: Rome, Italy, 2002; pp. 141–156.
25. Löffler, M.; Schön, C.C.; Miedaner, T. Revealing the genetic architecture of FHB resistance in hexaploid wheat (*Triticum aestivum* L.) by QTL meta-analysis. *Mol. Breed.* **2009**, *23*, 473–488. [CrossRef]
26. Zhao, M.; Leng, Y.; Chao, S.; Xu, S.S.; Zhong, S. Molecular mapping of QTL for Fusarium head blight resistance introgressed into durum wheat. *Theor. Appl. Genet.* **2018**, *131*, 1939–1951. [CrossRef] [PubMed]
27. Bai, G.; Kolb, F.L.; Shaner, G.; Domier, L.L. Amplified fragment length polymorphism markers linked to a major quantitative trait locus controlling scab resistance in wheat. *Phytopathology* **1999**, *89*, 343–348. [CrossRef] [PubMed]
28. Anderson, J.A.; Stack, R.W.; Liu, S.; Waldron, B.L.; Fjeld, A.D.; Coyne, C.; Moreno-Sevilla, B.; Fetch, J.M.; Song, Q.J.; Cregan, P.B.; et al. DNA markers for Fusarium head blight resistance QTLs in two wheat populations. *Theor. Appl. Genet.* **2001**, *102*, 164–1168. [CrossRef]
29. Buerstmayr, H.; Lemmens, M.; Hartl, L.; Doldi, L.; Steiner, B.; Stierschneider, M.; Ruckenbauer, P. Molecular mapping of QTLs for Fusarium head blight resistance in spring wheat. I. Resistance to fungal spread (Type II resistance). *Theor. Appl. Genet.* **2002**, *104*, 84–91. [CrossRef]
30. Buerstmayr, H.; Steiner, B.; Hartl, L.; Griesser, M.; Angerer, N.; Lengauer, D.; Miedaner, T.; Schneider, B.; Lemmens, M. Molecular mapping of QTLs for Fusarium head blight resistance in spring wheat. II. Resistance to fungal penetration and spread. *Theor. Appl. Genet.* **2003**, *107*, 503–508. [CrossRef]
31. Buerstmayr, H.; Stierschneider, M.; Steiner, B.; Lemmens, M.; Griesser, M.; Nevo, E.; Fahima, T. Variation for resistance to head blight caused by *Fusarium graminearum* in wild emmer (*Triticum dicoccoides*) originating from Israel. *Euphytica* **2003**, *130*, 17–23. [CrossRef]
32. Steiner, B.; Buerstmayr, M.; Wagner, C.; Danler, A.; Eshonkulov, B.; Ehn, M.; Buerstmayr, H. Fine-mapping of the Fusarium head blight resistance QTL Qfhs.ifa-5A identifies two resistance QTL associated with anther extrusion. *Theor. Appl. Genet.* **2019**, *132*, 2039–2053. [CrossRef]

33. Lemmens, M.; Scholz, U.; Berthiller, F.; Dall'Asta, C.; Koutnik, A.; Schuhmacher, R.; Adam, G.; Buerstmayr, H.; Mesterházy, Á.; Krska, R.; et al. The ability to detoxify the mycotoxin deoxynivalenol colocalizes with a major quantitative trait locus for Fusarium head blight resistance in wheat. *Mol. Plant-Microbe Interact.* **2005**, *18*, 1318–1324. [CrossRef]
34. Brown, D.W.; Butchko, R.A.E.; Proctor, R.H. Fusarium genomic resources: Tools to limit crop diseases and mycotoxin contamination. *Mycopathologia* **2006**, *162*, 191–199. [CrossRef]
35. Mesterházy, Á. Role of deoxynivalenol in aggressiveness of *Fusarium graminearum* and *F. culmorum* and in resistance to Fusarium head blight. In *Mycotoxins in Plant Disease*; Springer: Dordrecht, The Netherlands, 2002; pp. 675–684. [CrossRef]
36. Mendes, G.D.R.L.; Ponte, E.M.D.; Feltrin, A.C.; Badiale-Furlong, E.; Oliveira, A.C.D. Common resistance to Fusarium head blight in Brazilian wheat cultivars. *Sci. Agric.* **2018**, *75*, 426–431. [CrossRef]
37. Mesterhazy, A.; Bartók, T.; Kászonyi, G.; Varga, M.; Tóth, B.; Varga, J. Common resistance to different Fusarium spp. causing Fusarium head blight in wheat. *Eur. J. Plant Pathol.* **2005**, *112*, 267–281. [CrossRef]
38. Miedaner, T.; Heinrich, N.; Schneider, B.; Oettler, G.; Rohde, S.; Rabenstein, F. Estimation of deoxynivalenol (DON) content by symptom rating and exoantigen content for resistance selection in wheat and triticale. *Euphytica* **2004**, *139*, 123–132. [CrossRef]
39. Proctor, R.H.; Desjardins, A.E.; McCornick, S.P.; Plattner, R.D.; Alexander, N.J.; Brown, D.W. Genetic analysis of the role of Trichothecene and Fumonisin mycotoxins in the virulence of Fusarium. *Eur. J. Plant Pathol.* **2002**, *108*, 691–698. [CrossRef]
40. Lancova, K.; Hajslova, J.; Poustka, J.; Krplova, A.; Zachariasova, M.; Dostálek, P.; Sachambula, L. Transfer of Fusarium mycotoxins and 'masked'deoxynivalenol (deoxynivalenol-3-glucoside) from field barley through malt to beer. *Food Addit. Contam.* **2008**, *25*, 732–744. [CrossRef] [PubMed]
41. Janick, J. (Ed.) *Plant Breeding Reviews*; John Wiley & Sons: Hoboken, NJ, USA, 2008; Volume 18, pp. 177–250.
42. Steiner, B.; Buerstmayr, M.; Michel, S.; Schweiger, W.; Lemmens, M.; Buerstmayr, H. Breeding strategies and advances in line selection for Fusarium head blight resistance in wheat. *Trop. Plant Pathol.* **2017**, *42*, 165–174. [CrossRef]
43. Majumdar, R.; Rajasekaran, K.; Cary, J.W. RNA interference (RNAi) as a potential tool for control of mycotoxin contamination in crop plants: Concepts and considerations. *Front. Plant Sci.* **2017**, *8*, 200. [CrossRef]
44. Baulcombe, D. RNA silencing in plants. *Nature* **2004**, *431*, 356. [CrossRef]
45. Watson, J.M.; Fusaro, A.F.; Wang, M.; Waterhouse, P.M. RNA silencing platforms in plants. *Febs Lett.* **2005**, *579*, 5982–5987. [CrossRef]
46. Small, I. RNAi for revealing and engineering plant gene functions. *Curr. Opin. Biotechnol.* **2007**, *18*, 148–153. [CrossRef]
47. Koch, A.; Kogel, K.H. New wind in the sails: Improving the agronomic value of crop plants through RNAi-mediated gene silencing. *Plant Biotechnol. J.* **2014**, *12*, 821–831. [CrossRef]
48. Nunes, C.C.; Dean, R.A. Host-induced gene silencing: A tool for understanding fungal host interaction and for developing novel disease control strategies. *Mol. Plant Pathol.* **2012**, *13*, 519–529. [CrossRef] [PubMed]
49. Yin, C.; Scot, H.H. Host-induced gene silencing (HIGS) for elucidating Puccinia gene function in wheat. In *Plant Pathogenic Fungi and Oomycetes*; Humana Press: New York, NY, USA, 2018; pp. 139–150. [CrossRef]
50. Machado, A.K.; Brown, N.A.; Urban, M.; Kanyuka, K.; Hammond-Kosack, K.E. RNAi as an emerging approach to control Fusarium head blight disease and mycotoxin contamination in cereals. *Pest Manag. Sci.* **2018**, *74*, 790–799. [CrossRef] [PubMed]
51. Yin, C.; Jurgenson, J.E.; Hulbert, S.H. Development of a host-induced RNAi system in the wheat stripe rust fungus *Puccinia striiformis* f. sp. tritici. *Mol. Plant-Microbe Interact.* **2011**, *24*, 554–561. [CrossRef]
52. Cheng, W.; Song, X.S.; Li, H.P.; Cao, L.H.; Sun, K.; Qiu, X.L.; Xu, Y.B.; Yang, P.; Huang, T.; Zhang, J.B.; et al. Host-induced gene silencing of an essential chitin synthase gene confers durable resistance to Fusarium head blight and seedling blight in wheat. *Plant Biotechnol. J.* **2015**, *13*, 1335–1345. [CrossRef] [PubMed]
53. Low, L.Y.; Yang, S.K.; Andrew Kok, D.X.; Ong-Abdullah, J.; Tan, N.P.; Lai, K.S. Transgenic plants: Gene constructs, vector and transformation method. *New Vis. Plant Sci.* **2018**, 41–61. [CrossRef]
54. Harwood, W.A.; Hardon, J.; Ross, S.M.; Fish, L.; Smith, J.; Snape, J.W. Analysis of transgenic barley in a small scale field trial. *John Innes Cent. Sainsbury Lab. Annu. Rep.* **2000**, *29*.
55. Romano, A.; Raemakers, K.; Visser, R.; Mooibroek, H. Transformation of potato (*Solanum tuberosum*) using particle bombardment. *Plant Cell Rep.* **2001**, *20*, 198–204. [CrossRef]
56. Morikawa, H.; Sakamoto, A.; Hokazono, H.; Irifune, K.; Takahashi, M. Mechanism of transgene integration into a host genome by particle bombardment. *Plant Biotechnol.* **2002**, *19*, 219–228. [CrossRef]
57. Bode, J.; Benham, C.; Knopp, A.; Mielke, C. Transcriptional augmentation: Modulation of gene expression by scaffold/matrix-attached regions (S/MAR elements). *Crit. Rev. Eukaryot. Gene. Expr.* **2000**, *10*, 73–90. [CrossRef]
58. Haber, J.E. Partners and pathways: Repairing a double-strand break. *Trends Genet.* **2000**, *16*, 259–264. [CrossRef]
59. Gelvin, S.B. *Agrobacterium* and plant genes involved in T-DNA transfer and integration. *Annu. Rev. Plant Physiol. Plant Mol. Biol.* **2000**, *51*, 223–256. [CrossRef] [PubMed]
60. Ward, D.V.; Zambryski, P.C. The six functions of *Agrobacterium* VirE2. *Proc. Natl. Acad. Sci. USA* **2001**, *98*, 385–386. [CrossRef] [PubMed]
61. Gelvin, S.B. *Agrobacterium*-mediated transformation of germinating seeds of *Arabidopsis thaliana*: A non-tissue culture approach. *Mol. Gen. Genet.* **2003**, *208*, 1–9. [CrossRef]

62. Zhang, Z.; Liu, X.; Wang, X.; Zhou, M.; Zhou, X.; Ye, X.; Wei, X. An R2R3 MYB transcription factor in wheat, Ta PIMP 1, mediates host resistance to *Bipolaris sorokiniana* and drought stresses through regulation of defense-and stress-related genes. *New Phytol.* **2012**, *196*, 1155–1170. [CrossRef] [PubMed]
63. Liu, W.; Frick, M.; Huel, R.; Nykiforuk, C.L.; Wang, X.; Gaudet, D.A.; Eudes, F.; Conner, R.L.; Kuzyk, A.; Chen, Q.; et al. The stripe rust resistance gene Yr10 encodes an evolutionary-conserved and unique CC–NBS–LRR sequence in wheat. *Mol. Plant* **2014**, *7*, 1740–1755. [CrossRef]
64. Gao, L.; Wang, S.; Li, X.Y.; Wei, X.J.; Zhang, Y.J.; Wang, H.Y.; Liu, D.Q. Expression and functional analysis of a pathogenesis-related protein 1 gene, TcLr19PR1, involved in wheat resistance against leaf rust fungus. *Plant Mol. Biol. Report.* **2015**, *33*, 797–805. [CrossRef]
65. Jha, S.; Chattoo, B.B. Expression of a plant defensin in rice confers resistance to fungal phytopathogens. *Transgenic Res.* **2010**, *19*, 373–384. [CrossRef]
66. Li, X.; Zhong, S.; Chen, W.; Fatima, S.; Huang, Q.; Li, Q.; Tan, F.; Luo, P. Transcriptome analysis identifies a 140 kb region of chromosome 3B containing genes specific to Fusarium Head Blight resistance in wheat. *Int. J. Mol. Sci.* **2018**, *19*, 852. [CrossRef]
67. Ye, X.; Cheng, H.; Xu, H.; Du, L.; Lu, W.; Huang, Y. Development of transgenic wheat plants with chitinase and β-1, 3-glucosanase genes and their resistance to fusarium head blight. *Zuo Wu Xue Bao* **2005**, *31*, 583–586.
68. Ishida, Y.; Tsunashima, M.; Hiei, Y.; Komari, T. Wheat (*Triticum aestivum* L.) transformation using immature embryos. In *Agrobacterium Protocols*; Springer: New York, NY, USA, 2015; pp. 189–198. [CrossRef]
69. San Filippo, J.; Sung, P.; Klein, H. Mechanism of eukaryotic homologous recombination. *Annu. Rev. Biochem.* **2008**, *77*, 229–257. [CrossRef]
70. Lieber, M.R. The mechanism of double-strand DNA break repair by the nonhomologous DNA end-joining pathway. *Annu. Rev. Biochem.* **2010**, *79*, 81–211. [CrossRef] [PubMed]
71. Chapman, J.R.; Taylor, M.R.; Boulton, S.J. Playing the end game: DNA double-strand break repair pathway choice. *Mol. Cell* **2012**, *47*, 497–510. [CrossRef] [PubMed]
72. Jiang, F.; Doudna, J.A. CRISPR–Cas9 structures and mechanisms. *Annu. Rev. Biophys.* **2017**, *46*, 505–529. [CrossRef] [PubMed]
73. Matsu-ura, T.; Baek, M.; Kwon, J.; Hong, C. Efficient gene editing in *Neurospora crassa* with CRISPR technology. *Fungal Biol. Biotechnol.* **2015**, *2*, 4. [CrossRef]
74. Fuller, K.K.; Chen, S.; Loros, J.J.; Dunlap, J.C. Development of the CRISPR/Cas9 system for targeted gene disruption in *Aspergillus fumigatus*. *Eukaryot. Cell* **2015**, *14*, 1073–1080. [CrossRef] [PubMed]
75. Nødvig, C.S.; Nielsen, J.B.; Kogle, M.E.; Mortensen, U.H. A CRISPR-Cas9 system for genetic engineering of filamentous fungi. *PLoS ONE* **2015**, *10*, e0133085. [CrossRef] [PubMed]
76. Pohl, C.; Kiel, J.A.K.W.; Driessen, A.J.M.; Bovenberg, R.A.L.; Nygard, Y. CRISPR/Cas9 based genome editing of *Penicillium chrysogenum*. *ACS Synth. Biol.* **2016**, *5*, 754–764. [CrossRef]
77. Wenderoth, M.; Pinecker, C.; Voß, B.; Fischer, R. Establishment of CRISPR/Cas9 in *Alternaria alternata*. *Fungal Genet. Biol.* **2017**, *101*, 55–60. [CrossRef]
78. Arazoe, T.; Miyoshi, K.; Yamato, T.; Ogawa, T.; Ohsato, S.; Arie, T.; Kuwata, S. Tailor-made CRISPR/Cas system for highly efficient targeted gene replacement in the rice blast fungus. *Biotechnol. Bioeng.* **2015**, *112*, 2543–2549. [CrossRef]
79. Schuster, M.; Schweizer, G.; Reissmann, S.; Kahmann, R. Genome editing in *Ustilago maydis* using the CRISPR–Cas system. *Fungal Genet. Biol.* **2016**, *89*, 3–9. [CrossRef]
80. Gardiner, D.M.; Kazan, K. Selection is required for efficient Cas9-mediated genome editing in *Fusarium graminearum*. *Fungal Biol.* **2018**, *122*, 131–137. [CrossRef] [PubMed]
81. Koch, A.; Hofle, L.; Werner, B.T.; Imani, J.; Schmidt, A.; Jelonek, L.; Kogel, K.-H. SIGS vs. HIGS: A study on the efficacy of two dsRNA delivery strategies to silence *Fusarium* FgCYP51 genes in infected host and non-host plants. *Mol. Plant Pathol.* **2019**, *20*, 1636–1644. [CrossRef] [PubMed]
82. He, F.; Zhang, R.; Zhao, J.; Qi, T.; Kang, Z.; Guo, J. Host-induced silencing of *Fusarium graminearum* genes enhances the resistance of *Brachypodium distachyon* to Fusarium head blight. *Front. Plant Sci.* **2019**, *10*, 1362. [CrossRef] [PubMed]
83. Su, Z.; Amy, B.; Bin, T.; Hui, C.; Shan, W.; Hongxiang, M.; Shibin, C.; Liu, D.; Zhang, D.; Li, T.; et al. A deletion mutation in TaHRC confers Fhb1 resistance to Fusarium head blight in wheat. *Nat. Gen.* **2019**, *51*, 1099–1105. [CrossRef]
84. Mandalà, G.; Tundo, S.; Francesconi, S.; Gevi, F.; Zolla, L.; Ceoloni, C.; D'Ovidio, R. Deoxynivalenol detoxification in transgenic wheat confers resistance to Fusarium head blight and crown rot diseases. *Mol. Plant Microbe Interact.* **2019**, *32*, 583–592. [CrossRef]
85. Koch, A.; Kumar, N.; Weber, L.; Keller, H.; Imani, J.; Kogel, K.H. Host-induced gene silencing of cytochrome P450 lanosterol C14 alpha-demethylase-encoding genes confers strong resistance to *Fusarium* species. *Proc. Natl. Acad. Sci. USA* **2013**, *110*, 19324–19329. [CrossRef]
86. McLaughlin, J.E.; Darwish, N.I.; Garcia-Sanchez, J.; Tyagi, N.; Trick, H.N.; McCormick, S.; Dill-Macky, R.; Tumer, N.E. A lipid transfer protein has antifungal and antioxidant activity and suppresses Fusarium head blight disease and DON accumulation in transgenic wheat. *Phytopathology* **2021**, *4*, 671–683. [CrossRef]
87. Wang, M.; Wu, L.; Mei, Y.; Zhao, Y.; Ma, Z.; Zhang, X.; Chen, Y. Host-induced gene silencing of multiple genes of *Fusarium graminearum* enhances resistance to Fusarium head blight in wheat. *Plant Biotechnol. J.* **2020**, *18*, 2373. [CrossRef]
88. Kumar, J.; Rai, K.M.; Pirseyedi, S.; Elias, E.M.; Xu, S.; Dill-Macky, R.; Kianian, S.F. Epigenetic regulation of gene expression improves Fusarium head blight resistance in durum wheat. *Sci. Rep.* **2020**, *10*, 1–15. [CrossRef]

89. Gatti, M.; Florence, C.; Caroline, T.; Catherine, M.; Florence, G.; Thierry, L.; Marie, D. The *Brachypodium distachyon* UGT Bradi5gUGT03300 confers type II fusarium head blight resistance in wheat. *Plant Pathol.* **2019**, *68*, 334–343. [CrossRef]
90. Li, G.; Jiyang, Z.; Haiyan, J.; Zhongxia, G.; Min, F.; Yanjun, L.; Panting, Z.; Xue, S.; Li, N.; Yuan, Y.; et al. Mutation of a histidine-rich calcium-binding-protein gene in wheat confers resistance to Fusarium head blight. *Nat. Gen.* **2019**, *51*, 1106–1112. [CrossRef] [PubMed]
91. Mandalà, G.; Carla, C.; Isabella, B.; Francesco, F.; Silvio, T. Transgene pyramiding in wheat: Combination of deoxynivalenol detoxification with inhibition of cell wall degrading enzymes to contrast Fusarium Head Blight and Crown Rot. *Plant Sci.* **2021**, *313*, 111059. [CrossRef] [PubMed]
92. Hu, W.; Gao, D.; Wu, H.; Liu, J.; Zhang, C.; Wang, J.; Jiang, Z.; Liu, Y.; Li, D.; Zhang, Y.; et al. Genome-wide association mapping revealed syntenic loci QFhb-4AL and QFhb-5DL for Fusarium head blight resistance in common wheat (*Triticum aestivum* L.). *BMC Plant Biol.* **2020**, *20*, 1–13. [CrossRef] [PubMed]
93. Słomińska-Durdasiak, K.M.; Sonja, K.; Viktor, K.; Daniela, N.; Patrick, S.; Armin, D.; Jochen, C.R. Association mapping of wheat Fusarium head blight resistance-related regions using a candidate-gene approach and their verification in a biparental population. *Theor. Appl. Genet.* **2020**, *133*, 341–351. [CrossRef]
94. Tessmann, E.W.; Dong, Y.; Van Sanford, D.A. GWAS for Fusarium head blight traits in a soft red winter wheat mapping panel. *Crop Sci.* **2019**, *59*, 1823–1837. [CrossRef]
95. Sonia, E.; Dorothée, S.; Sandrine, G.; Corinne, C.; Christian, L.; Henri, L.M.; Valérie, L. Optimized real time QPCR assays for detection and quantification of *Fusarium* and *Microdochium species* involved in wheat head blight as defined by MIQE guidelines. *BioRxiv* **2018**. [CrossRef]
96. Burlakoti, R.R.; Estrada Jr, R.; Rivera, V.V.; Boddeda, A.; Secor, G.A.; Adhikari, T.B. Real-time PCR quantification and mycotoxin production of *Fusarium graminearum* in wheat inoculated with isolates collected from potato, sugar beet, and wheat. *Phytopathology* **2007**, *97*, 835–841. [CrossRef]
97. Munis, M.F.H.; Xu, S.; Hakim, H.J.C.; Masood, S.; Farooq, A.B.U. Diagnosis of *Fusarium graminearum* in Soil and Plant Samples of Wheat by Real-Time PCR. *Rom. Biotechnol. Lett.* **2018**, *23*, 14035–14042. [CrossRef]
98. Tralamazza, S.M.; Braghini, R.; Corrêa, B. Trichothecene genotypes of the *Fusarium graminearum* species complex isolated from Brazilian wheat grains by conventional and quantitative PCR. *Front. Microbiol.* **2016**, *7*, 246. [CrossRef]
99. Gatti, M.; Choulet, F.; Macadré, C.; Guérard, F.; Seng, J.M.; Langin, T.; Dufresne, M. Identification, molecular cloning and functional characterization of a wheat UDP-glucosyltransferase involved in resistance to Fusarium Head Blight and to mycotoxin accumulation. *Front. Plant Sci.* **2018**, *9*, 1853. [CrossRef]

plants

MDPI

Article

Evaluation of Two Predictive Models for Forecasting Olive Leaf Spot in Northern Greece

Thomas Thomidis [1,*], Konstantinos Michos [2], Fotis Chatzipapadopoulos [2] and Amalia Tampaki [2]

1 Department of Nutritional Science and Diabetics, International Hellenic University, Sindos, 57400 Thessaloniki, Greece
2 Neuropublic S.A., Information Technologies & Smart Farming Services, Piraeus, 18545 Attica, Greece; k_michos@neuropublic.gr (K.M.); f_chatzipapadopoulos@neuropublic.gr (F.C.); a_tampaki@neuropublic.gr (A.T.)
* Correspondence: thomidis@cp.teithe.gr; Tel.: +30-2310013342

Abstract: Olive leaf spot (*Venturia oleaginea*) is a very important disease in olive trees worldwide. The introduction of predictive models for forecasting the appearance of a disease can lead to improved disease management. One of the aims of this study was to investigate the effect of temperature and leaf wetness on conidial germination of local isolates of *V. oleaginea*. The results showed that a temperature range of 5 to 25 °C was appropriate for conidial germination, with 20 °C being the optimum. It was also found that at least 12 h of leaf wetness was required to start the germination of *V. oleaginea* conidia at the optimum temperature. The second aim of this study was to validate the above generic model and a polynomial model for forecasting olive leaf spot disease under the field conditions of Potidea Chalkidiki, Northern Greece. The results showed that both models correctly predicted infection periods. However, there were differences in the severity of the infection, as demonstrated by the goodness-of-fit for the data collected on leaves of olive trees in 2016, 2017 and 2018. Specifically, the generic model predicted lower severity, which fits well with the incidence of the disease symptoms on unsprayed trees. In contrast, the polynomial model predicted high severity levels of infection, but these did not fit well with the incidence of disease symptoms.

Keywords: leaf wetness; temperatures; validation; *Venturia oleaginea*

Citation: Thomidis, T.; Michos, K.; Chatzipapadopoulos, F.; Tampaki, A. Evaluation of Two Predictive Models for Forecasting Olive Leaf Spot in Northern Greece. *Plants* **2021**, *10*, 1200. https://doi.org/10.3390/plants10061200

Academic Editor: Alessandro Vitale

Received: 17 March 2021
Accepted: 10 June 2021
Published: 12 June 2021

1. Introduction

Olive leaf spot (*Venturia oleaginea* (Castagne) Rossman & Crous, comb. nov.) is the cause of a very important disease in olive trees worldwide. According to Trapero and Blanco [1] and Viruega et al. [2], *V. oleaginea* over summers as mycelium in infected leaves that remain in the tree canopy or fallen to the soil surface, while in autumn, mycelia resume growth from the latent infections caused during the last spring or from old lesions, and new conidia are produced, which are dispersed by rain splash and run-off. The main symptoms of the disease are dark sooty spots (commonly known as peacock spots) which appear on the upper surface of leaves, mainly in the low canopy. Rarely, similar spots may also appear on the stem and fruit [3]. Heavy premature defoliation, which sometimes leads to twig death of olive (*Olea europaea* L.), can been caused by this pathogen [4] when no preventive or curative sprays are applied. According to Prota [5], in the Mediterranean region, fungicides are usually applied in the main shoot-growth seasons (spring and/or autumn).

Meteorological factors play a key role in infection by the olive leaf spot fungus. Temperature and moisture are the main climate factors influencing the development of *V. oleaginea* in olive trees. Relatively mild to low temperatures and free moisture on the leaves favor infections during the rainy periods in fall, winter, and spring [6–8]. Previous works have shown that the minimum temperature for conidia germination of the fungus is 5 °C, the optimum 20 °C, and the maximum 30 °C [6,9,10]; this pathogen is able to sporulate at temperatures from 5 to 25 °C [11]. In Greece, these temperatures occur mainly in the period of September to May [12]. Saad and Masri [13] also demonstrated the relationship between conidial germination and leaf wetness duration. They found that a minimum of 42 h leaf wetness was required for *V. oleaginea* conidia to germinate at 12 °C, while at 20 °C, 18 h was required. Infection occurred from 5 to 25 °C, and disease severity was the greatest at ~20 °C for wetness durations of 12 to 24 h and at ~15 °C for longer durations, while the optimum temperature and minimum wetness durations for infection were 15.5 °C and 11.9 h [6,12,14]. Obanor et al. [15] found that temperature affected olive leaf spot severity, with the lesion numbers increasing gradually from 5 °C to a maximum at 15 °C, and then declining to a minimum at 25 °C, while the numbers of lesions increased with increasing leaf wetness period at all temperatures tested. The minimum leaf wetness periods for infection at 5, 10, 15, 20 and 25 °C were 18, 12, 12, 12 and 24 h, respectively.

Several forecasting models to predict the infection for specific plant diseases have been developed. Each of the developed models has their strengths and weakness, so choosing the right one is based on many factors. A widely used generic infection model is that developed by Magarey et al. [16], which described pathosystems in which the parameters of temperature and wetness duration were supplied for each of the studied pathogens. In addition to the generic model, an infection model using regression equations, such as those based on polynomials, logistic equations can be developed by conducting combinations of multiple temperature and wetness. A forecasting model to predict olive leaf spot infection was developed by Obanor et al. [15] based on a polynomial equation with linear and quadratic terms of temperature, wetness and leaf age. However, this model has not been validated under field conditions.

The successful development of a plant disease forecasting system also requires the proper validation of a developed model. There are a large number of predictive models for the many important plant diseases in the international [17] literature. However, the accuracy of the predictions of each model must be tested under field conditions in order to reduce: (a) erroneous indications of high risk in cases where, in fact, no disease was observed, and (b) erroneous indications of no risk where, in fact, the disease was observed [18]. Although the effect of temperature and leaf wetness on the conidial germination of *V. oleaginea* has been previous studied [15], repetition to validate previous results with local isolates of the pathogen is essential to fit these parameters in the generic model developed by Magarey et al. [16] for local uses. Thus, one of the main aims of this study was to investigate the minimum, maximum and optimum temperatures and leaf wetness durations for conidia germination of local isolates of *V. oleaginea*. Because validation of prediction models for forecasting plant disease under field conditions is important prior to commercial use, a second aim of this study was to validate the above generic model and the polynomial model developed by Obanor et al. [15] under the field conditions in Potidea Chalkidiki, Northern Greece.

2. Results

2.1. Effect of Temperature and Leaf Wetness on Conidial Germination

There was no significant difference between repeated trials ($p = 0.201$), so the data from the two trials were pooled. Temperature significantly influenced ($p < 0.001$, SE = 0.687) conidial germination. Under continuous wetness, the optimum temperature for conidial germination was 20 °C, whereas conidial germination was inhibited at 30 and 0 °C. Conidial germination was significantly less at 15 and 25 °C than at 20 °C. The percentage of conidial germination was significant higher at 15 and 25 °C than at 10 °C. Conidial germination was significantly less at 5 than 10 °C. The estimates of the parameters from the quadratic function of temperature ($R^2 = 0.739$; $Y = 7.46 + 5.22 \times X - 0.16 \times X^2$) and leaf wetness ($R^2 = 0.946$; $Y = 23.88 + 4.43 \times X - 0.06 \times X^2$) are presented in Figure 1.

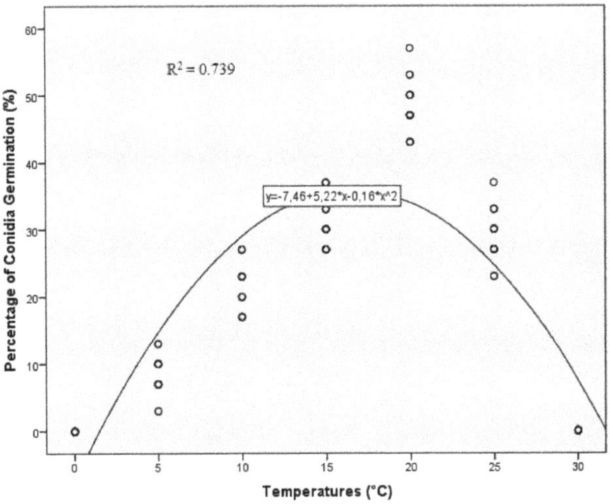

Figure 1. Effect of temperature on conidial germination of *Venturia oleaginea.* The parameters of minimum, maximum and optimum temperatures were fit to the generic model.

There was no significant difference between repeated trials ($p = 0.186$), so the data from the two trials were pooled. Leaf wetness also significantly influenced ($p < 0.001$, SE = 0.073) conidial germination. Under constant temperature at 20 °C, the conidial germination started after 12 h of continues wetness. In contrast, no conidial germination was observed after 6 h of wetness. The percentage of conidial germination after 18 h of wetness was significantly higher than 12 h, but significantly less than 24 h. No significant difference was observed in the percentage of conidial germination after 24, 36 and 48 h of leaf wetness. The estimates of the parameters from the quadratic function are presented in Figure 2.

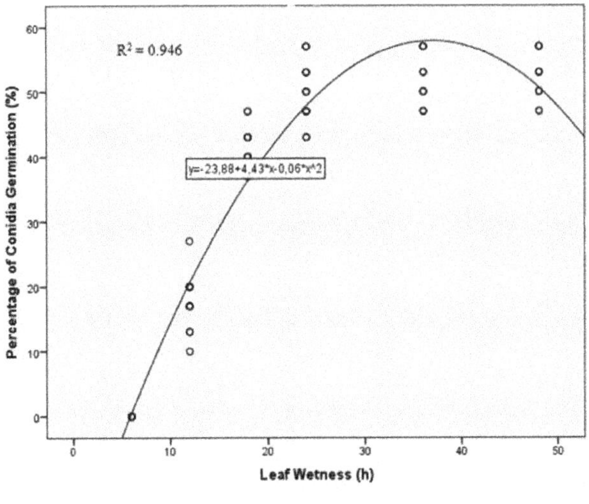

Figure 2. Effect of leaf wetness on the conidia germination of *Venturia oleaginea.* The parameters of minimum and maximum leaf wetness were fit to the generic model.

2.2. *Evaluation of Model Accuracy*

The average temperature, rainfall, and leaf wetness for the period May to December for each of these three years is presented in Figure 3. Figure 4 presents the predictions of the generic and polynomial models for the period of May to December for three consecutive years (2016, 2017, 2018).

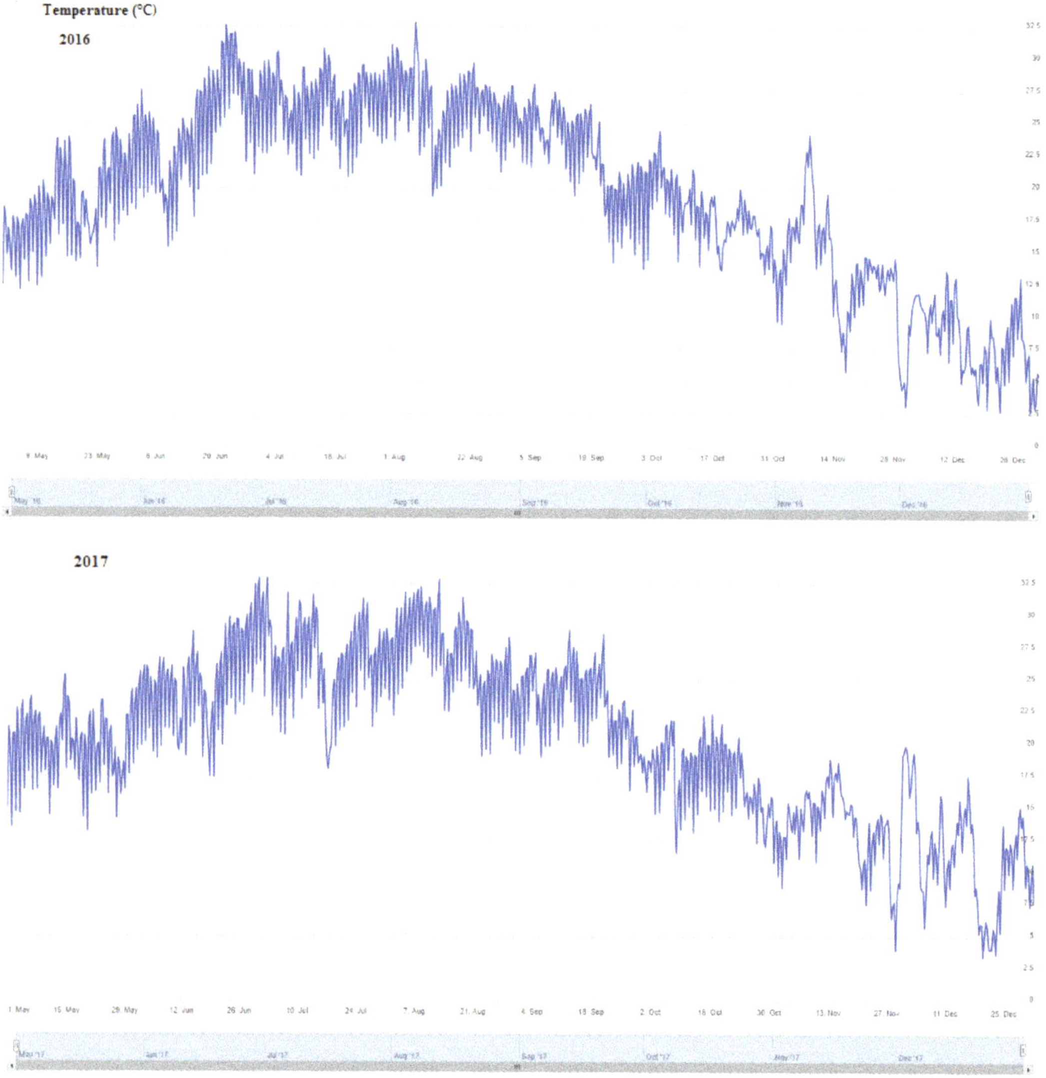

Figure 3. *Cont.*

2018

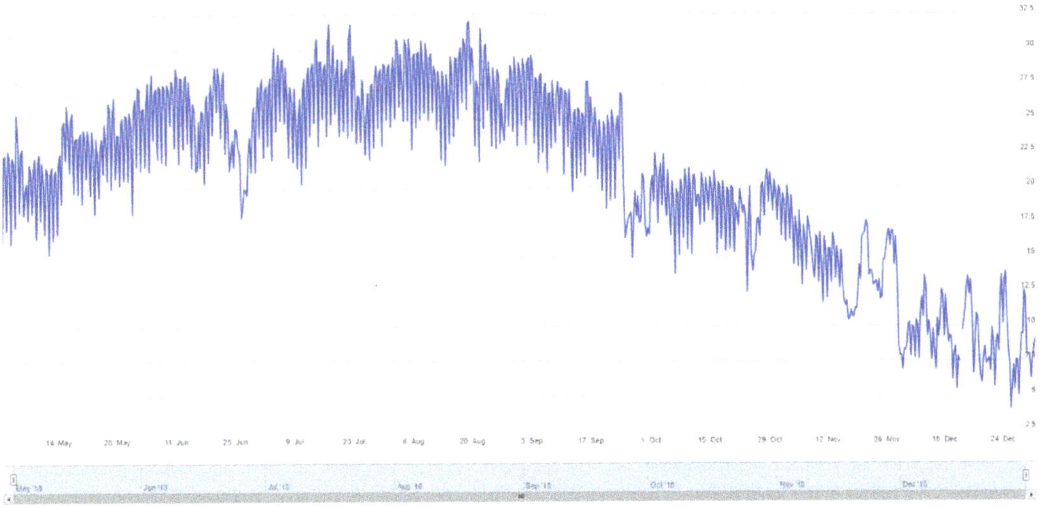

Leaf Wetness (h)

2016

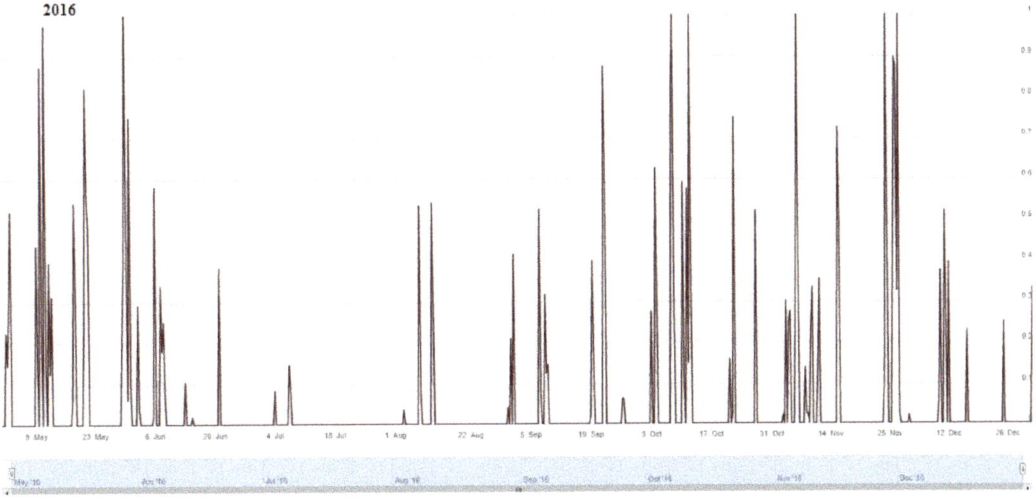

Figure 3. *Cont.*

2017

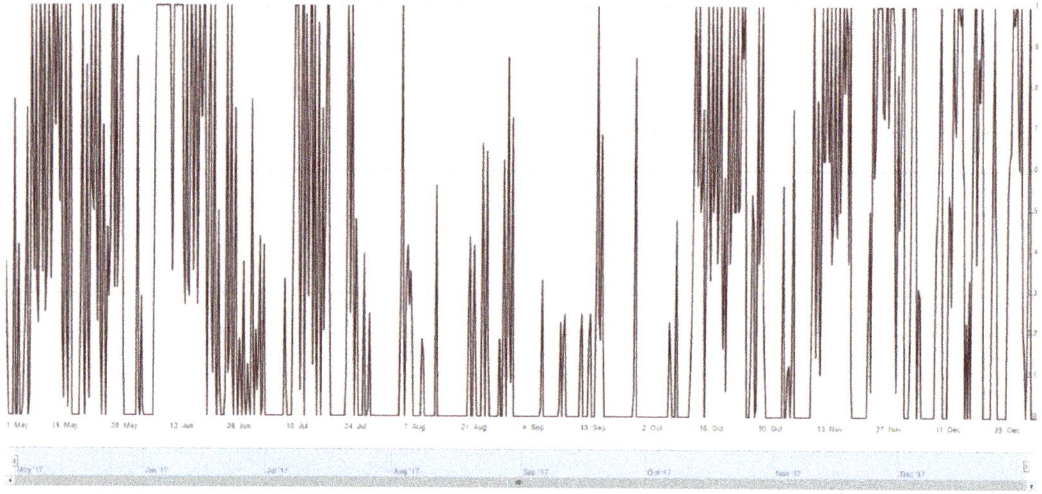

2018

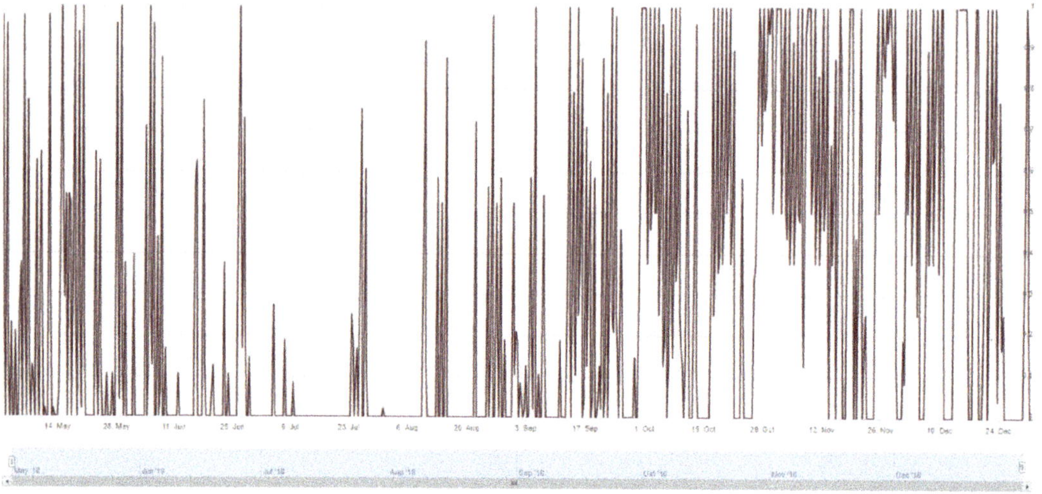

Figure 3. *Cont.*

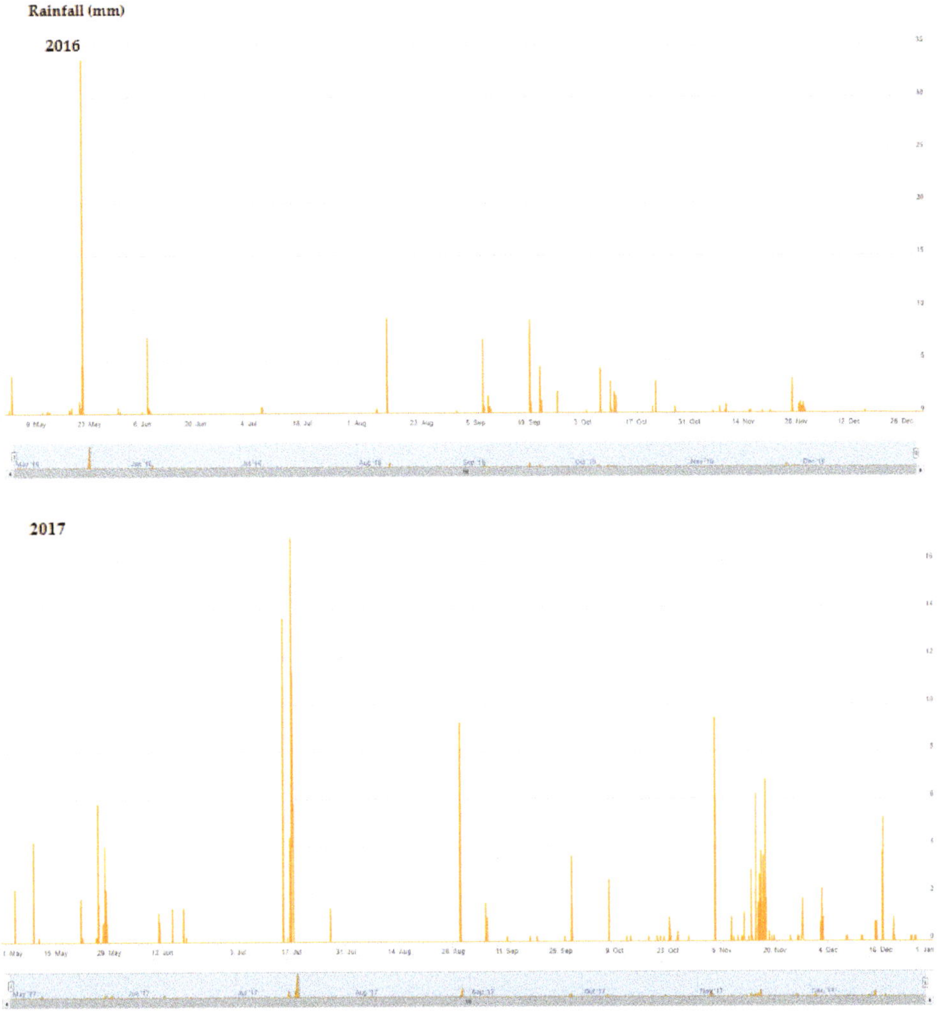

Figure 3. *Cont.*

2018

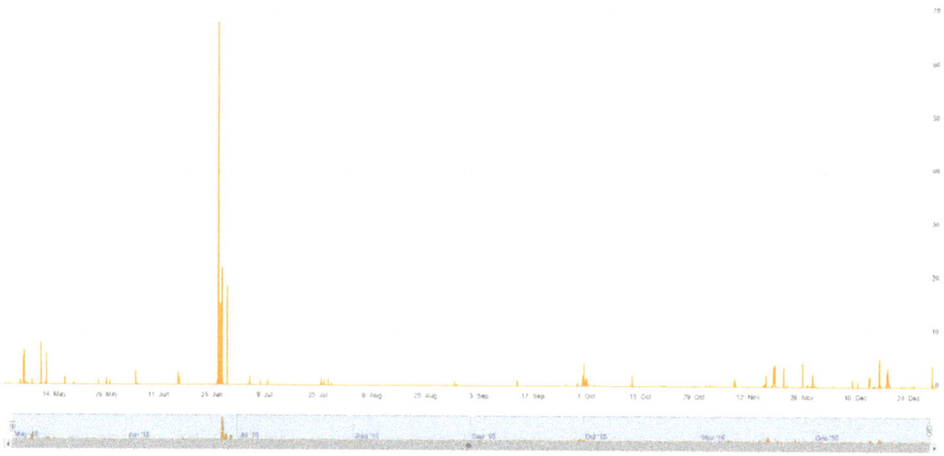

Figure 3. Average temperature, rainfall, and leaf wetness for the period May to December in 2016, 2017 and 2018.

In 2016, the polynomial model predicted risk >29 on the 15th, 20th and 31st of May, 23th September, 10th and 15 October, 8th November and 1st December (Table 1). In contrast, no prediction of risk >29 was given from the generic model in the period April–December 2016. Very low incidence symptoms (percentage of diseased leaves < 5%) of the disease were observed in the unsprayed control trees at the same period. The mean temperature was 18.6 °C in May, which increased to 24.8 °C, 26.6 °C and 26.2 °C in June, July and August, and decreased to 22.7 °C, 17.8 °C, 13.8 °C in September, October and November respectively. The above temperatures were not a limiting factor for infections of the olive trees by the fungus *V. oleaginea*. In contrast, the total degree of hourly leaf wetness was very low in the period of May to December and not favorable for the development of the disease.

Table 1. Dates of the first seasonal infection of *Venturia oleaginea* predicted by the Generic and Polynomial models in 2016, 2017 and 2018; corresponding values of the risk index calculated by the model; times of actual disease onset; and percentage of diseased leaves.

Year	Date of First Seasonal Infection Predicted by Model (>29) [a]	Risk Value [b]	Time (Days) of Actual Disease Onset [c]	Percentage of Diseased Leaves
		Generic Model		
2016	-	-	-	<5%
2017	15th May	59	14	18%
2018	3rd October	66	23	36%
		Polynomial Model		
2016	15th May [d]	56	-	<5%
2017	5th May [d]	86	24	18%
2018	2rd October [d]	64	25	36%

[a] Date when model prediction value is higher to 29. [b] Risk was calculated on a scale from 0 (no risk) to 100 (maximum risk) based on weather data and tree growth stage. [c] Days after predicted infection–incubation period. [d] Model predicted risk >29 was also observed on other dates of the year without correlated with actual disease onset (see Figure 4).

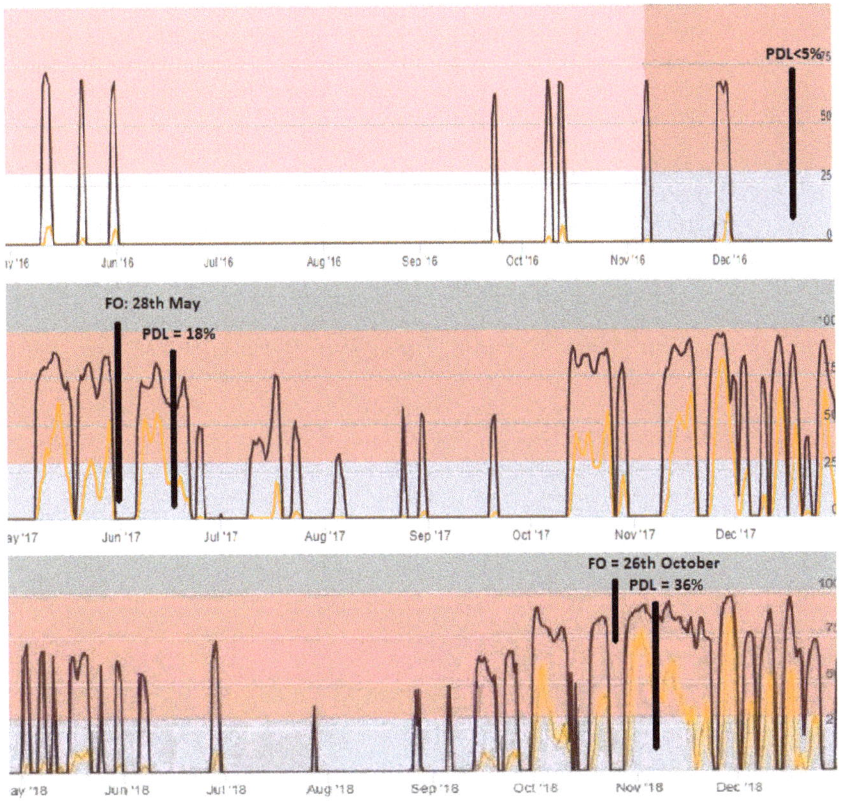

Figure 4. Predictions (as presented on the screen of the computer) of the generic model (orange line) and polynomial models (black line) to forecast infections from the fungus *Venturia oleaginea* on olive trees for the period May to December of three consecutive years (2016, 2017, 2018) (FO = First Observation; PDL = Percentage of the Diseased Leaves).

In 2017, the polynomial model predicted nearly continuously risk >29 between the 5th and 31st of May. The generic model predicted risk >29 at 15th and 31st May (Table 1). The first symptoms were observed at 28th May, and the final incidence of the symptoms was moderate when recorded 15 days later (percentage of diseased leaves in unsprayed trees was 18%). The mean temperature was 19.8 °C in May, while the total degree of hourly leaf wetness per day was about 20 in the same period. Those climate conditions were favorable for infections of olive trees by the fungus *V. oleaginea*. The optimum temperature and leaf wetness for growth of *V. oleaginea* occurred from 15th to 28th May, justifying the short incubating period of 14 days.

In 2018, the polynomial model predicted risk >29 at 2–5th, 10th, 20th, 25th and 30th May, 8th and 30th June, 25 July, 28th September (Table 1), while the predicted risk >29 was nearly always the period October-December. In contrast, no prediction of risk >29 was given from the generic model in the period April–September 2018. The generic model predicted risk >29 at the 3rd, 6th, and 20th October, 26th October, 16th November, 22th and 28th November, 3rd, 8th and 15th December. The first symptoms of the disease were observed at the 26th October, while the incidence of the symptoms was relatively high at the 10th November (percentage of diseased leaves <36%). No other results were collected after 10 November. The mean temperature was 21.7 °C in May, which increased to 24.6 °C, 26.6 °C and 27.4 °C in June, July and August, and decreased to 23.4 °C, 18.8 °C, 14.3 °C, 9.2 °C in September, October, November, December, respectively. The above temperatures were not a limiting factor for infections of the olive trees by the fungus *V. oleaginea*. The total degree of hourly leaf wetness was very low in the period of May to September, and not favorable for the development of the disease. In contrast, there was a high number of hourly leaf wetness in October, making the climate conditions favorable for the development of the disease.

The estimates of the parameters from the linear regression analysis to find the relationship between model predictions and level of the disease (Generic Model: $R^2 = 0.917$, $Y = 4.74 + 0.47 \times X$, Beta Value = 0.958; Polynomial Model: $R^2 = 0.578$, $Y = 7.86 + 0.34 \times X$, Beta Value = 0.76) are presented in Figure 5.

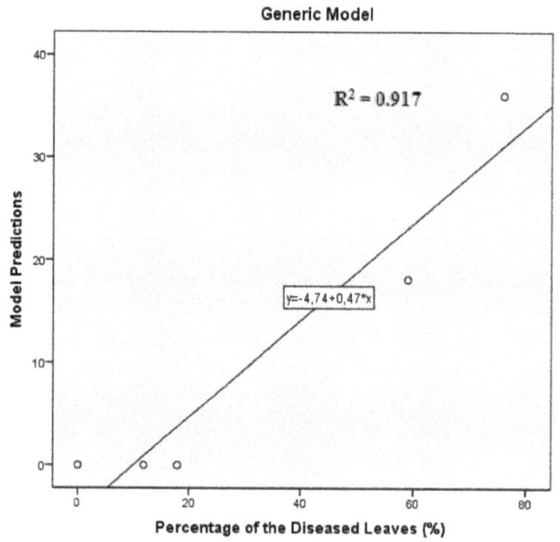

Beta Value = 0.958

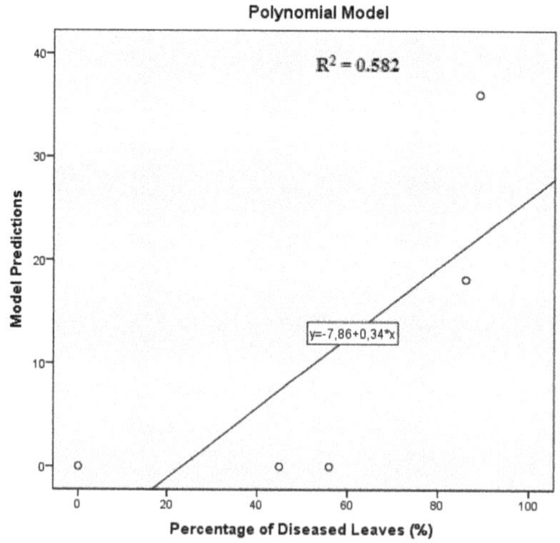

Beta Value = 0.76

Figure 5. Relationship between model (generic and polynomial) predictions and percentage of diseased leaves.

3. Discussion

So far, control of olive leaf spot has been based on prognosis. This method is adequate, but possesses disadvantages including inopportune and unnecessary spray applications. It increases the cost of production, and also the risk of environmental pollution. The introduction of predictive models to forecast the appearance of a disease could improve crop management by reducing the number of spray applications and improving the effectiveness of spray applications conducted.

Magarey et al. [16] developed a generic model appropriate for predicting the appearance of a high number of plant diseases. This model requires some climate parameters such as the minimum, maximum and optimum temperatures, as wells as the minimum and maximum numbers of hours of leaf wetness. The results of this study showed that a temperature range of 5 to 25 °C was appropriate for the conidial germination on detached olive leaves with 20 °C being the optimum. Previous works showed that conidial germination of *S. oleagina* (synonym of *V. oleaginea*) on agar was at its minimum at 5 °C, with an optimum at 20 °C, and a maximum at 30 °C [9,19,20], while Saad and Masri [13] found that conidial germination of *S. oleagina* could be observed in temperatures ranging from 5 to 25 °C. The above range of temperatures at which *V. oleaginea* conidia germinate suggests that infection may occur mainly throughout the period of September–June in olive growing regions of Northern Greece. This study also showed that at least 12 h of leaf wetness was required to start the conidial germination of *V. oleaginea* at optimum temperatures. According to Obanor et al. [15], the minimum leaf wetness periods for infection of olive trees from *S. oleagina* were 18, 12, 12, 12 and 24 h at 5, 10, 15, 20 and 25 °C, respectively, while the minimum leaf wetness periods required for conidial germination at 5, 10, 15, 20 and 25 °C were 24, 12, 9, 9 and 12 h, respectively [6].

Specific factors, such as pathogen biology, host phenology, and host variety in a specific area may significantly affect the input variables for a predictive model. As the predictive model could contain assumptions about site specific conditions, each model must be validated for a specific location by testing for one or more seasons under local conditions to verify that it works with precision in this location. Obanor et al. [15] developed a regression model to predict the infections of olive trees by the fungus *S. oleagina*. However, this model was not evaluated and validated under field conditions. In this study, the generic model developed by Magarey et al. [16] and the polynomial model developed by Obanor et al. [15] were evaluated to predict infection of olive trees by the fungus *V. oleaginea* under the climate conditions of Potidea Chalkidiki, Northern Greece. Because the purpose of the model was to be part of a warning system for olive leaf spot management, the ability to correctly predict infection periods is crucial. The results showed that both models correctly predicted infection periods. However, there was difference in the severity of the infection, as demonstrated by the goodness-of-fit for the data collected on leaves of olive trees in 2016, 2017 and 2018. Specifically, the generic model predicted lower severity of the infection which fits very well with the incidence of the symptoms of the disease. In contrast, the model developed by Obanor et al. [15] predicted high severity of the infection, but these did not fit as well with the incidence of the symptoms. Based on the above results, the polynomial model gave false positive predictions and did not generate proper spray recommendations increasing the fungicide applications with indirect results the increase of cost production and possible environmental pollution. It is recommended that the polynomial model be calibrated and re-validated under field conditions before commercial use. In contrast, the generic model gave a correct prediction for the appearance of the disease, and it seems to fit better in computer-assisted Decision Support Systems (DSSs).

Considering that this study did not include olive cultivars with different levels of susceptibility, it was not possible to evaluate whether each of the above predictive model can be fitted better to specific olive cultivars as the same climate conditions could favor different incidence of the symptoms depending on the level of cultivar susceptibility.

4. Materials and Methods

4.1. Effect of Temperature and Leaf Wetness on Conidial Germination

4.1.1. Effect of Temperature

The effect of temperature on conidial germination of *V. oleaginea* was investigated by using the methodology described by Obanor et al. [6]. Leaves (cv. Chondrolia Chalkidikis) with symptoms of olive leaf spot were collected from a commercial field established in Potidea Chalkidiki (40.1939° N, 23.3301° E) in November 2015. The leaves were agitated in distilled water and the conidial suspension filtered through a double layer of cheesecloth to remove leaf debris. Inoculum suspensions were adjusted to 6×10^4 conidia mL^{-1} by using a hemocytometer. Seven temperatures (0, 5, 10, 15, 20, 25, and 30 °C) were tested to find the upper and lower limits of spore germination. Fully expanded leaves (4 weeks old) without symptoms of the disease were excised from olive trees grown in a commercial field by cutting at the stem end of the petiole. The leaves, before being inoculated, were disinfested by dipping them in 10% vol/vol domestic bleach solution (4.85% NaOCl) for 5 min, washed three times with sterile distilled water and left to dry at room temperature. The leaves were inoculated with two drops (10 μL) of the conidial suspension deposited on the upper leaf surface. After inoculation, the leaves were placed in 9-cm petri dishes (wet paper towel was placed onto bottom and leaves was placed on plastic sticks so that to avoid any contact) arranged randomly in the growth chamber (97–100% RH) described below. Results were collected by recording the germination of 30 conidia/leaf (10 leaves for each treatment) 24h later. A conidium was considered germinated when the germ tube was equal to the greatest diameter of the swollen conidia (1 to 1.5×) [21,22].

4.1.2. Effect of Leaf Wetness

Similarly, fully expanded leaves of olive trees (cv. Chondrolia Chlakidiki) without symptoms of disease were inoculated with two drops (10 μL) of the conidia suspension as described above. After inoculation, the leaves were placed in 9-cm petri dishes (wet paper towel was placed onto the bottom, and leaves were placed on plastic sticks so as to avoid any contact) arranged randomly in the growth chamber (Emmanuel E. Chryssagis, Growth Plant Chambers—GRW 500/CMP2) (97% ± 3 Relative Humidity) under continuous wetness at the 20 °C (optimal temperature identified above) and incubated for 6, 12, 18, 24, 36 and 48 h. Results were collected by recording the germination of 30 conidia as described above.

Both experiments were repeated once. General linear regression analysis was performed (SPSS Grad Pack 23, SPSS Inc., Chicago, IL, USA) in order to determine the relationship between leaf wetness, temperatures and conidia germination.

4.2. Model Development and Validation

4.2.1. The Models

The generic model developed by Magarey et al. [16] was used. The parameters were used to run the model based on the results produced in the above experiments: Minimum Temperature (Tmin) = 5 °C, Maximum Temperature (Tmax) = 25 °C, Optimum Temperature (Topt) = 20 °C, Minimum Leaf Wetness (Wmin) = 12 h, Maximum Leaf Wetness (Wmax) = 24 h. In addition to the above, the predictive model (polynomial model; $Y = \beta 0 + \beta 1A + \beta 2T + \beta 3W + \beta 4(A \times W) + \beta 5T2 + \beta 6W2$), where Y is $\sqrt{}$(disease severity), A is the leaf age (weeks), T is temperature (°C), W is wetness period (h), and $\beta 0 \ldots \beta 6$ are determined parameters)to forecast the appearance of olive leaf spot developed by Obanor et al. [15] was simultaneously evaluated under the field conditions of Potidea Chalkidiki.

The leaf wetness was estimated from the hourly data: if an (i) hour is wet, it is counted as 1, or when it is dry it is designated as 0 (so the dry hours are not counted and are not taken into account). Continuous wet hours are summed to determine leaf wetness. However, if there is an interruption of fewer than or equal to 20 dry hours and low relative humidity (<70%) (based on the result published by Villalta et al. [23] for the fungus *Venturia pirina*), the summation of hours is continued. In contrast, if the interruption of dry hours is longer than 20 dry hours, a new summation of hours is started. Temperature is the event average temperature during each wet period. Cultivar susceptibility and inoculum level were not considered because insufficient information was available about their effects on the occurrence of infection.

4.2.2. Evaluation of Model Accuracy under Field Conditions

Model accuracy in predicting the day of infection was evaluated by comparing actual and predicted times of symptom appearance. In the Potidea Chalkidiki, which is one of the most important olive production areas in Greece, a telemetric meteorological station (NEUROPUBLIC S. A., Information Technologies & Smart Farming Services, Piraeus, 18545, Attica, Greece) was established to record weather data, which were used to run the models. The model was operated hourly, starting from the 1st of May (aiming to include both periods favorable for the development of the disease (May to June) and unfavorable periods for the development of the disease (July to August)), 00.01, and ending at 31st December using hourly leaf wetness and hourly temperatures as driving variables for calculation. The date of the first observation of the symptoms (in young leaves) was used to verify the prediction of the models, while the final incidence of the symptoms was recorded 15–20 days later by calculating the percentage of diseased leaves to a sample of 100 leaves randomly selected from each of 10 trees in total. The period of possible appearance of the disease was calculated on each day when Risk (LW, T) > 30, which was considered to be the incubation period. According to Bakarić [24], an incubation period depends on the environmental conditions, and it lasts 15 days, but can be extended from three to eight or more months. The model predictions were ranged from 0 (when Risk = 0) to 100 (when Risk = the highest possible value).It was calculated with 0 being the minimum value that could be given by the model, and 100 the maximum value. All the other values were distributed between 0 and 100. Previous preliminary work under field conditions to find the threshold for the model predictions showed that no symptoms or very light-sporadic symptoms (percentage of diseased leaves < 5%) of the disease could be observed when the model predictions were in the range 0–29 (indicating that a spray application against olive leaf spot disease would not be financially justifiable; the spray program usually includes copper-based fungicides applied before the onset of the main infection periods, which often coincide with the main shoot-growth seasons (spring and autumn)).

A commercial olive field (cv. Chondrolia Chalkidiki, 7- to 10-year-old trees), located in Potidea, Chalkidiki, was chosen to record the appearance of olive leaf spot symptoms. Trees were pruned to a vase shape by hand pruning. Five to six irrigations were provided yearly. Nitrogen was applied yearly as $(NH_4)_2SO_4$ at 100 N units per hectare. Selected trees did not show any symptom of the disease before starting the trial. The trees (kept unsprayed) were inspected twice per week to determine the time of symptom onset. The trees were carefully inspected for the appearance of the first symptoms, which are dark sooty spots (commonly known as peacock spots) appear on the upper surface of leaves, mainly in the low canopy. When the symptoms were unclear, the leaves were marked and observed during the following surveys. To assess the severity of the disease, 100 random leaves were observed for the symptoms of the disease per tree (results were collected from 10 trees), and the disease incidence was calculated as the percentage of leaves with leaf spot symptoms. The predicted period of disease onset was then compared with the actual one. The model was judged to have provided an accurate prediction when the observed symptom onset coincided with the time interval predicted by the model [25].

This experiment was conducted for three consecutive years (2016, 2017, and 2018). General linear regression analysis was performed (SPSSGradPack23, SPSS Inc., Chicago, IL, USA) in order to determine the relationship between model predictions and level of the disease (observations).

5. Conclusions

The effects of the air temperature and leaf wetness on *V. oleaginea* infection of olive leaves were clarified. Based on those results, disease forecasting systems were developed to find the proper timing of foliar fungicidal sprays. The generic model predicted lower severity, which fits well with the incidence of the symptoms of the disease in unsprayed trees, and this model could be used to schedule the spray applications against olive spot disease. In contrast, the polynomial model predicted high severity levels of infection, but this did not fit well with the incidence of the symptoms. This study could help pest managers and researchers predict the risk of olive leaf spot in different regions or under different crop management practices.

Author Contributions: Conceptualization, T.T. and A.T. methodology, F.C. software, K.M.; A.T. validation, T.T. formal analysis, T.T.; writing—original draft preparation, T.T.; K.M., F.C., A.T. writing—review and editing, T.T.; supervision. All authors have read and agreed to the published version of the manuscript.

Funding: This research was fund by NEUROPUBLIC S. A., Information Technologies & Smart Farming Services, Piraeus, 18545, Attica, Greece.

Institutional Review Board Statement: Not applicable.

Informed Consent Statement: Not applicable.

Data Availability Statement: The data presented in this study are available on request from the corresponding author. The data are not publicly available due to the agreement with the NEUROPUBLIC S. A.

Acknowledgments: In this section, you can acknowledge any support given which is not covered by the author contribution or funding sections. This may include administrative and technical support, or donations in kind (e.g., materials used for experiments).

Conflicts of Interest: The authors declare no conflict of interest.

References

1. Trapero, A.; Blanco, M.A. Diseases. In *Olive Growing*; Barranco, D., Fernández-Escobar, R., Rallo, L., Eds.; Junta de Andalucía/Mundi-Prensa/RIRDC/AOA: Pendle Hill, NSW, Australia, 2010; pp. 521–578.
2. Viruega, J.R.; Moral, J.; Roca, L.F.; Navarro, N.; Trapero, A. *Spilocaea oleagina* in olive groves of southern Spain: Survival, inoculum production, and dispersal. *Plant Dis.* **2013**, *97*, 1549–1556. [CrossRef]
3. Sanei, S.J.; Razavi, S.E. Survey of olive fungal disease in North of Iran. *Annu. Rev. Res. Biol.* **2012**, *2*, 27–36.
4. Salman, M.; Hawamda, A.; Amarni, A.A.; Rahil, M.; Hajjeh, H.; Natsheh, B.; Abuamsha, R. Evaluation of the incidence and severity of Olive Leaf Spot caused by *Spilocaea oleagina* on olive trees in Palestine. *Am. J. Plant Sci.* **2011**, *2*, 457–460. [CrossRef]
5. Prota, U. Le malattiedell' olivo. *Inf. Fitopato.* **1995**, *45*, 16–26.
6. Obanor, F.O.; Walter, M.; Jones, E.E.; Jaspers, M.V. Effect of temperature, relative humidity, leaf wetness and leaf age on *Spilocaea oleagina* conidium germination on olive leaves. *Eur. J. Plant Pathol.* **2008**, *120*, 211–222. [CrossRef]
7. Roubal, C.; Regis, S.; Nicot, P.C. Field models for the prediction of leaf infection and latent period of *Fusicladium oleagineum* on olive based on rain, temperature and relative humidity. *Plant Pathol.* **2013**, *62*, 657–666. [CrossRef]
8. Viruega, J.R.; Roca, L.F.; Moral, J.; Trapero, A. Factors affecting infection and disease development on olive leaves inoculated with *Fusicladium oleagineum*. *Plant Dis.* **2011**, *95*, 1139–1146. [CrossRef]
9. Obanor, F.O.; Walter, M.; Jones, E.E.; Jaspers, M.V. Sources of variation in a field evaluation of the incidence and severity of olive leaf spot. *N. Z. Plant Prot.* **2005**, *58*, 273–277. [CrossRef]
10. Sanei, S.J.; Razavi, S.E. Survey of *Spilocaea oleagina*, causal agent of olive leaf spot, in North of Iran. *J. Yeast Fungal Res.* **2011**, *2*, 33–38. [CrossRef]
11. González-Domínguez, E.; Armengol, J.; Rossi, V. Biology and epidemiology of Venturia species affecting fruit crops: A review. *Front. Plant Sci.* **2017**, *8*, 1496. [CrossRef]
12. Hellenic National Meteorological Services. Available online: http://www.hnms.gr/emy/en/climatology/climatology_month?minas=08 (accessed on 15 May 2020).
13. Saad, A.; Masri, S. Epidemiological studies on olive leaf spot incited by *Spilocaea oleagina* (Cast.) Hugh. *Phytopathol. Mediter.* **1978**, *17*, 170–173.
14. Viruega, J.R.; Trapero, A. Effect of temperature, wetness duration and leaf age on infection and development of olive leaf spot. *Acta Hortic.* **2002**, *586*, 797–800. [CrossRef]
15. Obanor, F.O.; Walter, M.; Jones, E.E.; Jaspers, M.V. Effects of temperature, inoculum concentration, leaf age, and continuous and interrupted wetness on infection of olive plants by *Spilocaea oleagina*. *Plant Pathol.* **2011**, *60*, 190–199. [CrossRef]

16. Magarey, R.D.; Sutton, T.B.; Thayer, C.L. A simple generic infection model for foliar fungal plant pathogens. *Phytopathology* **2005**, *95*, 92–100. [CrossRef] [PubMed]
17. De Wolf, D.E.; Scott, A.I. Disease cycle approach to plant disease prediction. *Annu. Rev. Phytopathol.* **2007**, *45*, 203–220. [CrossRef] [PubMed]
18. Esker, P.D.; Sparks, A.H.; Campbell, L.; Guo, Z.; Rouse, M.; Silwal, S.D.; Tolos, S.; Van Allen, B.; Garrett, K.A. Ecology and Epidemiology in R: Disease forecasting and validation. *Plant Health Instr.* **2008**. [CrossRef]
19. Dzaganiya, A.M. Data on the characteristics of development of the pathogen of olive leaf spot in Georgia. *Rev. Appl. Mycol.* **1996**, *46*, 2788.
20. Wilson, E.E.; Miller, H.N. Olive leaf spot and its control with fungicides. *Hilgardia* **1949**, *19*, 1–24. [CrossRef]
21. Dantigny, P.; Bensoussan, M.; Vasseur, V.; Lebrihi, A.; Buchet, C.; Ismaili-Alaoui, M.; Devlieghere, F.; Roussos, S. Standardisation of methods for assessing mould germination: A workshop report. *Int. J. Food Microbiol.* **2006**, *108*, 286–291. [CrossRef]
22. Dhingra, O.D.; Sinclair, J.B. *Basic Plant Pathology Methods;* CRC Press: Boca Raton, FL, USA, 1985.
23. Villalta, O.; Washington, W.S.; Rimmington, G.M.; Taylor, P.A. Influence of spore dose and interrupted wet periods on the development of pear scab caused by *Venturia pirina* on pear (*Pyrus communis*) seedlings. *Australas. Plant Pathol.* **2000**, *29*, 255–262. [CrossRef]
24. Bakaric, P. *Peacock Olive Leaf Spot/Spilocaea oleagina (Cast) Hughes;* HCPHS—Zavod za Zaštitu Bilja: Dubrovnik, Croatia, 2004; pp. 1–65.
25. Giosuè, S.; Spada, G.; Rossi, V.; Carli, G.; Ponti, I. Forecasting infections of the leaf curl disease on peaches caused by *Taphrina deformans*. *Eur. J. Plant Pathol.* **2000**, *106*, 563–571. [CrossRef]

plants

Article

Effects of Temperature and Wetness Duration on Infection by *Coniella diplodiella*, the Fungus Causing White Rot of Grape Berries

Tao Ji [1], Luca Languasco [1], Ming Li [2,3] and Vittorio Rossi [1,*]

1 Department of Sustainable Crop Production, Università Cattolica del Sacro Cuore, Via E. Parmense 84, 29122 Piacenza, Italy; tao.ji@unicatt.it (T.J.); luca.languasco@unicatt.it (L.L.)
2 National Engineering Research Center for Information Technology in Agriculture (NERCITA)/National Meteorological Service Center for Urban Agriculture, China Meteorological Administration & Ministry of Agriculture and Rural Affairs, Beijing 100097, China; lim@nercita.org.cn
3 Collaborative Innovation Center for Green Prevention and Control of Forest and Fruit Diseases and Insect Pests, Beijing Academy of Agriculture and Forestry Sciences, Beijing 100097, China
* Correspondence: vittorio.rossi@unicatt.it

Abstract: Grapevine white rot, caused by *Coniella diplodiella*, can severely damage berries during ripening. The effects of temperature and wetness duration on the infection severity of *C. diplodiella* were investigated by artificially inoculating grape berries through via infection pathways (uninjured and injured berries, and through pedicels). The effect of temperature on incubation was also studied, as was that of inoculum dose. Injured berries were affected sooner than uninjured berries, even though 100% of the berries inoculated with *C. diplodiella* conidia became rotted whether injured or not; infection through pedicels was less severe. On injured berries, the disease increased as the inoculum dose increased. Irrespective of the infection pathway, 1 h of wetness was sufficient to cause infection at any temperature tested (10–35 °C); with the optimal temperature being 23.8 °C. The length of incubation was shorter for injured berries than for uninjured ones, and was shorter at 25–35 °C than at lower temperatures; the shortest incubation period was 14 h for injured berries at 30 °C. Mathematical equations were developed that fit the data, with R^2 = 0.93 for infection through any infection pathway, and R^2 = 0.98 for incubation on injured berries, which could be used to predict infection period and, therefore, to schedule fungicide applications.

Keywords: *Coniella diplodiella*; environmental factors; infection pathway; inoculum dose; incubation

Citation: Ji, T.; Languasco, L.; Li, M.; Rossi, V. Effects of Temperature and Wetness Duration on Infection by *Coniella diplodiella*, the Fungus Causing White Rot of Grape Berries. *Plants* **2021**, *10*, 1696. https://doi.org/10.3390/plants10081696

Academic Editor: Alessandro Vitale

Received: 22 July 2021
Accepted: 13 August 2021
Published: 18 August 2021

1. Introduction

White rot of grapevines is also known as "hail disease" because it frequently develops following hailstorms [1]; the disease, however, can develop even in the absence of hailstorms [2]. White rot, which is caused by the fungus *Coniella diplodiella* (Speg.) Petrak and Sydow (syn. *Coniothyrium diplodiella* [Speg.] Sacc. and *Pilidiella diplodiella* [Speg.] Crous and van Niekerk; [3]), was first described in 1878 in Italy [4], and is currently distributed worldwide [5–7]. In addition to *C. diplodiella*, other species of *Coniella*—including *C. vitis* [6] and *Pilidiella castaneicola* [8]—can naturally infect vines in the vineyard, or can infect grapevines in inoculation studies in the case of *C. petrakii* and *C. fragariae* [9].

Grapevine white rot affects all green tissues of the vine, but mainly damages ripening clusters. Yellowish and water-soaked lesions appear on the edges or tips of the leaves, which gradually expand inward, forming concentric, wheel-shaped lesions that eventually dry; grayish-white pycnidia can appear on the diseased tissue in moist weather [10,11]. Long, depressed, brownish necrotic areas may also occur around the nodes of green shoots that can evolve in cankers [2]. As previously noted, however, the typical symptoms appear on clusters. The pathogen forms yellowish-brown and water-soaked areas that rapidly enlarge to cover the whole berry, which finally becomes soft and rotten. In a later stage, the berry's surface is densely covered with small, brown-violet pustules consisting of immature pycnidia that rise from the epidermal layers of the cuticle without rupturing it; the layer of air between the cuticle and the epidermis plus the grayish-white color of the mature pycnidia make the infected berries appear white, which explains why the disease is named white rot [2,12].

C. diplodiella overwinters as mycelia or pycnidia in infected canes, rachises, and mummified berries that fall into the soil [13], where pycnidia can remain viable for more than 15 years, and can repeatedly discharge conidia under moist conditions [2,14]. Conidia (and soil particles containing conidia) are transported to plant surfaces by splashing rain or farm equipment. Conidia then germinate and cause infection through stomata [5,10], microcracks, or mechanical wounds [4,10], or via direct penetration [4].

White rot is often closely associated with hailstorms that occur during berry ripening, because hail causes wounds that facilitate the infection of berries. In areas where hailstorms are sufficiently frequent to cause a buildup of *C. diplodiella* inoculum in the vineyard, a single-spray application of a fungicide is recommended as soon as possible after the storm; the recommended fungicides include folpet, dichlofluanid, or thiram [15]; captan and dicarboximides [2]; chlorothalonil [13]; and tebuconazole, pyraclostrobin, carbendazol, and mancozeb [16]. In affected vineyards, however, summer rain followed by persistent moisture and temperatures of 24–27 °C can occasionally lead to disease outbreaks [2]. In some viticultural areas, the disease occurs almost every season. For instance, grape white rot has been reported as one of the main fungal diseases affecting grapes in China [11,17,18], where it causes an annual production loss of 16% [17]. In these cases, repeated fungicide applications are required to control the disease [19].

Knowledge of weather conditions leading to infection may be important for timing both single hail-related fungicide applications and repeated fungicide applications. Researchers have found that infection occurs over a wide temperature range, from <12 °C and >33 °C, with the optimum between 24 and 27 °C [2,13,20,21]. Infection also requires a minimum of 2 h of wetness [10].

Knowledge about the pathways and environmental conditions for infection by *C. diplodiella* is crucial for disease control, but the available information is fragmented, inconsistent, and based on only a few early studies. We therefore conducted a study with detached grapevine berries under controlled environmental conditions to determine (1) the temperature and moisture conditions required for different infection pathways, and their interactive effects on disease severity; (2) the effect of inoculum dose on berry infection; and (3) the effect of temperature on incubation length.

2. Results

2.1. *Effect of Temperature and Wetness Duration on Berry Infection (Experiment 1)*

The disease progressed very rapidly when berries were injured before they were inoculated with *C. diplodiella* conidia; 90% of the berries had initial water-soaked areas at 1 d after inoculation, and all of the injured berries were completely rotten after 7 d (Figure 1b). Disease progress was slower for uninjured berries, with approximately 80% of berries showing symptoms on the majority of the berry surface after 10 d (Figure 1a). Disease progress was slowest for pedicel inoculation, with less than 50% of the berries affected after 13 d (Figure 1c). The AUDPC values were then calculated at 4 d for injured berries, 10 d for uninjured berries, and 13 d for pedicel-inoculated berries, and then standardized as previously described (Figure 2).

The ANOVA carried out on standardized AUDPC data showed that all of the main sources of variation affected disease progress with $p < 0.001$. Infection pathway accounted for >50% of the variance, with the AUDPC highest for injured berries (overall average 0.476 ± 0.011), intermediate for uninjured berries (0.395 ± 0.014), and lowest for pedicel inoculation (0.249 ± 0.007). Duration of the wet period (WD) accounted for 27.9% of the variance, and the pathway × WD interaction accounted for 5.8% of the variance. For any infection pathway, the AUDPC for berries that were kept wet for only 1 h after inoculation was approximately 50% of the AUDPC for berries that were kept wet for 24 h after inoculation (Figure 2). The AUDPC then increased with increasing WD up to 24 h, except that the AUDPC for pedicel inoculation did not increase further after a 3 h WD (Figure 2). Temperature during the wet period accounted for 7.9% of the variance, and the infection pathway × T interaction was not significant (p = 0.086). For all infection pathways, the optimal temperatures for infection were therefore between 20 and 30 °C; at 10 and 35 °C, the AUDPC values were significantly lower than at the optimal temperature range (p < 0.001). The infection pathway × WD × T interaction was significant at p = 0.044, but accounted for only 4.1% of the variance.

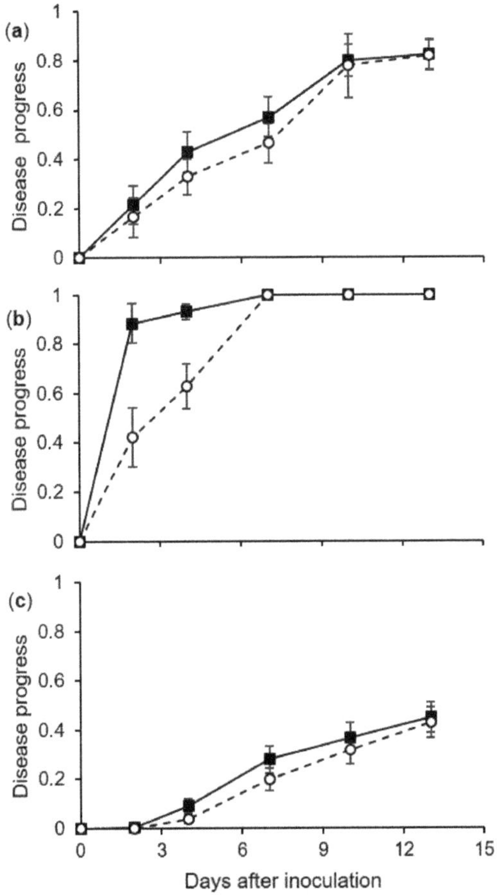

Figure 1. Progress of disease incidence (as the proportion of berries showing symptoms, on a 0–1 scale, where 1 means all of the berries are affected; black squares) and severity (as the proportion of the berry surface showing symptoms, on a 0–1 scale, where 1 means all of the berry surface is affected; white circles) on grapevine berries inoculated with conidia of *Coniella diplodiella* (experiment 1). After uninjured berries (**a**), injured berries (**b**), or pedicels of uninjured berries (**c**) were inoculated, the berries were kept at 30 combinations of 6 temperatures (T: 10, 15, 20, 25, 30, and 35 °C) and 5 wetness duration periods (WD; 1, 3, 7, 12, and 24 h); each combination was represented by 3 replicates, with 15 berries per replicate. In each panel, values are means (±SE, n = 90) of disease incidence or severity with the 6 temperatures and 5 wetness duration periods.

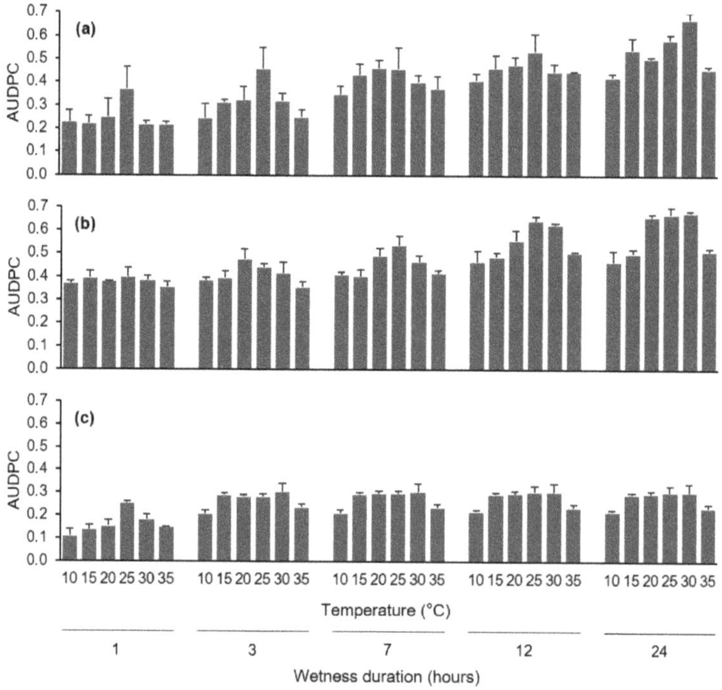

Figure 2. Values of the area under the disease progress curve (AUDPC, dimensionless) for grapevine berries inoculated with conidia of *Coniella diplodiella* (experiment 1). After uninjured berries (**a**), injured berries (**b**), or pedicels attached to uninjured berries (**c**) were inoculated, the berries were kept at 30 combinations of 6 temperatures and 5 wetness duration periods. Values are means (± SE, n = 45) of AUDPC data. AUDPC data were standardized by dividing the original AUDPC values by the time considered for their calculation, i.e., 10, 4, and 13 d for A, B, and C, respectively.

The estimated parameters for Equation (2) were Tmin = 5 °C, Tmax = 40 °C, $a = 5.067 \pm 1.777$, $b = 1.162 \pm 0.088$, $c = 0.350 \pm 0.054$, $m = 0.468 \pm 0.286$, and $n = 0.363 \pm 0.092$, which provided a three-dimensional representation of the changes in the relative infection severity based on the combined effect of temperature and wetness duration (Figure 3), with $R^2 = 0.93$, CCC = 0.96, RMSE = 0.10, and CRM < 0.0001. The plot of predicted versus observed values did not show systematic deviations (not shown). For the whole dataset, the intercept of the regression line of predicted versus observed data was $\alpha < 0.0001$ (not significantly different from $\alpha = 0$, at $p = 0.998$) and slope $\beta > 0.999$ (not significantly different from $\beta = 1$, at $p = 0.982$).

Based on Equation (3), the optimal temperature was Topt = 23.8 °C, with a 95% confidence interval from 22.4 to 25.0 °C.

2.2. *Effect of Inoculum Dose on Berry Infection (Experiment 2)*

The progress of disease severity on the berries that were injured and then drop-inoculated with conidia of *C. diplodiella* differed depending on the inoculum dose and the temperature at which the berries were incubated following inoculation (Figure 4). Specifically, disease severity increased very rapidly in the first 7 d following inoculation when the inoculum dose was high (10^4 and 10^5 conidia/mL), and with temperatures of 25 and 35 °C; in these treatments, the disease did not substantially increase in the next 7 d (Figure 4). No or very light disease was observed at low inoculum doses (10 or 10^2 conidia/mL, respectively), regardless of temperature (Figure 4).

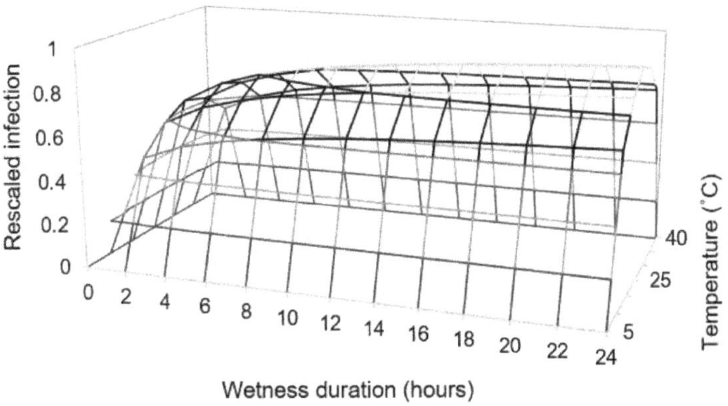

Figure 3. Effects of temperature and wetness duration on the rescaled infection severity in grapevine berries inoculated with conidia of *Coniella diplodiella*, as estimated by Equation (2) and with data from experiment 1. After uninjured berries, injured berries, or pedicels attached to uninjured berries were inoculated, the berries were kept at 5 wetness duration periods (1–24 h) and at 6 temperatures (10–35 °C). For fitting purposes, AUDPC data were rescaled on a 0–1 scale by dividing the original AUDPC values by the maximal value for each infection pathway (uninjured, injured, and pedicels).

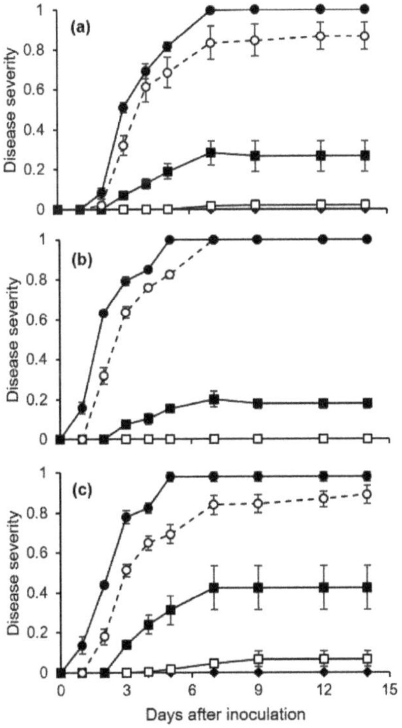

Figure 4. Progress of disease severity (as the proportion of berry surface showing symptoms, on a 0–1 scale, where 1 means that all of the berry surface is affected) on grapevine berries that were injured and then drop-inoculated with a suspension of *Coniella diplodiella* that contained 10 (black diamond), 10^2 (white squares), 10^3 (black squares), 10^4 (white circles), or 10^5 (black circles) conidia/mL (experiment 2). After inoculation, berries were incubated for a 24 h wet period at 15 (**a**), 25 (**b**), or 35 °C (**c**). Values are the means (± SE, n = 45).

The ANOVA of the normalized AUDPC data showed that the disease was significantly ($p < 0.001$) influenced by inoculum dose—which accounted for approximately 97% of the variance—and by temperature, but not by their interaction. Based on the ANOVA, the AUDPC decreased as the inoculum dose decreased (Figure 5), and the disease was more severe at 25 or 35 °C (average of normalized AUDPC = 0.38 ± 0.11 and 0.37 ± 0.09, respectively) than at 15 °C (0.30 ± 0.08).

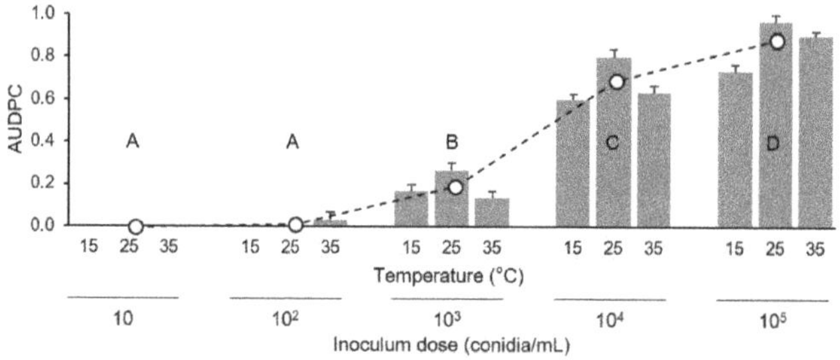

Figure 5. Values of the area under the disease progress curve (AUDPC, dimensionless) for grapevine berries that were injured and then drop-inoculated with conidia of *Coniella diplodiella* at 5 inoculum doses (10, 10^2, 10^3, 10^4, or 10^5 conidia/mL), and then incubated at 3 temperatures (T: 15, 25, or 35 °C) for 24 h (experiment 2). AUDPC data were standardized by dividing the original AUDPC values by the time considered for their calculation, i.e., 9 d. Bars indicate means (± SE) of 3 replicates, with 15 berries per replicate. White dots are the means (± SE, n = 45) of the 3 incubation temperatures for each inoculum dose; letters show comparisons of the latter means as determined by Tukey's HSD test, with $p = 0.05$.

2.3. *Effect of Temperature on Incubation Length (Experiment 3)*

Lesions on injured berries were first observed at 24 h after inoculation with *C. diplodiella* when the injured berries were kept at 20–35 °C (Figure 6a). The water-soaked lesions first appeared near the wound, and gradually extended to cover the entire berry surface, and the berries subsequently became soft and rotten (yellow to pink) in the next 1–2 weeks. On uninjured berries, symptoms first appeared after 48 h at 30 °C (Figure 6b). Disease symptoms developed more rapidly on injured than on uninjured berries (Figure 6a,b). For both injured and uninjured berries, the disease developed faster at 25–35 °C than at 10–20 °C.

The incubation period was longest at 10 °C, with IP_{50} = 236 h for injured berries and 416 h for uninjured berries. The length of the incubation period decreased as temperature increased; at 30 °C, IP_{50} = 14 h for injured berries and 101 h for uninjured berries (Figure 7). The following parameter estimates of Equation (5) fit the incubation data for injured clusters with R^2 = 0.98 and CCC= 0.98: $IP_{50}min$ = 16 h, Tmin = 5.5 °C, Topt = 30 °C, and Tmax = 41.5 °C (Figure 7a). Estimates for uninjured clusters (with R^2 = 0.80 and CCC = 0.85) were $IP_{50}min$ = 101 h, Tmin = 1 °C, Topt = 30 °C, and Tmax = 43.5 °C (Figure 7b).

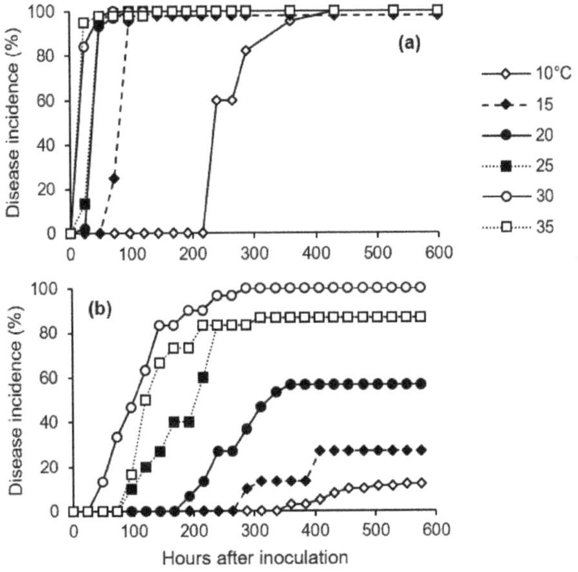

Figure 6. Percentage of injured (**a**) and uninjured (**b**) grape berries showing white rot symptoms following artificial infection with *Coniella diplodiella* conidia (experiment 3). After they were inoculated, berries were initially kept at 30 °C and with 100% relative humidity for 12 h to facilitate infection, and were then kept at one of the six indicated temperatures.

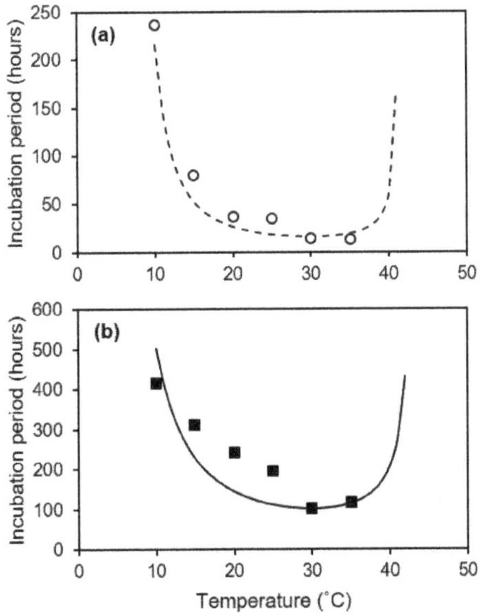

Figure 7. Effect of temperature on the length of the incubation period in grape berries that were injured (**a**) or uninjured (**b**) and artificially inoculated with *Coniella diplodiella* conidia (experiment 3). After they were inoculated, the berries were kept at 30 °C and with 100% relative humidity for 12 h to facilitate infection, and were subsequently kept at one of the six indicated temperatures. Period lengths were expressed as the number of hours required to reach 50% disease incidence (IP_{50}). The lines show the fit of the data with Equation (5).

3. Discussion

In this study, we inoculated injured grape berries, uninjured grape berries, or berry pedicels with *C. diplodiella* conidia. To our knowledge, this is the first time different methods of *C. diplodiella* inoculation have been compared and the subsequent disease progress observed. In previous studies, inoculations were performed with mechanically wounded and chemically dewaxed berries [4], or with damaged berries only [5]. The information obtained by comparing different inoculation methods that represent the possible infection pathways in the vineyard increases our understanding of white rot's epidemiology.

Berries that were injured before artificial inoculation with conidia of *C. diplodiella* were rapidly and severely affected, with some hyphae being visible on the inoculation site as early as 1 d after inoculation. The germination of *C. diplodiella* conidia, for example, was previously found to require external nutrients [14,22]. An early study found that germination was much more rapid in grape juice than in water, and that the minimal sugar concentration required for germination was 0.01% [5]. In another study, the germination rate increased linearly with sugar concentration [10]. Therefore, the presence of wounds on our mature berries likely provided nutrients that supported rapid germination of conidia, hyphal growth, and berry rotting. Our results are consistent with previous field reports that linked white rot outbreaks to hailstorms that occurred during berry ripening [1,2,5,11,23].

Uninjured berries were also severely affected in our experiments, but the disease developed more slowly on uninjured berries than on injured berries. Infection of uninjured berries may have occurred through microscopic cracks ("microcracks") [10] that were not visible at the time of berry sampling in the vineyard. Microcracks in the cuticular membrane of berries increase susceptibility to pathogens [24]. Microcracks occur naturally on ripening berries, but their causes are not well understood [25,26]. Because *C. diplodiella* requires a high sugar concentration for conidial germination [10], and does not grow on the berry surface [4], penetration through microcracks may occur for those conidia that deposit near microcracks. Under natural conditions, the probability that conidia are deposited near microcracks may be much lower than in the current study, because we sprayed the entire berry surface with *C. diplodiella* conidia, and this may have resulted in a higher probability that some conidia were located near microcracks. Under natural conditions, the probability that conidia deposit near to microcracks may depend on both the number of conidia depositing on the berry surface and the frequency of microcracks. Becker and Knoche [25] found that the number of microcracks per mm^2 on "Riesling" grape berries ranged from 0.02 to 1.35, and that the value depended on the orientation of the berry in the cluster and berry region (stylar scar, cheek, or pedicel end). The hypothesis that the probability of conidia landing near microcracks and infecting uninjured berries is low under natural conditions is supported by the steep decline in infection as the concentration of *C. diplodiella* conidia declined (Figures 4 and 5).

Berries inoculated via the pedicel were also infected, but much less rapidly and less severely than injured or uninjured berries that were directly sprayed with conidia. The pedicel is a site where nutrients accumulate, because the retention of water on the pedicels may lead to the leakage of nutrients from the inner tissue, which may favor conidial germination and infection [10]. The development of water-soaked lesions in our berries, in fact, began at the base of the pedicels and gradually extended into the entire berry. That the fungus can infect through the pedicel and also through rachises was also observed by Locci and Quaroni [4].

In addition to infection pathways, our research provided quantitative data on the effects of temperature and moisture on berry infection. Irrespective of the infection pathway, *C. diplodiella* was able to cause infection under a wide range of temperatures, from 10 to 35 °C, with an estimated optimum of 23.8 °C, with a 95% confidence interval from 22.4 to 25.0 °C. The time required for the first visible symptoms to appear on artificially inoculated berries was dependent on temperature after infection, and was 1–7 d for injured berries and 2–14 d for uninjured berries; the overall time required was shorter at 20–35 °C and longer at 15 and 10 °C; this result clarifies the effect of temperature on infection, and also helps explain inconsistencies in previously published books and compendia [2,20,21]. For instance, Bisiach [2] indicated that the optimal temperature ranged from 24 to 27 °C, that the disease was slight above 34 °C, and that when the temperature fell below 15 °C for 24–48 h following a hailstorm, infection was negligible. In the field, the incubation period was reported to be 3–12 d, with the shortest durations at 25–27 °C [1,2,20,22]. We also found that a 1 h wet period was sufficient for infection via any of the three infection pathways and at any of the temperatures tested, and that the disease progression slightly increased with a longer wet period. To our knowledge, these are the first experimental data on the moisture requirements for *C. diplodiella* infection. The only previous data we found were from Chen et al. [10], who observed that 2 h of wetness following wounding was sufficient for a 59% disease incidence under field conditions.

Our overall results demonstrated that the infection of ripe grape berries can occur over a wide range of temperatures, and requires only a very short wet period. Considering that both hyphal growth and sporulation also occur between 10 and 35 °C [27–30]—i.e., at temperatures that are typical during berry ripening in the areas where grapes are grown [31,32]—and considering that short wet periods are likely to occur when splashes from rain disperse the conidia from the soil to clusters, we infer that the *C. diplodiella* infection cycle is not substantially limited by environmental conditions.

This provides insight into disease control. In locations where white rot outbreaks commonly follow hailstorms, growers usually apply fungicides after hail, but with inconsistent results. For instance, the efficacy was approximately 75% when folpet or captan was applied within 12–18 h after hail, but was only 0–50% when the treatments were applied 21–24 h after hail [33]. Dicarboximides were also effective when applied within 12–18 h after a hailstorm if temperatures were below 20 °C [2]. In early studies, interventions within 2–20 h after hailstorm [23,33–36] were often effective, those applied as late as 21 h after a hailstorm provided adequate protection, and those applied 24 h after a hailstorm were ineffective [37]. In other studies, however, treatments after 18 h were ineffective [33]. Our results indicate that fungicides should be applied as early as possible to control white rot following a hailstorm.

For the infections that are not related to hailstorms, and occur through pedicels and through microcracks on berry surfaces, disease control should be preventative, and should be based on one or more of the following: (1) reduction of inoculum; (2) prevention of microcracks on berry surfaces; and (3) fungicide application. The sources of primary inoculum of *C. diplodiella* are diseased rachises and mummified berries that have remained in the soil from previous seasons [2,22,38,39]; such inocula can remain viable for many years [13]. Inoculum can be reduced by the timely removal of affected clusters from the vineyard [22], or by soil disinfestation [38]. Microcracks on berry surfaces can be prevented by the careful management of water and fertilizer, the monitoring of soil moisture, and the proper spraying of growth hormones and micronutrients [10,26].

The equations we developed in this work provide an estimation of the combined effect of temperature and wetness duration on infection severity, and of temperature on the length of incubation. To our knowledge, this is the first time such equations have been developed. These equations could be used as risk algorithms [40] to predict infection period and to schedule fungicides accordingly. Further studies in vineyards are needed, however, in order to verify the utility of the equations.

4. Materials and Method

4.1. Fungal Isolates and Grapevine Cultivars

Three strains of *Coniella diplodiella* were used in this study: strain CBS166.84 was provided by Westerdijk Fungal Biodiversity Institute, and had been isolated from *Vitis vinifera* L. in Germany; strains COD2D and COD5A were isolated for the purpose of this study from grapevine cv. Merlot (*Vitis vinifera* L.), in Castell'Arquato in northern Italy. The identification of these strains was confirmed at the molecular level (see Supplementary

Materials). Fungal colonies were maintained on water agar in tubes that were kept at 4 °C until used.

For the preparation of inoculum for the artificial inoculation of grape berries, 36-day-old cultures of the three fungal strains that had been grown on PDA at 20 °C were washed three times with 5 mL of sterile water and gently shaken. The resulting suspensions were passed through a double layer of sterile cheesecloth with sterile water. The suspensions of conidia were then adjusted to equal concentrations and were used in infection studies by mixing equal amounts of the three strains.

Two grapevine cultivars—Chardonnay and Ortrugo (*V. vinifera* L.)—were used; the vines had been planted in 2018 in a vineyard at Castell'Arquato, northern Italy (latitude 40°51′29″ N; longitude 9°51′17″ E, 258 m ASL). The vines were trained using a Guyot system (2.4 m between rows, 0.8 m between vines in rows) and were managed following the local practices, with soil management under the rows and natural grass growing in the inter-row space; no fungicides were applied starting from 45 d before the beginning of the study period. Thirty random clusters were sampled at ripening. For "Chardonnay", sugars were 18.1° Brix, pH was 3.2, and total acidity (as tartaric acid) was 6.4 g/L; for "Ortrugo", which is a local variety, sugars were 18.4° Brix, pH was 3.0, and total acidity was 6.0 g/L. The berries used for artificial inoculations were cut from clusters at their pedicels using scissors; all of the berries were turgid and visually disease-free with intact skins. In preliminary tests, the two cultivars showed high and similar susceptibility to *C. diplodiella*.

4.2. *Effects of Temperature and Wetness Duration on Berry Infection (Experiment 1)*

The first experiment assessed the effects of temperature (10, 15, 20, 25, 30, or 35 °C) and wetness duration (1, 3, 7, 12, or 24 h) on the infection of berries by *C. diplodiella*. Each of the 30 combinations of temperature × wetness duration was represented by 3 replicates, with 15 berries per replicate.

Berries with no visible damage and with their pedicels attached were washed in running tap water for 10 min, immersed in 75% ethanol for 1 min and then in 10% sodium hypochlorite for 1 min, and finally washed three times with distilled water. Berries were arranged over a metal mesh so that they did not touch one another, and the metal mesh was placed in an aluminum foil box with a wet paper towel on the bottom; a box was considered a replicate, and there were 15 berries per box. Berries were then inoculated as described in the next paragraph. After inoculation, boxes were sealed in a plastic bag to create a saturated atmosphere, and were kept in growth chambers at each of the 6 temperatures, under fluorescent light (12 h photoperiod).

At the end of each of the 5 wetness duration periods, berries were taken from the box and deposited on filter paper under a laminar air flow for 10 min to dry the berry surface; for injured berries (see below), the drop that exuded from the wound was dried by touching it with a piece of filter paper. The dried berries were then returned to the boxes, which were sealed and incubated at room temperature (25 ± 2 °C) with a 12 h photoperiod.

Berries were individually observed at 1–3-day intervals for 2 weeks to assess disease incidence and disease severity. Disease incidence was calculated as the percentage of berries showing yellowish-brown and water-soaked areas, and/or tissue softening and rot. Disease severity was assessed on individual berries as the percentage of the berry's surface area showing disease symptoms by using a standard diagram that included the following classes: 0 (disease-free), 5, 10, 20, 30, 50, 75, and 100% (totally rotten). Disease incidence and severity were averaged for each replicate box.

We conducted experiment 1 three times, and we used three methods to inoculate the berries; these methods corresponded to the main infection pathways of *C. diplodiella*: (1) infection of uninjured berries; (2) infection of pedicels still attached to berries; and (3) infection of injured berries. For inoculation of uninjured berries (cv. Ortrugo), a conidial suspension (3×10^5 conidia/mL) was uniformly distributed on berry surfaces with a hand sprayer (500 µL of suspension per box). For pedicel inoculation (cv. Ortrugo), a 20 µL drop of conidial suspension (3×10^5 conidia/mL) was placed on the pedicel of each berry. For inoculation of injured berries (cv. Chardonnay), a sterilized needle was inserted through the skin and into the pulp of the berry, resulting in one wound per berry; a sterile pipette was then used to deposit a 20 μL drop of a conidial suspension (3×10^5 conidia/mL) on the wound.

4.3. Effect of Inoculum Dose on Berry Infection (Experiment 2)

The second experiment assessed the effects of 5 inoculum concentrations (10, 10^2, 10^3, 10^4, and 10^5 conidia/mL) on injured berries. Injured berries (cv. Chardonnay) were inoculated as described previously (experiment 1), and were kept at 15, 25, or 35 °C for 24 h; afterwards, the drop that exuded from the wound was dried by touching it with a piece of filter paper. The dried berries were then returned to the boxes, which were sealed and incubated at room temperature (25 ± 2 °C) under fluorescent light (12 h photoperiod). The berries were then assessed each day for 15 days for disease incidence and severity, as described in the previous section. There were 15 combinations of inoculum concentration × temperature, and each combination was represented by 3 replicate boxes, with 15 berries per box. The experiment was repeated once.

4.4. Effect of Temperature on Incubation Period (Experiment 3)

Experiment 3 assessed the effect of temperature on the incubation period for injured or uninjured berries that were artificially inoculated with *C. diplodiella*. Berries were inoculated as described for experiment 1.

Experiment 3 was conducted twice, and we used two methods to inoculate the berries: In the first method, the conidial suspension was uniformly distributed on the surfaces of uninjured berries (cv. Chardonnay) using a hand sprayer (500 μL of suspension per box). In the second method, a sterilized needle was inserted into the pulp of the berry (one wound per berry) (cv. Ortrugo), and a 20 μL drop of the conidial suspension was deposited on the wound using a sterile pipette. After uninjured and injured berries were inoculated, boxes were sealed in a plastic bag to create a saturated atmosphere, and were incubated in growth chambers at 30 °C for 12 h to facilitate infection. The berries were then removed from the box and placed on filter paper under a laminar air flow for 10 min to dry the berries' surfaces. The berries were then returned to the boxes, which were sealed and incubated in growth chambers at 6 temperatures (10, 15, 20, 25, 30, and 35 °C) under fluorescent light (12 h photoperiod); there were 3 replicate boxes (with 15 berries per box, and with uninjured and injured berries in separate boxes) for each combination of inoculation method and temperature treatment.

To assess the effects of inoculation method and temperature on the length of incubation, both injured and uninjured berries were individually observed daily for 24 d to assess disease incidence as the percentage of berries showing symptoms (as yellowish-brown and water-soaked areas, tissue softening, and rot).

4.5. Data Analysis

Disease severity data assessed at different times following inoculation of uninjured berries, injured berries, and pedicels under different conditions of wetness duration (WD) and temperature (T) were used to calculate the area under the disease progress curve (AUDPC) [41] by using the trapezoidal integration as follows:

$$\text{AUDPC} = \sum (y_i + y_{i+1})/2 \times (t_{i+1} - t_i) \quad (1)$$

where y_i and y_{i+1} are the disease severity data (on a 0–1 scale) at two consecutive assessment times—i and i + 1, respectively [42]. Based on a preliminary data analysis of the disease curves (see Figures 1 and 4), the AUDPC in the first experiment was calculated based on data collected at 10, 4, and 13 d after inoculation of uninjured berries, injured berries, and pedicels, respectively, and was calculated based on data collected at 9 d after inoculation in the second experiment. The AUDPC data were then normalized by dividing them by the time considered for their calculation [43].

Normalized AUDPC data were subjected to a factorial analysis of variance (ANOVA) to test the effect of infection pathway, WD, T, and their interactions in the first experiment, and the effect of inoculum dose, T, and their interaction in the second experiment. The ANOVAs were carried out using the SPSS package v. 19 (SPSS Inc., Chicago, IL, USA).

To study the relationship between temperature, wetness duration, and infection severity, we calculated a rescaled infection severity (0–1) by dividing the original AUDPC values by the maximal value in each replicated experiment (experiment 1). These data were then fitted by using following equation:

$$Y = (a\ \mathrm{Teq}^{b}\ (1 - \mathrm{Teq}))^{c}\ (1 - \exp(-((m\ \mathrm{WD})^{n}))) \tag{2}$$

where Y is the rescaled infection severity; Teq is an equivalent of temperature, calculated as Teq = (T − Tmin)/(Tmax − Tmin), in which T is the mean temperature during the wet period (°C), and Tmin and Tmax are the minimum and maximum temperature, respectively; WD is the duration of the wet period (hours); and a, b, c, m, and n are the equation parameters. The parameter a is a proportionality factor, such that infection at the optimal temperature = 1; b is the parameter that regulates the infection severity increase between the minimal and the optimal temperature; c is the parameter that regulates the decrease from the optimal to the maximal temperature; m refers to the intrinsic rate of increase of the infection severity with respect to WD; and n is the growth rate. Tmin and Tmax were also estimated as model parameters.

The optimal temperature (Topt) was estimated as described by Analytis [44], based on the estimated values of b, Tmin, and Tmax in Equation (2), as follows:

$$\mathrm{Topt} = b/(b + 1)\ (\mathrm{Tmax} - \mathrm{Tmin}) + \mathrm{Tmin} \tag{3}$$

The parameters of Equation (2) were estimated using the non-linear regression procedure of Origin 8 Pro (OriginLab Corporation, MicroCal). Goodness-of-fit of equations (e.g., Equation (2)) were determined based on the adjusted R^2, the magnitude of the standard errors of the parameters, the root-mean-square error (RMSE), the coefficient of residual mass (CRM), and the concordance correlation coefficient (CCC) [45,46]. RMSE is the measure of the average distance between the real data and the fitted line [46]. CRM represents the tendency of the model toward over- or underestimation; more specifically, a negative CRM indicates that the model overestimates, while a positive CRM indicates that the model underestimates [46]. CCC estimates the difference between the fitted line and the perfect agreement line; a CCC value of 1 indicates that the fitted line is identical to the perfect agreement line [40,45]. The predicted values were then regressed against the observed values, and the null hypotheses that the intercept of the regression line was $\alpha = 0$ and that the slope was $\beta = 1$ were tested using a t-test; when this test was not significant, both null hypotheses were accepted, and the model was considered a statistically accurate predictor of the real data [47].

The length of the incubation was expressed as the number of hours needed to attain 50% disease incidence by day 24 (hereafter referred to as the IP_{50}). IP_{50} was calculated as follows:

$$IP_{50} = (d(x) \times 24) + ((Y_{d(24)}/2) - Y_{d(x)})/((Y_{d(x+1)} - Y_{d(x)})/24) \tag{4}$$

where $Y_{d(24)}$ is the final disease incidence (at day 24); $Y_{d(x)}$ is the disease incidence at d x, i.e., the last d in which $Y_{d(x)} < Y_{d(24)}/2$; and $Y_{d(x+1)}$ is disease incidence at d x + 1, i.e., the first d in which $Y_{d(x+1)} \geq Y_{d(24)}/2$. For instance, if $Y_{d(24)} = 100\%$, $Y_{d(9)} = 0\%$, and $Y_{d(10)} = 60\%$, then $IP_{50} = (9 \times 24) + ((100/2) - 0)/((60 - 0)/24) = 236$ h.

The relationship between temperature and incubation length was fitted by using the equation of Magarey et al. [48] in the following form:

$$IP_{50} = f(T)/IP_{50}\mathrm{min} \tag{5}$$

$$f(T) = ((\mathrm{Tmax} - T)/(\mathrm{Tmax} - \mathrm{Topt}))((T - \mathrm{Tmin})/(\mathrm{Topt} - \mathrm{Tmin}))^{(\mathrm{Topt}-\mathrm{Tmin})/(\mathrm{Tmax}-\mathrm{Topt})} \tag{6}$$

where $IP_{50}min$ is the shortest duration of incubation; and Tmin, Topt, and Tmax are the minimal, optimal, and maximal temperatures, respectively. When $T < Tmin$ or $T > Tmax$, $f(T) = 0$. $IP_{50}min$ and cardinal temperatures were estimated by evaluating the goodness-of-fit of a set of equations calculated by using an iterative procedure in which $IP_{50}min$ changed in 1 h steps and the cardinal temperatures changed in 0.5 °C steps.

Supplementary Materials: The following are available online at https://www.mdpi.com/article/10.3390/plants10081696/s1: Identification of *Coniella diplodiella* isolates.

Author Contributions: Conceptualization, V.R., M.L. and T.J.; methodology, V.R. and L.L.; experiments, T.J. and L.L.; resources, V.R. and M.L.; writing—original draft preparation, T.J.; writing—review and editing, V.R.; supervision, V.R. All authors have read and agreed to the published version of the manuscript.

Funding: This research received no external funding.

Institutional Review Board Statement: Not applicable.

Informed Consent Statement: Not applicable.

Data Availability Statement: Data is contained within the article and Supplementary Material.

Acknowledgments: T.J. conducted this study within the Doctoral School of the Agro-Food System (Agrisystem) at the Università Cattolica del Sacro Cuore, Piacenza, Italy, with the support of the China Scholarship Council (no. 201809505008).

Conflicts of Interest: The authors declare no conflict of interest.

References

1. Osterwalder, A. Severe outbreak of white rot. *Schweiz. Z. Obs. Weinbau* **1930**, *39*, 339–340.
2. Bisiach, M. White rot. In *Compendium of Grape Diseases*; Pearson, R.C., Goheen, A.C., Eds.; American Phytopathological Society: St. Paul, MN, USA, 1988; pp. 22–23.
3. Alvarez, L.V.; Groenewald, J.Z.; Crous, P.W. Revising the *Schizoparmaceae*: *Coniella* and its synonyms *Pilidiella* and *Schizoparme*. *Stud. Mycol.* **2016**, *85*, 1–34. [CrossRef]
4. Locci, R.; Quaroni, S. Studies on *Coniothyrium diplodiella* I. isolation, cultivation and identification of the fungus. *Riv. Patol. Veg.* **1972**, *8*, 59–82.
5. Faes, H.; Staehelin, M. The coitre (*Coniothyrium diplodiella*) or hail disease of the Vine. In Pamphlet. Stat. Fed. D'essais Vitic. de Lausanne; 1922; p. 14. Available online: https://www.cabdirect.org/cabdirect/abstract/19231100062 (accessed on 6 June 2021).
6. Chethana, K.W.T.; Zhou, Y.; Zhang, W.; Liu, M.; Xing, Q.K.; Li, X.H.; Yan, J.Y.; Chethana, K.W.T.; Hyde, K.D. *Coniella vitis* sp. nov. is the common pathogen of white rot in Chinese vineyards. *Plant Dis.* **2017**, *101*, 2123–2136. [CrossRef]
7. CABI. *Coniella diplodiella* (Grapevine White Rot). Invasive Species Compendium. Available online: https://www.cabi.org/isc/datasheet/15183#toDistributionMaps (accessed on 6 June 2021).
8. He, Z.; Cui, C.Y.; Jiang, J.X. First Report of White Rot of Grape Caused by *Pilidiella castaneicola* in China. *Plant Dis.* **2017**, *101*, 1673. [CrossRef]
9. Tiedemann, A. *Untersuchungen zur Pathogenitat des Erregers der Weisfaule (Coniella Petrakii SUTT.) an Amerikaner-und Europarreben und Verbreitung und Bedeutung des Pilzes in den Deutschen Weinbaugebieten*; Gottingen University: Gottingen, Germany, 1985.
10. Chen, T.W.; Ye, Y.F.; Peng, F.Y.; Xiao, C.M. Studies on the white rot (*Coniothyrium diplodiella* (speg.) sacc.) of grapevine II. The environment conditions and paths of white rot infection. *J. Plant Prot.* **1980**, *7*, 27–34.
11. Kong, F.F.; Wang, Z.Y. Grapevine white rot. In *Crop Diseases and Insect Pests in China*; Guo, Y.Y., Wu, K.M., Chen, W.Q., Eds.; China Agricultural Press: Beijing, China, 2015; pp. 813–816.
12. Liu, C.Y.; Zhao, K.H.; Wang, K.; Bai, J.K. Redesignation of the causal agent of grape white rot in China. *Acta Phytopathol. Sin.* **1999**, *29*, 174–176.
13. Cortesi, P. White rot. In *Compendium of Grape Diseases, Disorders and Pests*; Wilcox, W.F., Gubler, W.D., Uyemoto, J.K., Eds.; American Phytopathological Society: St. Paul, MN, USA, 2015; pp. 92–94.
14. Liu, S.J.; Wang, P. The biology of white rot (*Coniothyrium diplodiella*) and the control of the fungus. *J. Southwest Agric. Univ.* **1992**, *14*, 504–506.
15. OEPP/EPPO. EPPO Standard PP 2/23 (1) Principles of good plant protection practice. *Bull. OEPP/EPPO Bull.* **2002**, *32*, 367–369. [CrossRef]
16. Li, B.Y.; Luan, B.H.; Shi, J.; Wang, S.L.; Tian, Y.Y.; Nie, L.X.; Wang, Y.Z. Sensitivity of *Coniella diplodiella* to pyraclostrobin in Jiaodong Area and comparison with four other fungicides. *Chin. J. Pestic. Sci.* **2020**, *22*, 959–966.
17. He, P.C. *Viticulture*; China Agriculture Press: Beijing, China, 1999.

18. Li, D.; Wan, Y.Z.; Wang, Y.J.; He, P.C. Relatedness of resistance to anthracnose and to white rot in Chinese wild grapes. *Vitis* **2008**, *47*, 213–215.
19. Zhang, Y.; Yao, J.L.; Feng, H.; Jiang, J.; Fan, X.; Jia, Y.F.; Wang, R.; Liu, C. Identification of the defense-related gene *VdWRKY53* from the wild grapevine *Vitis davidii* using RNA sequencing and ectopic expression analysis in Arabidopsis. *Hereditas* **2019**, *156*, 14. [CrossRef]
20. Storozhenko, E.M. White rot of Vine and control measures. Results of the sci. In Res. Work of the N. Caucasus Regional Sci.—Res. Inst. for Horticulture and Viticulture; 1959; pp. 296–336. Available online: https://www.cabdirect.org/cabdirect/abstract/19611102790 (accessed on 6 June 2021).
21. Kassemeyer, H.H. Fungi of Grapes. In *Biology of Microorganisms on Grapes, in Must and in Wine*; König, H., Unden, G., Fröhlich, J., Eds.; Springer: Cham, Switzerland, 2017; pp. 103–132.
22. Liu, F.C.; Wang, H.Y.; Shi, X.Q. Preliminary report on grape white rot research. *China Fruits* **1960**, *03*, 37–44.
23. Faes, H.; Staehelin, M. "Goitre" of the Vine (*Coniothyrium diplodiella*). In Terre Vaud; 1935; pp. 133, 154, 172, 193. Available online: https://www.cabdirect.org/cabdirect/abstract/19361100428 (accessed on 6 June 2021).
24. Considine, J.A. Physical aspects of fruit growth: Cuticular fracture and fracture patterns in relation to fruit structure in *Vitis vinifera*. *J. Hortic. Sci.* **1982**, *57*, 79–91. [CrossRef]
25. Becker, T.; Knoche, M. Water induces microcracks in the grape berry cuticle. *Vitis J. Grapevine Res.* **2012**, *51*, 141–142.
26. Ramteke, S.D.; Urkude, V.; Parhe, S.D.; Bhagwat, S.R. Berry cracking; its causes and remedies in grapes-a review. *Trends Biosci.* **2017**, *10*, 549–556.
27. Chen, Y.; Liu, C.Y.; Zhao, K.H.; Miao, Z.Y.; Liang, C.H. Biological Characters of the Pathogen of Grape White Rot. *J. Shenyang Agric. Univ.* **2006**, *37*, 840–844.
28. Dong, Y.H.; Xu, P.J.; Wang, Y.P.; Qian, J.R.; He, T.H.; Zhu, C.L.; Chen, B.Z.; Wang, H.X.; Zhang, J.P. Study on White Rot of Grape. *Acta Agric. Jiangxi* **2011**, *23*, 107–110.
29. Wang, Y.Z.; Zhang, W.; Zhao, M.; Liu, X.Q. Study on Mycelia Growth and Sporulation of the Pathogen of Grape White Rot. *J. Anhui Agric. Sci.* **2013**, *41*, 4828–4829, 4831.
30. Liu, W.Y.; Liu, L.; Li, H.Y. Biological characters of the pathogen of grapevine white rot. *Chin. Hortic. Abstr.* **2016**, *05*, 44–45.
31. Chen, T.W.; Ye, Y.F.; Peng, F.Y.; Xiao, C.M. Studies on the white rot (*Coniothyrium diplodiella* (speg.) sacc.) of grapevine I. The occurrence of white rot of grapevine in relation to the microclimate of vine yard. *J. Plant Prot.* **1979**, *6*, 9–18.
32. Wang, H.Y. Study on the epidemic temporal dynamics of grape clusters white rot in the Yellow River's old flooded area. *J. Fruit Sci.* **1999**, *16*, 123–125.
33. Terrier, C. Problems in the control of hail disease of grapes caused by *Coniella diplodiella* Pet. et Syd. *Rev Rom. Agric. VIticult Arboric.* **1949**, *5*, 89–91.
34. Faes, H.; Staehelin, M. New contribution to the study of coitre of the Vine (*Coniothyrium diplodiella*) or hail disease. In Annuaire Agric, at la Suisse; 1923; p. 10. Available online: https://www.cabdirect.org/cabdirect/abstract/19241100265 (accessed on 6 June 2021).
35. Staehelin, M. Control of coitre or hail disease of vines, *Coniothyrium diplodiella*. *Rev. Rom. Agric. Vit Arbor.* **1946**, *2*, 51–53.
36. Turian, G.; Leyvraz, H. Une nouvelle arme chimique contre le coître (*Coniella*) de la vigne. *Rev. Romande Agric. Vitic. Arboric.* **1954**, *10*, 97–98.
37. Bolay, A.; Corbaz, R. Nouvelles recherchies dans le domaine de la lutte contre le coître de la vigne (*Conilla diplodiella* [Speg.] Pet. et Syd.). *Ann. Agric. Suisse* **1961**, *62*, 239–248.
38. Chen, Z.W. Infection, occurrence and control of grape white rot. In Proceedings of the Annual Meeting of Hebei Fruit Tree Society, Changli, China, 18 December 1962; pp. 79–94.
39. Ye, Y.G.; Chen, X.T. Study on the Infection and Control of the Pathogen of Grape White Rot. *J. Fruit Sci.* **1986**, *03*, 41–45. [CrossRef]
40. Madden, L.V.; Hughes, G.; Van Den Bosch, F. *The Study of Plant Disease Epidemics*; American Phytopathological Society: St. Paul, MN, USA, 2007.
41. Berger, R.D. The Analysis of Effects of Control Measures on the Development of Epidemics. In *Experimental Techniques in Plant Disease Epidemiology*; Kranz, J., Rotem, J., Eds.; Springer: Berlin/Heidelberg, Germany, 1988; pp. 137–151.
42. Shaner, G.; Finney, R.E. The Effect of Nitrogen Fertilization on the Expression of Slow-Mildewing Resistance in Knox Wheat. *Phytopathology* **1977**, *67*, 1051–1056. [CrossRef]
43. Fry, W.E. Quantification of General Resistance of Potato Cultivars and Fungicide Effects for Integrated Control of Potato Late Blight. *Phytopathology* **1978**, *68*, 1650–1655. [CrossRef]
44. Analytis, S. Obtaining of sub-models for modeling the entire life cycle of a pathogen. *Z. Pflanzenkrankh. Pflanzenschutz* **1980**, *87*, 371–382.
45. Lin, L. A Concordance Correlation Coefficient to Evaluate Reproducibility. *Biometrics* **1989**, *45*, 255–268. [CrossRef] [PubMed]
46. Nash, J.E.; Sutcliffe, J.V. River flow forecasting through conceptual models part I—A discussion of principles. *J. Hydrol.* **1970**, *10*, 282–290. [CrossRef]
47. Teng, P.S. Validation of computer models of plant disease epidemics: A review of philosophy and methodology. *J. Plant Dis. Prot.* **1981**, *88*, 49–63.
48. Magarey, R.D.; Sutton, T.B.; Thayer, C.L. A simple generic infection model for foliar fungal plant pathogens. *Phytopathology* **2005**, *95*, 92–100. [CrossRef]

plants

Article

Possible Overestimation of Seed Transmission in the Spread of Pospiviroids in Commercial Pepper and Tomato Crops Based on Large-Scale Grow-Out Trials and Systematic Literature Review

Jacobus T. J. Verhoeven [1], Marleen Botermans [1], Ruben Schoen [1], Harrie Koenraadt [2] and Johanna W. Roenhorst [1,*]

[1] National Plant Protection Organization of the Netherlands, P.O. Box 9102, 6700 HC Wageningen, The Netherlands; koverhoeven@gmail.com (J.T.J.V.); m.botermans@nvwa.nl (M.B.); r.schoen@nvwa.nl (R.S.)

[2] Naktuinbouw Research and Development, P.O. Box 40, 2370 AA Roelofarendsveen, The Netherlands; h.koenraadt@naktuinbouw.nl

* Correspondence: j.w.roenhorst@nvwa.nl

Abstract: Several outbreaks of pospiviroids have been reported in pepper and tomato crops worldwide. Tracing back the origin of the infections has led to different sources. In some cases, the infections were considered to result from seed transmission. Other outbreaks were related to transmission from ornamental crops and weeds. Pospiviroids, in particular potato spindle tuber viroid, are regulated by many countries because they can be harmful to potatoes and tomatoes. Seed transmission has been considered an important pathway of introduction and spread. However, the importance of this pathway can be questioned. This paper presents data on seed transmission from large-scale grow-out trials of infested pepper and tomato seed lots produced under standard seed-industry conditions. In addition, it presents the results of a systematic review of published data on seed transmission and outbreaks in commercial pepper and tomato crops. Based on the results of the grow-out trials and review of the literature, it was concluded that the role of seed transmission in the spread of pospiviroids in practice is possibly overestimated.

Keywords: commercial seed lots; *Capsicum annuum*; epidemiology; source of infection; outbreak; pathway; plant disease management; *Solanum lycopersicum*

Citation: Verhoeven, J.T.J.; Botermans, M.; Schoen, R.; Koenraadt, H.; Roenhorst, J.W. Possible Overestimation of Seed Transmission in the Spread of Pospiviroids in Commercial Pepper and Tomato Crops Based on Large-Scale Grow-Out Trials and Systematic Literature Review. *Plants* **2021**, *10*, 1707. https://doi.org/10.3390/plants10081707

Academic Editor: Alessandro Vitale

Received: 21 July 2021
Accepted: 13 August 2021
Published: 19 August 2021

1. Introduction

Over the last few decades, several outbreaks of pospiviroids in pepper (*Capsicum annuum*) and tomato (*Solanum lycopersicum*) have been reported worldwide (reviewed by Candresse et al. [1] and Hammond [2]). Pospiviroids, and in particular potato spindle tuber viroid (PSTVd), are regulated in many countries because of their harmful effects on potato (*Solanum tuberosum*) and tomato crops [2–4]. Phytosanitary measures are taken to prevent and eradicate outbreaks, and efforts are made to trace the sources of infection. With regard to the outbreaks in pepper and tomato crops, both the introduction of pospiviroids, via infested seed lots, and transfer from infected ornamental crops and weeds, have been reported [2]. However, the relative importance of each of these sources as a pathway for introduction in pepper and tomato crops worldwide is unclear.

Pospiviroids are single-stranded circular RNA molecules of ca. 360 nucleotides able to infect plants. The genus *Pospiviroid*, one of the five genera within the family *Pospiviroidae*, includes nine species, of which *Potato spindle tuber viroid* is the type member [5,6]. All pospiviroid species, except iresine viroid 1, are able to infect at least one of the main solanaceous crops, i.e., pepper, potato, and tomato. In addition, pospiviroid infections have been reported from several ornamental crops and weeds, including many solanaceous species. However, pospiviroid infections in ornamentals often remain symptomless, which implies that these plants may serve as unnoticed sources of infection [7].

Pospiviroids can be spread by vegetative propagation and transmission via mechanical transfer (contact), insects, pollen, and seeds [1,4]. The role of vegetative propagation for spreading in potato and ornamental crops such as *Solanum jasminoides* is evident. Under experimental conditions, pospiviroids have been shown to be easily transferred mechanically [8,9]. In addition, in practice, the occurrence of spread by mechanical transfer is clear. Infections in tomato crops were found to spread rapidly within a row [10–12] and evidence was obtained that outbreaks in pepper and tomato crops were the result of mechanical transfer from symptomless-infected ornamentals and weeds, e.g., [13,14]. Regarding the role of insects in the spread of pospiviroids, aphids have been found to transmit PSTVd in mixed infections with potato leaf roll virus (PLRV) in potatoes [15,16], and tomato chlorotic dwarf viroid (TCDVd) with PLRV in tomatoes [17]. In addition, pollination by bumblebees has been associated with the spread of tomato apical stunt viroid (TASVd) [18,19] but was not observed for PSTVd [20]. The latter results [20] also indicated that the role of pollen in the spreading of pospiviroids is limited. Finally, seeds have been reported as a source of infection for seed-propagated crops such as pepper and tomato. However, reports on seed transmission in these crops are contradictory. Under experimental conditions, substantial transmission rates, as well as the complete absence of seed transmission, have been reported, e.g., [21,22]. For outbreaks in commercial crops, some were considered to result from infested seeds [23,24], whereas for others, no evidence was found for seed as a source of infection [12]. The rationale behind these different observations is not understood, and possible explanations include viroid load, 'age' and/or treatment of seeds, and climatic conditions (temperature) during germination and cultivation.

Several countries consider seeds as an important pathway for the introduction of pospiviroids in commercial pepper and tomato crops. Because production and processing occur at a global scale, this has led to installing strenuous seed-testing requirements for seed lots to be traded. However, the importance of seeds as a pathway for the introduction of pospiviroids in commercial crops has never been evaluated. Which part of the outbreaks can be attributed to the introduction via seed versus mechanical transfer from (symptomless) infected ornamentals and weeds, and what is the evidence?

To determine the role of seeds as a pathway for the introduction of pospiviroid infections in pepper and tomato crops, grow-out trials under standard conditions were performed using seed lots naturally infested by columnea latent viroid (CLVd), pepper chat fruit viroid (PCFVd), and PSTVd. In addition, published data were reviewed on the evidence of seed transmission under experimental conditions and in practice. The results of the grow-out trials, as well as the reviews of published data, will be discussed in view of the role of seeds as a pathway for the introduction of pospiviroids in commercial pepper and tomato crops.

2. Materials and Methods

2.1. Grow-Out Trials

2.1.1. Selection of Seed Lots

For the grow-out trials, commercially produced seed lots were selected that tested positive for pospiviroids during routine screening by real-time RT-PCR according to the EPPO standard PM7/138 [25], Appendix 4. To identify the seed lots with the highest viroid loads for each of these lots, three subsamples of 1000 seeds and eight subsamples of 100, 10, and 1 seed(s) were tested. The data was used to estimate the infestation rate using the software package SeedCalc8 (version 8.1.0; Geves, France). The lots with the highest viroid loads and the most consistent test results were selected for the grow-out trials. The selected seed lots concerned seeds from different cultivars and seed companies. They had been produced and initially processed in either Africa or Asia and were further processed in Europe, according to standard procedures. Except for a 1%-HCl treatment during the extraction of the tomato seeds, which is compulsory in the European Union, no additional treatments were applied for disinfection.

2.1.2. Confirmation of the Identity and Viability of the Viroids in the Selected Seed Lots

To confirm the presence and identity of the viroids detected during routine screening, conventional RT-PCRs were performed on RNA extracts using the following primers: pCLV4/pCLVR4 for CLVd [26]; Pospi1FW/RE and AP-FW1/RE2 for PCFVd [12,27], Pospi1FW/RE and Pospi2FW/RE for PSTVd lot 3, and Pospi1FW/RE for PSTVd lot 5 [12,28]. Amplicons were bi-directionally sequenced and analysed [29].

To determine the viability of the viroids, for each seed lot RNA extracts were mechanically inoculated onto four plants of *C. annuum* cv. Westlandse Grote Zoete and four plants of *S. lycopersicum* cv. Moneymaker according to Verhoeven et al. [28]. After six weeks, the inoculated pepper and tomato plants were tested in groups of four by real-time RT-PCR [30].

2.1.3. Conditions Grow-Out Trials

The grow-out trials of both the infested pepper and tomato seed lots were performed at seven different locations in the Netherlands. Three seed lots were grown in dedicated greenhouses of six seed companies and two seed lots in the quarantine greenhouse of the National Plant Protection Organization (NPPO) of the Netherlands. To prevent viroid transmission to and from external hosts, all treatments were performed under containment conditions.

Between 12 and 17 days after germination, when the first fully developed leaf had appeared in most seedlings, plants were manually transplanted in rock-wool blocks or potting soil. Plants were grown for 56–59 days at a minimum temperature of 25 °C and a day length of at least 14 h. Common cultivation practices such as irrigation and insect control were applied, but pruning was limited and aberrant plants were not removed.

2.1.4. Detection of Pospiviroids in Leaf Samples from Seedlings

At the end of the grow-out trials, the top leaflet of a young (50–90% full-grown) leaf of each seedling was collected. Equal parts of leaves were pooled up to 25 and tested for pospiviroids at the laboratories of the participating seed companies or Naktuinbouw, using routine procedures adapted from Botermans et al. [30].

2.2. Review of Published Data on Seed Transmission and Outbreaks

Data on seed transmission and outbreaks of pospiviroids in pepper and tomato crops were extracted from publications in scientific journals and the EPPO Reporting Service, the latter only when data on outbreaks were not published elsewhere. In addition, data of the grow-out trials reported in this paper was included for seed transmission. For reviewing, publications were grouped per crop and considered per viroid species. Regarding seed transmission, the following data were considered: type of seed (produced experimentally or commercially), number of infected/raised seedlings, and specific conditions if applicable. For reviewing the data on outbreaks, the term outbreak had been defined as the report of a single viroid per crop and country. For each outbreak, the evidence for the indicated source of infection was considered in relation to the stage of the crop at the time of viroid detection, the genome sequence of the isolate, information on other crops grown from the same seed lot, environmental conditions, etc. Based on an 'unbiased' review of these data, conclusions were drawn on the most probable source(s) of infection. These conclusions were compared with those given by the authors.

3. Results

3.1. Grow-Out Trials

3.1.1. Confirmation of the Identity and Viability of the Viroids in the Selected Seed Lots

For the selected seed lots, the presence and identity of the pospiviroids were confirmed by sequencing and analysis of the amplicons produced by RT-PCR. Table 1 shows the identified viroids and accession numbers in NCBI GenBank. For seed lots 1 to 4, complete genome sequences of the viroid isolates were obtained, except for the primer positions. For lot 5, only a partial sequence of 112 nt of PSTVd was obtained.

Table 1. Seed lots used for grow-out trials to determine the transmission from seeds to seedlings.

Lot Number	Crop [1]	Origin	Viroid(s) Identified	Genbank Accession Number	Estimated Infestation Rate [2]	Number of Plants Raised
1	pepper	Asia	PCFVd	MW422288	13% (CL_{95} 0–52%)	27,735
2	pepper	Asia	CLVd	MW422289	9% (CL_{95} 3–22%)	27,703
			PCFVd	MW422290	13% (CL_{95} 4–29%)	
3	pepper	Africa	PSTVd	MW422291	nd	2500
4	tomato	Asia	PCFVd	MW422292	63% (CL_{95} 25–91%)	47,528
5	tomato	Asia	PSTVd	MW422293	nd	2500

[1] Seed lots concerned distinct pepper and tomato cultivars from different seed companies. [2] Based on SeedCalc8 (version 8.1.0; Geves, France); CL_{95}: confidence level 95%; nd: not determined.

For the viroid isolates from the three pepper seed lots, the viability was shown by successful infection of *C. annuum* cv. Westlandse Grote Zoete and *S. lycopersicum* cv. Moneymaker after mechanical inoculation (results not shown). For the viroid isolates from the two tomato seed lots, the viability could not be confirmed. These results show that at least for pepper, the viability of the viroids was maintained during the processing and storage of the seeds, whereas for tomatoes such a conclusion could not be drawn.

3.1.2. Detection of Pospiviroids in Leaf Samples from Seedlings

Table 2 shows the number of pepper and tomato seedlings included in the grow-out trials and their distribution over the seven locations. Up to 56–59 days after sowing, no viroid symptoms were observed and testing of all plants at the end of the growing period did not reveal any viroid infection. This means that in both crops, CLVd, PCFVd, and PSTVd were not transmitted from seeds to seedlings/plants even though the environmental conditions were favourable for viroid transmission and replication.

Table 2. The number of pepper and tomato plants raised from infested seed lots and distributed over the different locations.

Location [1]	Pepper			Tomato	
	Lot 1	Lot 2	Lot 3	Lot 4	Lot 5
	PCFVd	CLVd and PCFVd	PSTVd	PCFVd	PSTVd
1	5575	5500		8256	
2	5805	6045		9985	
3	6000	6000		10,000	
4	5605	5358			
5	4750	4800		10,215	
6				9072	
7			2500		2500
Total	27,735	27,703	2500	47,528	2500

[1] Locations 1–6: dedicated greenhouses at seed companies; location 7: quarantine greenhouse at the NPPO in the Netherlands.

3.2. Review of Published Data on Seed Transmission and Data from This Study

A total of 27 publications on seed transmission of pospiviroids were considered and reviewed per crop and type of seed production. The results are summarised in Tables 3–6 and further detailed in Supplementary Materials Table S1 (A–D).

3.2.1. Pepper

For experimentally produced pepper seeds, seed transmission has been reported for PCFVd and PSTVd (Table 3). Verhoeven et al. [27] reported transmission of PCFVd for 11 out of 59 seedlings. However, Verhoeven et al. [31] were not able to confirm the previously reported seed transmission for PCFVd in 179 seedlings of the same pepper variety and 158 seedlings of another variety. In addition, no seed transmission of PCFVd was reported in 46 seedlings by Yanagisawa and Matsushita [32]. For PSTVd, Matsushita and Tsuda [33] reported transmission for seven out of 2230 seedlings. In contrast, no seed transmission was reported by Lebas et al. [34] and by Verhoeven et al. [31] for 25 and 222 pepper seedlings. With regard to the other pospiviroids, no seed transmission was reported for columnea latent viroid (CLVd) [31], TASVd [31], TCDVd [33], and tomato planta macho viroid (TPMVd) [32], based on examination of 179, 217, 1105, and 46 seedlings, respectively.

Table 3. The number of infected/raised pepper seedlings in ten separate grow-out trials of experimentally produced seeds reported in the literature.

CLVd		PCFVd		PSTVd		TASVd		TCDVd		TPMVd	
0	179 [a]	11	59 [b]	0	25 [d]	0	217 [a]	0	1105 [e]	0	46 [c]
		0	337 [1a]	7	2230 [2e]						
		0	46 [c]	0	222 [a]						

[1] Total of two varieties; [2] total of three varieties; [a] [31]; [b] [27]; [c] [32]; [d] [34]; [e] [33].

For commercially produced pepper seeds, the grow-out trials reported in this paper did not show seed transmission for CLVd, PCFVd, and PSTVd in 27,703, 55,438, and 2500 seedlings, respectively. Furthermore, Verhoeven et al. [28] did not observe seed transmission for TASVd in 1200 seedlings (Table 4).

Table 4. The number of infected/raised pepper seedlings in four separate grow-out trials of commercially produced seeds reported in the literature and in this paper.

CLVd		PCFVd		PSTVd		TASVd	
0	27,703 [a]	0	55,438 [1a]	0	2500 [a]	0	1200 [b]

[1] Total of two varieties; [a] this study; [b] [28].

3.2.2. Tomato

Regarding experimentally produced tomato seeds, seed transmission has been reported for all seven pospiviroids found in commercial tomato crops (Table 5). However, for most pospiviroids results were contradictory. For citrus exocortis viroid (CEVd), Semancik [35] reported seed transmission in tomato without further details, whereas Faggioli et al. [21] did not find transmission for 1849 seedlings. For CLVd, Matsushita and Tsuda [33] reported transmission for 46 out of 793 seedlings, whereas no seed transmission was found by Fox and Monger [36] and Faggioli et al. [21] for 200 and 1599 seedlings. For PCFVd, Yanagisawa and Matsushita [32] reported three infected seedlings out of 941. For PSTVd, again, contradictory results have been reported. Successful seed transmission was reported by Khoury et al. [37], Singh and Dilworth [38], Simmons et al. [22] and Matsushita and Tsuda [33] for 459 out of 1933 seedlings in total. In addition, seed transmission without further details was reported by Benson and Singh [39], Kryczynski et al. [40], Menzel and Winter [41], and Batuman et al. [42]. In contrast, Lebas et al. [34] and Faggioli et al. [21] did not find seed transmission of PSTVd for a total of 47 seedlings, neither did McClean [43] for tomato bunchy top virus (synonym PSTVd) in 92 seedlings. For TASVd seed transmission was reported for 24 out of 30 seedlings by Antignus et al. [18] and without further details by Batuman et al. [42], but no transmission was found by Faggioli et al. [21] and Matsushita and Tsuda [33] for 2575 seedlings. For TCDVd, Singh and Dilworth [38] reported seed transmission for 209 out of 280 seedlings, whereas no transmission was found in over 4251 seedlings by Singh et al. [44], Koenraadt et al. [45] and Matsushita and Tsuda [33].

Finally, for TPMVd, seed transmission was reported by Yanagisawa and Matsushita [32] for 13 out of 1039 seedlings, but not by Belalcazar and Galindo-Alonso [46] for 425 seedlings. Overall, 724 seedlings were found infected among a total of 16,054 raised plants, the efficiency of the seed transmission varying from 0 to 80%.

Table 5. The number of infected/raised tomato seedlings in 22 separate grow-out trials of experimentally produced seeds reported in the literature.

CLVd		CEVd		PCFVd		PSTVd		TASVd		TCDVd		TPMVd	
0	1599 [1,a]	0	1849 [1,a]	3	941 [d]	0	22 [a]	24	30 [k]	0	4000 [l]	0	425 [n]
0	200 [b]					3	60 [e]	0	1232 [1,a]	0	251 [c]	13	1039 [d]
46	793 [2c]					0	25 [f]	0	1343 [3,c]	0	? [m]		
						111	285 [4,c]			209	280 [j]		
						0	92 [5,g]						
						178	350 [h]						
						107	1192 [i]						
						30	46 [j]						

[1] Total of two varieties; [2] total of four varieties; [3] total of three varieties; [4] total of five varieties; [5] results for tomato bunchy top virus (synonym PSTVd [43]); [a] [21]; [b] [36]; [c] [33]; [d] [32]; [e] [37]; [f] [34]; [g] [43]; [h] [22]; [i] [47]; [j] [38]; [k] [18]; [l] [45]; [m] [44]; [n] [46]. Seed transmission without further details was reported by Benson and Singh [39] (PSTVd), Semancik [35] (CEVd), Kryczynski et al. [40] (PSTVd), Menzel and Winter [41] (PSTVd), and Batuman et al. [42] (PSTVd and TASVd).

For commercially produced tomato seeds, also successful and unsuccessful seed transmission have been reported (Table 6). In comparison to experimentally produced seeds, both frequency and the rate of seed transmission were low, i.e., one out of 370 seedlings for PSTVd by Van Brunschot et al. [24] and 2–20 out of 2500 seedlings for TCDVd by Candresse et al. [23]. In the latter case, the viroid was detected in two out of 250 samples by testing bulked samples of 10 seedlings, explaining the range of 2–20 seedlings. In the grow-out trials reported in this paper, no seed transmission of PSTVd was found for 2500 seedlings. Similarly, no transmission was found for 1000 seedlings in a previous trial, where the seed was harvested from an infected crop according to commercial practices [48]. For CLVd and PCFVd, no seed transmission was found for 25,500 [36] and 47,477 tomato seedlings (this study). Overall, for a total of 79,398 tomato seedlings only in two cases was seed transmission reported.

Table 6. The number of infected/raised tomato seedlings in six separate grow-out trials of commercially produced seeds reported in the literature and in this paper.

CLVd		PCFVd		PSTVd		TCDVd	
0	25,500 [a]	0	47,528 [b]	1	370 [c]	2–20 [1]	2500 [e]
				0	1000 [d]		
				0	2500 [b]		

[1] Two out of 250 bulked samples of 10 seedlings tested positive; [a] [36]; [b] this study; [c] [24]; [d] [48]; [e] [23].

3.3. *Review of Published Data on Outbreaks*

3.3.1. Pepper

For pepper, six publications on at least 14 outbreaks of pospiviroids were reviewed with regard to the source of infection (Table 7). In five cases, the current review resulted in the same conclusion as drawn by the authors, i.e., the most probable origin being other host plants or unknown (Table 7; Supplementary Materials Table S2A). Only in one case in New Zealand, where the authors assumed seeds as the source of the PSTVd infection [34], the conclusion could be questioned. Firstly, none of the 25 seedlings grown from seeds of the infected pepper plants was found to be infected. Secondly, highly similar PSTVd sequences were found in four pepper varieties in five different glasshouses, as well as in tomato and cape gooseberry crops grown in this country [34,49,50]. Together, these observations are reasons to consider locally infected host plants as a more probable source of infection. For similar reasons, Mackie et al. [51] considered local wild plants as the source of infection of repeated PSTVd outbreaks in pepper crops in Western Australia.

Table 7. Conclusions on the most probable source of pospiviroid infections in pepper crops as reported and drawn upon in the current review of the provided data.

Source of Infection	Reported in Publication	Based on Review
Seeds	1 [a]	0
Plants [1]	4 [b]	5
Unknown	1 [c]	1

[1] Plants include: plants for planting of the same crop and other hosts; [a] [34]; [b] [27,51–53]; [c] [54].

3.3.2. Tomato

For tomato, 43 publications on outbreaks of pospiviroids were reviewed (Table 8; Supplementary Materials Table S2B). Thirty-nine publications concerned one viroid, three reported on two viroids [42,55,56], and one publication reported on four pospiviroids, of which one was reported from two countries without obvious connection to each other [12]. This makes a total of 50 outbreaks. For 11 out of these 50 pospiviroid outbreaks in tomatoes, the authors indicated seeds as the most likely source of infection. Upon review, in nine cases the provided data were insufficient to substantiate this conclusion [18,49,57–60], [42] (two outbreaks), [55] (one outbreak), and seeds seemed to be indicated as a source of infection by default. For seven outbreaks too little information was provided to conclude on the source of infection and, therefore, upon review it was considered unknown. In two cases plants were considered as the most probable source of infection, instead of seeds. Firstly, in 2002 Antignus et al. [61] reported on the spread of TASVd from a few infected plants in a tomato crop and indicated that the origin of the infections in these plants was unknown. In 2007, the same authors reporting on the same outbreak suggested seeds as the most probable source of infection [18]. The latter conclusion was based on the demonstration of seed transmission under experimental conditions for seed batches not related to the outbreak. Secondly, Elliott et al. [49] indicated seeds as the most probable source of the PSTVd outbreak in a tomato crop in New Zealand in 2000, without providing data to support this suggestion. Since the first symptomatic plants were found near the entrance of the glasshouse, and the viroid sequence was almost similar to that of other PSTVd isolates reported in tomato, pepper and *Physalis peruviana* in New Zealand [34,50], infected host plants in the environment seem a more probable source of infection. In only two publications, data were provided indicating that seeds could have been the source of infection of the PSTVd and TCDVd infections in tomatoes [23,24]. For ten outbreaks, infected plants were considered as the most probable source of infection [12] (three outbreaks) [13,51,62–66]. Reviewing the data lead to the same conclusion in nine cases. For the outbreak in Germany [63], however, it was concluded that the data provided did not allow to conclude on the source of infection. In the remaining 29 outbreaks in tomato (27 publications), no conclusion was drawn on the source of infection. For three of these publications, reviewing the data led to the conclusion that infected host plants were the most probable source of infection. Firstly, Ling and Sfetcu [67] reported the source of the PSTVd outbreak in tomatoes in California (USA) in 2009 to be unknown, although tomato plants in the affected glasshouse and its surroundings were already known to be infected by PSTVd for several years. Therefore, these infected plants were considered the most probable source of the infections in 2010 (Verhoeven, unpublished data). Secondly, for the outbreak reported by Mackie et al. [10], the genome sequence of the PSTVd isolate was found to show the highest identities with PSTVd sequences from the *P. peruviana* cluster, which also includes other isolates from wild plants in Australia [51,68]. The publication of Mackie et al. [51] provided further evidence that local plants most likely served as the source of infection. Thirdly, in Japan, infected petunia plants appeared to be the most probable source of infection of the TCDVd outbreak in tomatoes in 2008, since the genome sequence of the isolate identified in tomato (AB329668) [11] was reported from many petunia selections originating in Japan [69,70]. The latter conclusion was further supported by the observation that most TCDVd sequences from tomato crops outside Japan do not group in the cluster of petunia sequences [69]. For the remaining 26 outbreaks, no source of infection could be indicated, based on the data provided.

Table 8. Conclusions on the most probable source of pospiviroid infections in tomato crops as reported and drawn upon in the current review of the provided data.

Source of Infection	Reported in Publication	Based on Review
Seeds	11 [a]	2
Plants [1]	10 [b]	14
Unknown	29 [c]	34

[1] Plants include: plants for planting of the same crop and other hosts. [a] Single outbreaks [18,23,24,49,55,57–60], two outbreaks [42]; [b] Single outbreaks [13,51,62–66], three outbreaks [12]; [c] Single outbreaks [10,11,44,46,55,61,67,71–88], two outbreaks [12,56].

4. Discussion

The results of the grow-out trials of commercially produced pepper and tomato seeds, combined with the results of the reviews of data from publications on seed transmission and outbreaks, indicate that the role of seed transmission in the spread of pospiviroids in these crops may have been overestimated.

The results presented here show that none of the 57,938 pepper and 50,028 tomato seedlings raised from commercially produced infested seed lots was found infected, neither by symptom observation nor by testing. For pepper, the lack of seed transmission could not be ascribed to the absence of viable viroid, given the successful infection of healthy pepper and tomato plants after mechanical inoculation. For tomatoes, the viability of the viroids could not be established. The lack of successful infection of the inoculated plants does not seem to relate to the amount of viroid, since the estimated viroid load of one of the tomato seed lots was substantially higher than that of the pepper seed lots. An effect of the seed processing procedure on the viability of the viroid cannot be excluded, such as the HCl treatment, neither an effect of the matrix on the success of mechanical transmission. A matrix effect has been observed, e.g., for the transmission of PSTVd from leaf material of *Brugmansia* sp. in comparison to *Solanum jasminoides* [9,14,64]. In conclusion, these results indicate that for commercially produced seeds the chance is low that pospiviroids are transmitted from seed to seedling, even in the case that viable viroids are present as shown for pepper.

The lack of seed transmission in grow-out trials of commercially produced seed lots was further supported by a review of publications on seed transmission of pospiviroids in pepper (Table 4) and tomato (Table 6). Although the rate of seed transmission may be affected by the viroid variant, (a cultivar of) the host plant and environmental conditions, the substantially higher seed-transmission rates reported for experimentally produced seeds (Tables 3 and 5) raised the question of whether the industrial seed processing also may affect the transmission from seed to seedling. For pepper seeds, Verhoeven et al. [31] did not find such an effect, when simulating industrial processing by postponing the date of sowing. Neither does industrial processing seem to abolish the viability of the viroids of infested seeds. For tomato seeds, the effect of seed processing on transmission cannot be excluded, since the viability of the viroid on infested tomato seeds could not be established. Nevertheless, the question remains why the figures of seed transmission of pospiviroids are so different for experimentally and commercially produced seeds. Could cross-contamination have occurred under experimental conditions where research with these highly contagious viroids is performed by staff working on various aspects of these pathogens while using the same facilities? For example, Verhoeven et al. [31] could not confirm the earlier seed transmission of PCFVd in pepper [27], despite using the same isolate and cultivar under the same growing conditions. Therefore, they questioned whether these different results could be attributed to the occurrence of cross-contamination in the previous experiments. More generally, it raises the question of whether cross-contamination could also have accounted for the establishment of infections in seedlings in other experiments? In relation to the TCDVd outbreak reported by Candresse et al. [23], in 4000 seedlings grown from seeds harvested from the affected crops, no infections were found by Koenraadt et al. [45]. In addition, for the single positive seedling reported by van Brunschot et al. [24], the identified genome sequence of PSTVd was highly similar to the majority of PSTVd genotypes reported in the region [10,51,68] and, therefore, does not exclude local plants as a source of infection. Nevertheless, the results of Candresse et al. [23] and Van Brunschot et al. [24], indicate that exceptional seed transmission of pospiviroids in tomatoes cannot be excluded.

Overall, the review of published data on pospiviroid outbreaks in pepper and tomato crops indicate that designating seed as a source of infection is often by default rather than evidence-based. Reviewing the publications in which seeds were suggested as a source of infection resulted in different conclusions in the only case for pepper and nine from eleven cases for tomatoes (Tables 7 and 8). Regarding the two remaining publications suggesting seeds as the source of infection in tomato crops, at least some questions can be raised from this conclusion, as discussed before. In the majority of the publications, the source of infection was not specified or unknown, which did not change after review.

Finally, the conclusion that the role of seed transmission in the global spread of pospiviroids may be overestimated, is supported by the fact that the number of outbreaks in tomato and pepper crops has been limited even before the testing of seed lots was implemented. Constable et al. [89] reported the detection of pospiviroids in 36 out of 553 (6.5%) imported pepper seed lots and in 91 out of 1562 (5.8%) tomato seed lots in Australia since 2008. These seed lots were reported to originate from Africa, Asia (eastern and southern regions), Europe, the Middle East, as well as North and South America. In addition, in the Netherlands, pospiviroids were detected in 70 out of 2997 (2.3%) pepper and 140 out of 5874 (2.4%) tomato seed lots from various origins between 2015 and 2018 (Koenraadt, unpublished data). The recent detection of TASVd in a pepper seed lot produced in 1992, however, indicated that pospiviroids already occurred in these crops before testing started [28]. Therefore, far more reports of outbreaks would have been expected when seeds are a substantial source of infection. Nevertheless, in Australia, the number of reported pospiviroid outbreaks was limited, also before testing of imported seed lots started in 2008. For none of the reported cases from this period [10,59,68] were seeds considered the most probable source of infection in the current review (Table S2). For the two outbreaks in Australia since then, Mackie et al. [51] reported plants and Van Brunschot et al. [24] reported seeds as the most likely sources of infection. Moreover, in other countries, the number of reported pospiviroid outbreaks in tomato and pepper crops was low, and outbreaks were only occasionally assumed to be related to infested seed lots (Tables 7 and 8, Table S2). Taken together, both the substantial number of infested seed lots found by Constable et al. [89] and Koenraadt (unpublished data) and the low number of pospiviroid outbreaks reported, support the conclusion that the role of seed transmission in the spread of pospiviroids in pepper and tomato crops has been overestimated.

5. Conclusions

In conclusion, this systematic review of published data on seed transmission and outbreaks in pepper and tomato, including the results of the grow-out trials described in this paper, sheds new light on the contradictory views on the contribution of seeds to the spread of posiviroids. These new insights can be used to reassess the role of seeds as a pathway for the spread of pospiviroids in these crops, and to develop and substantiate alternative disease management strategies at lower costs by avoiding unnecessary destruction of 'infested' seed lots.

Supplementary Materials: The following are available online at https://www.mdpi.com/article/10.3390/plants10081707/s1, Table S1: Pospiviroid seed transmission experiments in pepper and tomato, Table S2: Pospiviroid outbreaks in pepper and tomato crops.

Author Contributions: Conceptualization, J.T.J.V. and J.W.R.; data curation, J.T.J.V.; methodology, J.T.J.V., M.B. and H.K.; writing—original draft, J.T.J.V. and J.W.R.; writing—review and editing, M.B. and R.S. All authors have read and agreed to the published version of the manuscript.

Funding: This study was funded by the NVWA and Naktuinbouw, with in kind funding from BASF Vegetable Seeds (Nunhems), Bejo Zaden B.V., ENZA Zaden B.V., Monsanto Vegetable Seeds, Rijk Zwaan, and Syngenta.

Institutional Review Board Statement: Not applicable.

Informed Consent Statement: Not applicable.

Data Availability Statement: Data is contained within the article and the Supplementary Materials.

Acknowledgments: Authors would like to acknowledge the (technical) support by staff of the molecular and virology groups of the National Reference Centre of the Netherlands Food and Consumer Product Safety Authority (NVWA), Naktuinbouw, BASF Vegetable Seeds (Nunhems), Bejo Zaden B.V., ENZA Zaden B.V., Monsanto Vegetable Seeds, Rijk Zwaan and Syngenta.

Conflicts of Interest: The authors declare no conflict of interest.

References

1. Candresse, T.; Verhoeven, J.T.J.; Stancanelli, G.; Hammond, R.W.; Winter, S. Other pospiviroids infecting solanaceous plants. In *Viroids and Satellites*; Hadidi, A., Flores, R., Palukaitis, P., Randles, J., Eds.; Elsevier Academic Press: London, UK, 2017; pp. 159–168.
2. Hammond, R.W. Economic significance of viroids in vegetable and field crops. In *Viroids and Satellites*; Hadidi, A., Flores, R., Palukaitis, P., Randles, J., Eds.; Elsevier Academic Press: London, UK, 2017; pp. 5–13.
3. Barba, M.; James, D. Quarantine and certification for viroids and viroid diseases. In *Viroids and Satellites*; Hadidi, A., Flores, R., Palukaitis, P., Randles, J., Eds.; Elsevier Academic Press: London, UK, 2017; pp. 415–424.
4. Owens, R.A.; Verhoeven, J.T.J. Potato spindle tuber viroid. In *Viroids and Satellites*; Hadidi, A., Flores, R., Palukaitis, P., Randles, J., Eds.; Elsevier Academic Press: London, UK, 2017; pp. 149–158.
5. Di Serio, F.; Li, S.-F.; Pallás, V.; Owens, R.A.; Randles, J.W.; Sano, T.; Verhoeven, J.T.J.; Vidalakis, G.; Flores, R. Viroid taxonomy. In *Viroids and Satellites*; Hadidi, A., Flores, R., Palukaitis, P., Randles, J., Eds.; Elsevier Academic Press: London, UK, 2017; pp. 135–146.
6. Di Serio, F.; Owens, R.A.; Li, S.-F.; Matoušek, J.; Pallás, V.; Randles, J.W.; Sano, T.; Verhoeven, J.T.J.; Vidalakis, G.; Flores, R. ICTV Virus Taxonomy Profile: *Pospiviroidae*. *J. Gen. Virol.* **2020**, *102*, 001543.
7. Verhoeven, J.T.J.; Hammond, R.W.; Stancanelli, G. Economic significance of viroids in ornamental crops. In *Viroids and Satellites*; Hadidi, A., Flores, R., Palukaitis, P., Randles, J., Eds.; Elsevier Academic Press: London, UK, 2017; pp. 27–38.
8. Manzer, F.E.; Merriam, D. Field transmission of the potato spindle tuber virus and virus X by cultivating and hilling equipment. *Am. Potato J.* **1961**, *38*, 346–352. [CrossRef]
9. Verhoeven, J.T.J.; Huner, L.; Virscek Marn, M.; Mavric Plesko, I.; Roenhorst, J.W. Mechanical transmission of *Potato spindle tuber viroid* between plants of *Brugmansia suaveolens*, *Solanum jasminoides* and potatoes and tomatoes. *Eur. J. Plant Pathol.* **2010**, *128*, 417–421. [CrossRef]
10. Mackie, A.E.; McKirdy, S.J.; Rodoni, B.; Kumar, S. *Potato spindle tuber viroid* eradicated in Western Australia. *Australas. Plant Pathol.* **2002**, *31*, 311–312. [CrossRef]
11. Matsushita, Y.; Kanda, A.; Usugi, T.; Tsuda, S. First report of *Tomato chlorotic dwarf viroid* disease on tomato plants in Japan. *J. Gen. Plant Pathol.* **2008**, *74*, 182–184. [CrossRef]
12. Verhoeven, J.T.J.; Jansen, C.C.C.; Willemen, T.M.; Kox, L.F.F.; Owens, R.A.; Roenhorst, J.W. Natural infections of tomato by *Citrus exocortis viroid*, *Columnea latent viroid*, *Potato spindle tuber viroid* and *Tomato chlorotic dwarf viroid*. *Eur. J. Plant Pathol.* **2004**, *110*, 823–831. [CrossRef]
13. Navarro, B.; Silletti, M.R.; Trisciuzzi, V.N.; Di Serio, F. Identification and characterization of *Potato spindle tuber viroid* infecting tomato in Italy. *J. Plant Pathol.* **2009**, *91*, 723–726.
14. Verhoeven, J.T.J.; Jansen, C.C.C.; Botermans, M.; Roenhorst, J.W. Epidemiological evidence that vegetatively propagated, solanaceous plant species act as sources of *Potato spindle tuber viroid* inoculum for tomato. *Plant Pathol.* **2010**, *59*, 3–12. [CrossRef]
15. Querci, M.; Owens, R.A.; Bartoli, I.; Lazarte, V.; Salazar, L.F. Evidence for heterologous encapsidation of potato spindle tuber viroid in particles of potato leafroll virus. *J. Gen. Virol.* **1997**, *78*, 1207–1211. [CrossRef] [PubMed]
16. Singh, R.; Kurz, J. RT-PCR analysis of PSTVd aphid transmission in association with PLRV. *Can. J. Plant Pathol.* **1997**, *19*, 418–424. [CrossRef]
17. Vo, T.T.; Dehne, H.W.; Hamacher, J. Transmission of *Tomato chlorotic dwarf viroid* by *Myzus persicae* assisted by *Potato leafroll virus*. *J. Plant Dis. Prot.* **2018**, *125*, 259–266. [CrossRef]
18. Antignus, Y.; Lachman, O.; Pearlsman, M. Spread of *Tomato apical stunt viroid* (TASVd) in greenhouse tomato crops is associated with seed transmission and bumble bee activity. *Plant Dis.* **2007**, *91*, 47–50. [CrossRef]

19. Matsuura, S.; Matsushita, Y.; Kozuka, R.; Shimizu, S.; Tsuda, S. Transmission of *Tomato chlorotic dwarf viroid* by bumblebees (*Bombus ignitus*) in tomato plants. *Eur. J. Plant Pathol.* **2010**, *126*, 111–115. [CrossRef]
20. Nielsen, S.L.; Enkegaard, A.; Nicolaisen, M.; Kryger, P.; Virscek Marn, M.; Mavric Plesko, I.; Kahrer, A.; Gottsberger, R.A. No transmission of *Potato spindle tuber viroid* shown in experiments with thrips (*Frankliniella occidentalis, Thrips tabaci*), honey bees (*Apis mellifera*) and bumblebees (*Bombus terrestris*). *Eur. J. Plant Pathol.* **2012**, *133*, 505–509. [CrossRef]
21. Faggioli, F.; Luigi, M.; Sveikauskas, V.; Olivier, T.; Virscek Marn, M.; Mavric Plesko, I.; De Jonghe, K.; Van Bogaert, N.; Grausgruber-Gröger, S. An assessment of the transmission rate of four pospiviroid species through tomato seeds. *Eur. J. Plant Pathol.* **2015**, *143*, 613–617. [CrossRef]
22. Simmons, H.E.; Ruchti, T.B.; Munkvold, G.P. Frequencies of seed infection and transmission to seedlings by potato spindle tuber viroid (a Pospiviroid) in tomato. *J. Plant Pathol. Microbiol.* **2015**, *6*, 1000275.
23. Candresse, T.; Marais, A.; Tassus, X.; Suhard, P.; Renaudin, I.; Leguay, A.; Poliakoff, F.; Blancard, D. First report of *Tomato chlorotic dwarf viroid* in tomato in France. *Plant Dis.* **2010**, *94*, 633. [CrossRef]
24. Van Brunschot, S.L.; Verhoeven, J.T.J.; Persley, D.M.; Geering, A.D.W.; Drenth, A.; Thomas, J.E. An outbreak of *Potato spindle tuber viroid* in tomato is linked to imported seed. *Eur. J. Plant Pathol.* **2014**, *139*, 1–7. [CrossRef]
25. EPPO. PM7/138 Detection and identification of Pospiviroids. *Bull. OEPP* **2021**, *51*, 144–177.
26. Spieker, R.L. A viroid from *Brunfelsia undulata* closely related to the *Columnea latent viroid*. *Arch. Virol.* **1996**, *141*, 1823–1832. [CrossRef]
27. Verhoeven, J.T.J.; Jansen, C.C.C.; Roenhorst, J.W.; Flores, R.; De la Peña, M. *Pepper chat fruit viroid*: Biological and molecular properties of a proposed new species of the genus *Pospiviroid*. *Virus Res.* **2009**, *144*, 209–214. [CrossRef]
28. Verhoeven, J.T.J.; Koenraadt, H.M.S.; Westenberg, M.; Roenhorst, J.W. Characterization of tomato apical stunt viroid isolated from a 24-year old seed lot of *Capsicum annuum*. *Arch. Virol.* **2017**, *162*, 1741–1744. [CrossRef] [PubMed]
29. Van de Vossenberg, B.T.L.H.; Van der Straten, M.J. Development and validation of real-time PCR tests for the identification of four *Spodoptera* species: *Spodoptera eridania, Spodoptera frugiperda, Spodoptera littoralis*, and *Spodoptera litura* (Lepidoptera: Noctuidae). *J. Econ. Entomol.* **2014**, *107*, 1643–1654. [CrossRef] [PubMed]
30. Botermans, M.; Van de Vossenberg, B.T.L.H.; Verhoeven, J.T.J.; Roenhorst, J.W.; Hooftman, M.; Dekter, R.; Meekes, E.T.M. Development and validation of a real-time RT-PCR assay for generic detection of pospiviroids. *J. Virol. Meth.* **2013**, *187*, 43–50. [CrossRef] [PubMed]
31. Verhoeven, J.T.J.; Koenraadt, H.M.S.; Jodlowska, A.; Huner, L.; Roenhorst, J.W. Pospiviroids infections in *Capsicum annuum*: Disease symptoms and lack of seed transmission. *Eur. J. Plant Pathol.* **2020**, *156*, 21–29. [CrossRef]
32. Yanagisawa, H.; Matsushita, Y. Host ranges and seed transmission of *Tomato planta macho viroid* and *Pepper chat fruit viroid*. *Eur. J. Plant Pathol.* **2017**, *149*, 211–217. [CrossRef]
33. Matsushita, Y.; Tsuda, S. Seed transmission of *potato spindle tuber viroid, tomato chlorotic dwarf viroid, tomato apical stunt viroid*, and *Columnea latent viroid* in horticultural plants. *Eur. J. Plant Pathol.* **2016**, *145*, 1007–1011. [CrossRef]
34. Lebas, B.S.M.; Clover, G.R.G.; Ochoa-Corona, F.M.; Elliott, D.R.; Tang, Z.; Alexander, B.J.R. Distribution of *Potato spindle tuber viroid* in New Zealand glasshouse crops of capsicum and tomato. *Australas. Plant Pathol.* **2005**, *34*, 129–133. [CrossRef]
35. Semancik, J.S. Citrus exocortis viroid. *Descr. Plant Viruses* **1980**, *226*. Available online: https://www.dpvweb.net/dpv/showdpv/?dpvno=226 (accessed on 23 May 2021).
36. Fox, A.; Monger, W. *Detection and Elimination of Solanaceous Viroids in Tomato Seeds and Seedlings*; Final Report of the Project PC294, December 2010; Horticultural Development Company: Kenilworth, UK, 2011; 24p, Available online: https://ahdb.org.uk/pc-294-detection-and-elimination-of-solanaceous-viroids-in-tomato-seeds-and-seedlings (accessed on 23 May 2021).
37. Khoury, J.; Singh, R.P.; Boucher, A.; Coombs, D.H. Concentration and distribution of mild and severe strains of potato spindle tuber viroid in cross-protected tomato plants. *Phytopathology* **1988**, *78*, 1331–1336. [CrossRef]
38. Singh, R.P.; Dilworth, A.D. *Tomato chlorotic dwarf viroid* in the ornamental plant *Vinca minor* and its transmission through tomato seed. *Eur. J. Plant Pathol.* **2009**, *123*, 111–116. [CrossRef]
39. Benson, A.P.; Singh, R.P. Seed transmission of Potato Spindle Tuber Virus in tomato. *Am. Potato J.* **1964**, *41*, 294.
40. Kryczynski, S.; Paduch-Cichal, E.; Skrzeczkowsk, L.R. Transmission of three viroids through seed and pollen of tomato plants. *J. Phytopathol.* **1988**, *121*, 51–57. [CrossRef]
41. Menzel, W.; Winter, S. Investigations on seed- and aphid-transmissibility of Potato spindle tuber viroid. *Julius-Kühn-Archiv* **2010**, *428*, 392–393.
42. Batuman, O.; Çiftçi, O.C.; Osei, M.K.; Miller, S.A.; Rojas, M.R.; Gilbertson, R.L. Rasta disease of tomato in Ghana is caused by the pospiviroids *potato spindle tuber viroid* and *tomato apical stunt viroid*. *Plant Dis.* **2019**, *103*, 1525–1535. [CrossRef]
43. McClean, A.P.D. Bunchy-top disease of the tomato: Additional host plants, and the transmission of the virus through seed of infected plants. *Union S. Afr. Dept. Agric. Sci. Bull.* **1948**, *256*, 1–28.
44. Singh, R.P.; Nie, X.; Singh, M. Tomato chlorotic dwarf viroid: An evolutionary link in the origin of pospiviroids. *J. Gen. Virol.* **1999**, *80*, 2823–2828. [CrossRef]
45. Koenraadt, H.; Jodlowska, A.; Van Vliet, A.; Verhoeven, K. Detection of TCDVd and PSTVd in seeds of tomato. *Phytopathology* **2009**, *99*, S66.
46. Belalcazar, C.S.; Galindo, A.J. Estudio sobre el virus de la "planta macho" de; jitomate (*Lycopersicon sculentum* Mill). *Agrocienc. Urug.* **1974**, *18*, 79–88.

47. Singh, R.P. Seed transmission of Potato spindle tuber virus in Tomato and Potato. *Am. Potato J.* **1970**, *47*, 225–227. [CrossRef]
48. Verhoeven, J.T.J.; Roenhorst, J.W.; Van Vliet, A.C.A.; Ebskamp, M.J.M.; Koenraadt, H.M.S. A potato spindle tuber viroid-positive tested seed lot of tomato may produce a viroid-free crop. In Proceedings of the Abstracts of the 5th Conference of the International Working Group of Legume and Vegetable Viruses, Haarlem, The Netherlands, 30 August–3 September 2015; p. 95.
49. Elliott, D.R.; Alexander, B.J.R.; Smales, T.E.; Tang, Z.; Clover, G.R.G. First report of *Potato spindle tuber viroid* in tomato in New Zealand. *Plant Dis.* **2001**, *85*, 1027. [CrossRef]
50. Ward, L.I.; Tang, J.; Veerakone, S.; Quinn, B.D.; Harper, S.J.; Delmiglio, C.; Clover, G.R.G. First Report of *Potato spindle tuber viroid* in cape gooseberry (*Physalis peruviana*) in New Zealand. *Plant Dis.* **2010**, *94*, 479. [CrossRef]
51. Mackie, A.E.; Rodoni, B.C.; Barbetti, M.J.; McKirdy, S.J.; Jones, R.A.C. *Potato spindle tuber viroid*: Alternative host reservoirs and strain found in a remote subtropical irrigation area. *Eur. J. Plant Pathol.* **2016**, *145*, 433–446. [CrossRef]
52. EPPO. First confirmed report of Potato spindle tuber viroid in Switzerland. *EPPO Report. Serv.* **2016**, *4*, 084.
53. Verhoeven, J.T.J.; Voogd, J.G.B.; Strik, N.; Roenhorst, J.W. First report of *Potato spindle tuber viroid* in vegetatively propagated plants of *Capsicum annuum* in The Netherlands. *N. Dis. Rep.* **2016**, *34*, 12. [CrossRef]
54. Verhoeven, J.T.J.; Botermans, M.; Janse, C.C.C.; Roenhorst, J.W. First report of *Pepper chat fruit viroid* in capsicum pepper in Canada. *N. Dis. Rep.* **2011**, *23*, 15. [CrossRef]
55. Ling, K.-S.; Zhang, W. First report of a natural infection *Mexican papita viroid* and *Tomato chlorotic dwarf viroid* on greenhouse tomatoes in Mexico. *Plant Dis.* **2009**, *93*, 1216. [CrossRef]
56. Matsushita, Y.; Usugi, T.; Tsuda, S. Development of a multiplex RT-PCR detection and identification system for *Potato spindle tuber viroid* and *Tomato chlorotic dwarf viroid*. *Eur. J. Plant Pathol.* **2010**, *128*, 165–170. [CrossRef]
57. Batuman, O.; Gilbertson, R.L. First Report of *Columnea latent viroid* (CLVd) in Tomato in Mali. *Plant Dis.* **2013**, *97*, 692. [CrossRef]
58. EPPO. Situation of Potato spindle tuber viroid in Austria in 2008. *EPPO Report. Serv.* **2008**, *9*, 177.
59. Hailstones, D.L.; Tesoriero, L.A.; Terras, M.A.; Dephoff, C. Detection and eradication of *Potato spindle tuber viroid* in tomatoes in commercial production in New South Wales, Australia. *Australas. Plant Pathol.* **2003**, *32*, 317–318. [CrossRef]
60. Nixon, T.; Glover, R.; Mathews-Berry, S.; Daly, M.; Hobden, E.; Lanbourne, C.; Harju, V.; Skelton, A. *Columnea latent viroid* (CLVd) in tomato: The first report in the United Kingdom. *N. Dis. Rep.* **2009**, *19*, 30. [CrossRef]
61. Antignus, Y.; Lachman, O.; Pearlsmand, M.; Gofman, R.; Bar-Joseph, M. A new disease of greenhouse tomatoes in Israel caused by a distinct strain to *Tomato apical stunt viroid* (TASVd). *Phytoparasitica* **2002**, *30*, 502–510. [CrossRef]
62. Fagoaga, C.; Duran-Vila, N. Naturally occurring variants of citrus exocortis viroid in vegetable crops. *Plant Pathol.* **1996**, *45*, 45–53. [CrossRef]
63. EPPO. Occurrence of Potato spindle tuber pospiviroid (Pstvd) in tomato plants in Germany. *EPPO Report. Serv.* **2004**, *1*, 006.
64. Verhoeven, J.T.J.; Botermans, M.; Meekes, E.T.M.; Roenhorst, J.W. *Tomato apical stunt viroid* in the Netherlands: Most prevalent pospiviroid in ornamentals and first outbreak in tomatoes. *Eur. J. Plant Pathol.* **2012**, *133*, 803–810. [CrossRef]
65. Fox, A.; Daly, M.; Nixon, T.; Brurberg, M.B.; Blystad, D.R.; Harju, V.; Skelton, A.; Adams, I.P. First report of *Tomato chlorotic dwarf viroid* (TCDVd) in tomato in Norway and subsequent eradication. *N. Dis. Rep.* **2013**, *27*, 8. [CrossRef]
66. Parrella, G.; Numitone, G. First report of *Tomato apical stunt viroid* in tomato in Italy. *Plant Dis.* **2014**, *98*, 1164. [CrossRef] [PubMed]
67. Ling, K.-S.; Sfetcu, D. First report of natural infection of greenhouse tomatoes by *Potato spindle tuber viroid* in the United States. *Plant Dis.* **2010**, *94*, 1376. [CrossRef]
68. Behjatnia, S.A.A.; Dry, I.B.; Krake, L.R.; Condé, B.D.; Connelly, M.I.; Randles, J.W.; Rezaian, M.A. New potato spindle tuber viroid and tomato leaf curl geminivirus strains from a wild Solanum sp. *Phytopathology* **1996**, *86*, 880–886. [CrossRef]
69. Shiraishi, T.; Maejima, K.; Komatsu, K.; Hashimoto, M.; Okano, Y.; Kitazawa, Y.; Yamaji, Y.; Namba, S. First report of tomato chlorotic dwarf viroid isolated from symptomless petunia plants (*Petunia* spp.) in Japan. *J. Gen. Plant Pathol.* **2013**, *79*, 214–216. [CrossRef]
70. Verhoeven, J.T.J.; Roenhorst, J.W. High stability of original predominant pospiviroid genotypes upon mechanical inoculation from ornamentals to potato and tomato. *Arch. Virol.* **2010**, *155*, 269–274. [CrossRef] [PubMed]
71. Parrella, G.; Crescenzi, A.; Pacella, R. First record of *Columnea latent viroid* (CLVd) in tomato in Italy. *Acta Hortic.* **2011**, *914*, 149–152. [CrossRef]
72. Candresse, T.; Smith, D.; Diener, T.O. Nucleotide sequence of a full-length infectious clone of the Indonesian strain of tomato apical stunt viroid (TASV). *Nucleic Acids Res.* **1987**, *15*, 10597. [CrossRef] [PubMed]
73. Candresse, T.; Marais, A.; Ollivier, F.; Verdin, E.; Blancard, D. First report of the presence of *Tomato apical stunt viroid* on tomato in Sénégal. *Plant Dis.* **2007**, *91*, 330. [CrossRef] [PubMed]
74. EPPO. Outbreak of Potato spindle tuber viroid in tomato in the United Kingdom. *EPPO Report. Serv.* **2011**, *9*, 202.
75. EPPO. Incidental finding of Potato spindle tuber viroid in tomatoes in the Netherlands. *EPPO Report. Serv.* **2013**, *7*, 148.
76. EPPO. First report of Tomato apical stunt viroid in France. *EPPO Report. Serv.* **2013**, *11*, 236.
77. Ling, K.-S.; Bledsoe, M.E. First report of *Mexican papita viroid* infecting greenhouse tomato in Canada. *Plant Dis.* **2009**, *93*, 839. [CrossRef]
78. Ling, K.-S.; Verhoeven, J.T.J.; Singh, R.P.; Brown, J.K. First report of *Tomato chlorotic dwarf viroid* in greenhouse tomatoes in Arizona. *Plant Dis.* **2009**, *93*, 1075. [CrossRef] [PubMed]
79. Ling, K.-S.; Li, R.; Panthee, D.R.; Gardner, R.G. First report of *Potato spindle tuber viroid* naturally infecting greenhouse tomatoes in North Carolina. *Plant Dis.* **2013**, *97*, 148. [CrossRef] [PubMed]

80. Ling, K.-S.; Li, R.; Groth-Helms, D.; Assis-Filho, F.M. First report of *Potato spindle tuber viroid* naturally infecting field tomatoes in the Dominican Republic. *Plant Dis.* **2014**, *98*, 701. [CrossRef]
81. Mishra, M.D.; Hammond, R.W.; Owens, R.A.; Smith, D.R.; Diener, T.O. Indian bunchy top disease of tomato plants is caused by a distinct strain of citrus exocortis viroid. *J. Gen. Virol.* **1991**, *72*, 1781–1785. [CrossRef]
82. Mumford, R.A.; Jarvis, B.; Skelton, A. The first report of *Potato spindle tuber viroid* (PSTVd) in commercial tomatoes in the UK. *Plant Pathol.* **2004**, *53*, 242. [CrossRef]
83. Puchta, H.; Herold, T.; Verhoeven, K.; Roenhorst, A.; Ramm, K.; Schmidt-Puchta, W.; Sänger, H.L. A new strain of potato spindle tuber viroid (PSTVd-N) exhibits major sequence differences as compared to all other PSTVd strains sequenced so far. *Plant Mol. Biol.* **1990**, *15*, 509–511. [CrossRef]
84. Reanwarakorn, K.; Klinkong, S.; Porsoongnurn, J. First report of natural infection of *Pepper chat fruit viroid* in tomato plants in Thailand. *N. Dis. Rep.* **2011**, *24*, 6. [CrossRef]
85. Steyer, S.; Olivier, T.; Skelton, A.; Nixon, T.; Hobden, E. *Columnea latent viroid* (CLVd): First report in tomato in France. *Plant Pathol.* **2009**, *59*, 794. [CrossRef]
86. Verhoeven, J.T.J.; Janse, C.C.C.; Roenhors, J.W. First report of *Tomato apical stunt viroid* in tomato in Tunisia. *Plant Dis.* **2006**, *90*, 528. [CrossRef]
87. Verhoeven, J.T.J.; Janse, C.C.C.; Roenhorst, J.W.; Steyer, S.; Michelante, D. First report of *Potato spindle tuber viroid* in tomato in Belgium. *Plant Dis.* **2007**, *91*, 1055. [CrossRef]
88. Walter, B.; Thouvenal, J.C.; Fauque, C. Les viruses de la tomate en Cote d'Ivore. *Ann. Phytopathol.* **1980**, *12*, 259–275.
89. Constable, F.; Chambers, G.; Penrose, L.; Daly, A.; Mackie, J.; Davis, K.; Rodoni, B.; Gibbs, M. Viroid-infected tomato and capsicum seed shipments to Australia. *Viruses* **2019**, *11*, 98. [CrossRef] [PubMed]

plants

Article

Development of a Real-Time Loop-Mediated Isothermal Amplification Assay for the Rapid Detection of Olea Europaea Geminivirus

Sofia Bertacca [1,†], Andrea Giovanni Caruso [1,2,†], Daniela Trippa [1], Annalisa Marchese [1], Antonio Giovino [2], Slavica Matic [3], Emanuela Noris [3], Maria Isabel Font San Ambrosio [4], Ana Alfaro [4], Stefano Panno [1,*] and Salvatore Davino [1,*]

1 Department of Agricultural, Food and Forest Sciences (SAAF), University of Palermo, Viale delle Scienze, 90128 Palermo, Italy; sofia.bertacca@unipa.it (S.B.); andreagiovanni.caruso@unipa.it (A.G.C.); danielaantonina.trippa@unipa.it (D.T.); annalisa.marchese@unipa.it (A.M.)
2 Research Centre for Plant Protection and Certification (CREA), Strada Statale, 113, 90011 Bagheria, Italy; antonio.giovino@crea.gov.it
3 Institute for Sustainable Plant Protection, National Research Council (IPSP-CNR), Strada delle Cacce, 73, 10135 Turin, Italy; slavica.matic@ipsp.cnr.it (S.M.); emanuela.noris@ipsp.cnr.it (E.N.)
4 Instituto Agroforestal Mediteráneo, Universitat Politécnica de València (IAM-UPV), Camino de Vera, s/n, 46022 Valencia, Spain; mafonsa@upv.edu.es (M.I.F.S.A.); analfer1@etsia.upv.es (A.A.)
* Correspondence: stefano.panno@unipa.it (S.P.); salvatore.davino@unipa.it (S.D.); Tel.: +39-09123896049 (S.D.)
† These authors contributed equally to this work.

Abstract: A real-time loop-mediated isothermal amplification (LAMP) assay was developed for simple, rapid and efficient detection of the Olea europaea geminivirus (OEGV), a virus recently reported in different olive cultivation areas worldwide. A preliminary screening by end-point PCR for OEGV detection was conducted to ascertain the presence of OEGV in Sicily. A set of six real-time LAMP primers, targeting a 209-nucleotide sequence elapsing the region encoding the coat protein (AV1) gene of OEGV, was designed for specific OEGV detection. The specificity, sensitivity, and accuracy of the diagnostic assay were determined. The LAMP assay showed no cross-reactivity with other geminiviruses and was allowed to detect OEGV with a 10-fold higher sensitivity than conventional end-point PCR. To enhance the potential of the LAMP assay for field diagnosis, a simplified sample preparation procedure was set up and used to monitor OEGV spread in different olive cultivars in Sicily. As a result of this survey, we observed that 30 out of 70 cultivars analyzed were positive to OEGV, demonstrating a relatively high OEGV incidence. The real-time LAMP assay developed in this study is suitable for phytopathological laboratories with limited facilities and resources, as well as for direct OEGV detection in the field, representing a reliable method for rapid screening of olive plant material.

Keywords: LAMP; OEGV; *Geminiviridae*; olive viruses

Citation: Bertacca, S.; Caruso, A.G.; Trippa, D.; Marchese, A.; Giovino, A.; Matic, S.; Noris, E.; Ambrosio, M.I.F.S.; Alfaro, A.; Panno, S.; et al. Development of a Real-Time Loop-Mediated Isothermal Amplification Assay for the Rapid Detection of Olea Europaea Geminivirus. *Plants* **2022**, *11*, 660. https://doi.org/10.3390/plants11050660

Academic Editor: Alessandro Vitale

Received: 7 February 2022
Accepted: 26 February 2022
Published: 28 February 2022

Publisher's Note: MDPI stays neutral with regard to jurisdictional claims in published maps and institutional affiliations.

1. Introduction

The olive tree (*Olea europaea* L.) belonging to the *Oleaceae* family is the most widely cultivated species of the *Olea* genus. Olive, providing edible fruits and storable oil, has been cultivated in the Mediterranean area since prehistoric times [1], and is regarded as the most economically important fruit tree in the Mediterranean basin [2]. Olive cultivation has, over the centuries, played an important role in the economic development of rural areas in the Mediterranean region, providing noteworthy sources of income and employment opportunities for the population in rainfed agricultural territories [3]. Even today, after thousands of years, the countries in this area produce about 90% of the olive fruits [4], while the olive cultivated area covers about 10 million hectares worldwide. According to the latest data available on FAOSTAT, among Mediterranean countries, Spain ranked the major olive producer in 2020 (2,623,720 ha; 8.137.810 tons), followed by Italy (1,145,520 ha; 2,207,150 tons), Morocco (1,068,895 ha; 1,409,266 tons), Greece (906,020 ha; 2,790,442 tons), and Turkey (887,077 ha; 1,316,626 tons). According to ISTAT data [5], among the Southern Italian regions, Sicily (161,661 ha; 255,798 tons) played a significant role in olive and olive oil production, industry and export in 2021, being the third largest producer after Apulia (379,960 ha; 708,400 tons) and Calabria (184,410 ha; 680,275 tons). In Sicily, the olive crop has been cultivated since ancient times, and it is characterized by many ancient landraces/cultivars of high organoleptic quality [6]; its germplasm is distinguished by a wide genetic diversity, possibly related to its past domestication and spread and to some reproductive biological peculiarities such as self-incompatibility [7]. The production of olive oil in Sicily is based mainly on the autochthonous cultivars (cvs) 'Biancolilla', 'Cerasuola', 'Moresca', 'Nocellara del Belice', 'Nocellara Etnea', 'Ogliarola Messinese', 'Santagatese' and 'Tonda Iblea' [8]. The table olive industry is also appreciable, accounting to 10% of the total production of this region [5], mainly relying on the cv. 'Nocellara del Belice' and, to a minor extent, 'Nocellara Etnea', 'Ogliarola Messinese', and 'Moresca', produce large-sized fruits of high commercial value [8]. The current tendency in olive tree cultivation is moving towards the use of local cvs for high quality oil production (such as DOP—protected designation of origin), which is typical of specific geographic areas. For this reason, the local administration currently supports studies and activities aimed at the characterization and recovery of local and ancient cvs, in order to establish germplasm collections that limit genetic erosion [9]. However, a large number of diseases and disorders affect this crop, mostly caused by fungi, such as *Arthrinium phaeospermum*, *Phoma cladoniicola* and *Ulocladium consortiale*, recently discovered as new olive pathogens in Italy [10], but also by systemic pathogens including bacteria and viruses, which provoke significant yield losses. Indeed, in the last decade, olive production has suffered an enormous decline due to the emergence of biotic agents that have significantly undermined the Mediterranean economy related to olive and the olive oil industry; a dramatic example being the *Xylella fastidiosa* epidemic in 2013, which decimated olive trees in Apulia [11] and created huge losses for the local olive economy and oil production outputs, posing critical challenges for its management, as well as dramatic changes in the landscape [12,13]. Furthermore, the vegetative propagation of olive trees using cuttings of semi-wood has contributed over the years to the spread of systemic-pathogens, particularly viruses [14]. Nevertheless, despite the difficulty of associating specific symptoms to a particular virus, many viruses are easily transmitted through infected propagation material [15], and many olive infecting viruses are symptomless. Therefore, it is essential to better elucidate the evolutionary aspects of latent viruses in olive crops. In the last year, thanks also to the application of new technologies such as high-throughput sequencing (HTS), a new geminivirus called Olea europaea geminivirus (OEGV) has been identified in the olive tree [16], but its spread and pathogenicity remain puzzling. Since its first identification in Apulia [16] in the "Ogliarola" and "Leccino" cvs, OEGV was reported in California and Texas [17], Portugal [15], and Spain [18]. OEGV is classified as a putative member within the *Geminiviridae* family [16], currently including 14 genera [19] and few other still unassigned geminiviruses [20]. The evolutionary relationship of OEGV with other geminiviruses indicated that OEGV has distinctive genome features, possibly representing a new genus [15–17]. OEGV is characterized by a bipartite genome containing DNA-A and DNA-B. DNA-A (2775 nucleotides, nts) includes four ORFs, three in the complementary-sense encoding the replication-associated protein Rep (AC1), the transcriptional activator protein TrAP (AC2), the replication enhancer protein Ren (AC3) and one in the virion-sense, (AV1), encoding the coat protein (CP). DNA-B (2763 nts) includes two ORFs, BC1 in the complementary sense, with an unknown function and lacking known conserved domains typical of geminiviral proteins, and BV1 on the virion sense, possibly encoding the movement protein (MP). In bipartite geminiviruses, AC4/C4 protein is a symptom determinant involved in cell-cycle control, and interacts with CP and/or MP in the replicated genome transport from nucleus to cytoplasm and from cell-to-cell [15]. Curiously, no genes encoding AC4/C4 were found on the OEGV genome.

In addition, the two DNA molecules present a common region (CR) of 403 nt that contains the TATA box and four replication-associated iterons with a unique arrangement compared to other geminiviruses [15–17]. In a recent survey, Alabi and co-workers [17] detected OEGV-positive olive trees originating from different locations, advancing the concept of a possible worldwide spread of this virus, likely due to the inadvertent movement of germplasms from clonally propagated infected but asymptomatic olive trees. As a matter of fact, OEGV does not appear to be clearly associated to any symptom in olive; moreover, a high degree of sequence conservation has been identified [18].

In this study, we aimed to investigate the presence of OEGV in Sicily and to develop a rapid detection protocol based on the LAMP methodology. In addition, an on-site olive sample homogenization procedure was developed replacing canonical DNA extraction methods, which is useful in evaluating the suitability of the LAMP assay for on-site OEGV testing.

2. Materials and Methods

2.1. *Plant Material Collection*

To investigate the presence of OEGV in Sicily, different surveys were carried out during spring 2021, focusing in particular on two olive producing sites in the Agrigento province (Sicily, Italy). The olive tree samples were randomly collected according to the hierarchical sampling scheme [21], with minor adaptations to olive plants. All samples were geo-referenced with the Planthology mobile application [22], collected from a total of 80 olive trees of 10 different cvs (40 trees randomly sampled for each site). Each sample consisted of 8 branches per plant (two for each plant cardinal point); samples were stored at 4 °C and processed within the next 24 h for subsequent molecular analyses.

2.2. *DNA Extraction and Sample Preparation*

Total DNA was extracted using the DNA extraction GenUP^{TM} Plant DNA kit (Biotechrabbit GmbH, Berlin, Germany), following manufacturer's instructions with slight modifications. In brief, 3 g of tissue were homogenized in an extraction bag (BIOREBA, Reinach, Switzerland) using the HOMEX 6 homogenizer (BIOREBA, Reinach, Switzerland), with 3 mL extraction buffer (1.3 g sodium sulphite anhydrous, 20 g polyvinylpyrrolidone MW 24–40,000, 2 g chicken egg chicken albumin Grade II, 20 g Tween-20 in one L of distilled water, pH 7.4). Aliquots of 400 µL of the extract were added to the same volume of lysis buffer. The eluted DNA was resuspended in 100 µL RNase-free water; following two measurements with a UV–Vis NanoDrop 1000 spectrophotometer (Thermo Fisher Scientific, Waltham, MA, USA), samples were adjusted to approximately 50 ng/µL and stored at −20 °C.

2.3. *Preliminary Screening of OEGV by End-Point PCR*

The end-point PCR was conducted using the primer pair A2for/A4rev [16], amplifying an 831 bp fragment within the AV1 gene. PCR was performed in a final volume of 25 µL, containing 1 µL of total DNA extract, 20 mM Tris-HCl (pH 8.4), 50 mM KCl, 3 mM $MgCl_2$, 0.4 mM dNTPs, 1 µM each primer, and 2 U Taq DNA polymerase (Thermo Fisher Scientific, Waltham, MA, USA) and RNase-free water to reach the final volume. Healthy olive plant DNA and water were used as control samples. The PCR was performed in a MultiGene OptiMax thermal cycler (Labnet International Inc., Edison, NJ, USA) with the following conditions: 95 °C for 5 min; 40 cycles of 95 °C for 30 s, 64 °C for 45 s, and 72 °C for 1 min; a final elongation at 72 °C for 10 min. PCR products were electrophoresed on 1.5% agarose gel, stained with SYBR^{TM} Safe (Thermo Fisher Scientific, Waltham, MA, USA) and visualized by UV light.

2.4. *LAMP Primers Design*

The OEGV DNA-A complete sequence (GenBank Acc. No. MW316657) was used to design LAMP primers by the PrimerExplorer version 5 software (http://primerexplorer.

jp/lampv5e/, accessed on 5 July 2021), choosing a 540-bp nucleotide sequence elapsing region within the AV1 gene. A set of six primers were selected, including two outer primers (forward and backward outer primer, F3 and B3, respectively), two inner primers (forward and backward inner primer, FIP and BIP, respectively), and two loop primers (forward and backward loop primer, LF and LB, respectively). The specificity of the primer set was tested in silico using the Nucleotide-BLAST algorithm (https://www.ncbi.nlm.nih.gov, accessed on 5 July 2021) available at the National Centre for Biotechnology Information (NCBI), in order to evaluate possible cross reactions with other viruses. This set of primers was also tested against the full genomic sequences of other geminiviruses reported in Italy using the Vector NTI Advance 11.5 software (Invitrogen, Carlsbad, CA, USA), in order to verify their affinity. The list included Tomato leaf curl New Delhi virus (ToLCNDV) (DNA-A: GenBank Acc. No. MK732932 and DNA-B: MK732933), Tomato yellow leaf curl Sardinia virus (TYLCSV) (GenBank Acc. No. GU951759), Tomato yellow leaf curl virus (TYLCV) (GenBank Acc. No. X15656), TYLCV-IL23 (GenBank Acc. No. MF405078), and TYLCV isolate 8-4/2004 (GenBank Acc. No. DQ144621).

2.5. OEGV Real-Time LAMP Assay Optimization

The real-time LAMP assay was performed in a 12 μL reaction mixture containing 1.6 mM each of FIP-OEGV and BIP-OEGV, 0.2 mM each of F3-OEGV and B3-OEGV, 0.4 mM each of forward loop primer (LF-OEGV) and backward loop primer (LB-OEGV), 6.25 μL WarmStart LAMP 2X Mastermix (New England Biolabs, Beverly, MA, USA), and 0.25 μL of LAMP Fluorescent dye (New England Biolabs, Beverly, MA, USA), 1 μL of total DNA as template and nuclease-free H_2O was added to reach the final volume. DNA extracted from ten samples previously analyzed by end-point PCR was used in the real-time LAMP assay, including a positive control (PC) and a healthy olive plant DNA as negative control (NC). Each sample was analyzed twice. The LAMP assay was conducted at 65 °C (according to manufacturer's instructions) for 60 min and fluorescence was acquired every 60 s, using a Rotor-Gene Q2plex HRM Platform Thermal Cycler (Qiagen, Hilden, Germany). A melting curve was calculated to record the fluorescence using the following protocol: 95 °C for 1 min, 40 °C for 1 min, 70 °C for 1 min and an increase of temperature at 0.5 °C/s up to 95 °C. During the amplification, the fluorescence data were obtained in the 6-carboxyfluorescein (FAM) channel (excitation at 450–495 nm and detection at 510–527 nm). The relative fluorescence units (RFU) threshold value was used, and the threshold time (Tt) was calculated as the time at which fluorescence was equal to the threshold value.

2.6. Features of Real-Time LAMP Assay: Sensitivity and Comparison to Conventional PCR, Reaction Time and Specificity

To set up the conditions of the LAMP assay, an amplicon obtained by subjecting an OEGV-positive sample to end-point PCR (see above) was purified from agarose gel using an UltraClean™ 15 DNA purification kit (MO-BIO Laboratories, Carlsbad, CA, USA), following manufacturer's instructions. The purified DNA (named pcr-DNA) was quantified using a UV–Vis NanoDrop 1000 spectrophotometer (Thermo Fisher Scientific, Waltham, MA, USA). The number of copies was determined as follows: [Number of copies = (amount of DNA in nanograms $\times$ 6.022 $\times$ 10^{23})/(length of DNA template in bp $\times$ 1 $\times$ 10^9 $\times$ 650)]. To determine the OEGV real-time LAMP optimal reaction time and sensitivity, ten-fold serial dilutions of the sample (named pcr-DNA) were used as a template for both real-time LAMP assay and end-point PCR. Moreover, to evaluate the specificity of the LAMP assay and to assess potential non-specific cross reactions with other geminiviruses, a LAMP assay was conducted with two OEGV-positive samples together with DNA extracts from other geminiviruses unrelated to OEGV; specifically, ToLCNDV (Acc. No. MK732932) [23], TYLCSV (Acc. No. GU951759) [24], TYLCV (Acc. No. DQ144621) [25], TYLCV-IS76 [26]. Each sample was analyzed in duplicate in two independent real-time LAMP assays. In each run, total DNA from a healthy olive plant (NC) was included. The assay was conducted as above described, including the melting curve steps.

2.7. Set up of a Rapid Sample Preparation Method Suitable for the Real-Time LAMP Assay

To set up a simple and inexpensive sample preparation procedure, a method that avoided DNA extraction named "membrane spot crude extract" was used. For this, 1.5 g of vegetable tissue was placed in an extraction bag and homogenized with 3 mL of extraction buffer (see above). Five µL of this extract was spotted on a 1 cm^2 Hybond®-N+ hybridization membrane (GE Healthcare, Chicago, IL, USA), dried at room temperature for 5 min, and placed in a 2 mL tube containing 0.5 mL of glycine buffer (0.1 M Glycine, 0.05 M NaCl, 1 mM EDTA). After 20-s vortexing, samples were heated at 95 °C for 10 min and 3 µL of the extract were used for the LAMP assay. Ten samples previously analyzed by end-point PCR were used in the real-time LAMP assay, including a positive control (PC) and a healthy olive plant DNA as negative control (NC).

2.8. Spread of OEGV in Different Cultivars

During autumn 2021, in different Sicilian areas, a second sampling was carried out to evaluate the OEGV spread in Sicily, this time sampling 10–15-year-old olive trees, belonging to 70 different cvs. A total of 560 samples were collected. For each cv, eight different trees were sampled and grouped, obtaining a total of 70 different batches. Sampling and geo-referencing were as described above. In this case, samples were prepared with the "membrane spot crude extract" method and subjected to real-time LAMP assays in a 12 µL final volume as described above. In the case of positive sample batches, they were re-sampled and analyzed individually to determine the effective number of positive plants for each cultivar.

3. Results

3.1. OEGV Detection by End-Point PCR

To ascertain the presence of OEGV in Sicily, a total of 80 samples representing 10 different cvs collected from two olive production sites in the Agrigento province were analyzed by end-point PCR. Overall, 44 of them were found to be positive to OEGV, demonstrating the presence of OEGV in Sicily also. However, OEGV was not equally distributed among the cvs tested, and some cvs tested negative for this virus, at least using the primer set mentioned in this manuscript (Table 1).

Table 1. Prevalence and cultivar distribution of OEGV analyzed by end-point PCR.

Cultivar	No. Positive/Tested Samples	Percentage of Positive Samples (%)
Cavalieri Standard	8/8	100
Cerasuola Nilo Paceco	8/8	100
Cerasuola Standard	8/8	100
Giarraffa	0/8	0
Nocellara del Belice Giafalione	8/8	100
Pizzutella	8/8	100
Salicina Vassallo	3/8	37.5
Uovo di piccione	1/8	12.5
Vaddara	0/8	0
Zaituna Florida	0/8	0
Total	**44/80**	**55**

3.2. OEGV Real-Time LAMP Primer Design

A real-time LAMP assay for the rapid detection of OEGV was developed using a set of six primers designed on the OEGV-AV1 coding region. The sequences and binding sites of the primers are reported in Table 2 and Figure 1, respectively.

Table 2. Primers used for OEGV detection by LAMP.

Primer Name	Sequence (5′-3′)	Amplicon Size (bp)
F3-OEGV	CGATACGAGACATACCCAG	209
B3-OEGV	TCCATGTTGATCATCCAAGT	
FIP-OEGV	CAGCCACTGCTTCATATTATGAACACGAATTGTGCTTAACGGTT	-
BIP-OEGV	GATGTGGCTCGTGTATGATAGACGTCTGGATCCCGACTTTCC	
LF-OEGV	GGCTTCGCTAGTCAACTTAACTG	-
LB-OEGV	TCCCGGTAATTCTAATCCCAGAG	

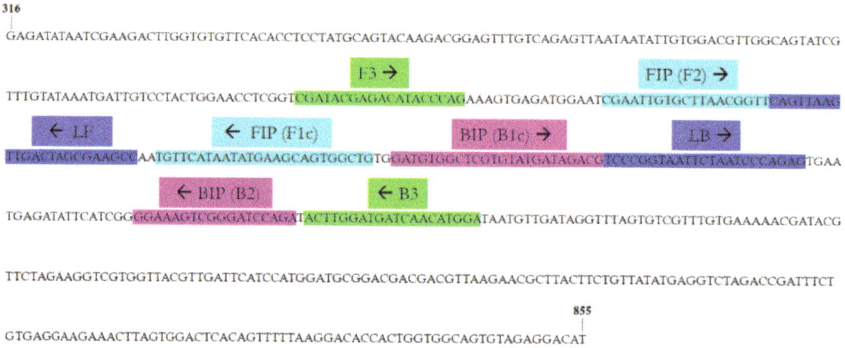

Figure 1. Location of loop-mediated isothermal amplification (LAMP) primer sets designed on the AV1 coding region of OEGV. F3 and B3 are shown in green, FIP (F1c-F2) in blue, BIP (B1c-B2) in pink, and the two loop primers LF and LB in brown. FIP is a hybrid primer consisting of the F1c and the F2 sequences, while BIP is a hybrid primer consisting of the B1c and B2 sequences. The arrows indicate the extension direction. The numbers at the beginning and end of the sequence represent the genomic position of the first and last nucleotide in the selected sequence (GenBank Acc. No. MW316657).

Both the in silico analysis of LAMP primers using Nucleotide-BLAST algorithm and the hybridisation analysis against other geminiviruses performed with the Vector NTI 11.5 program allowed for the exclusion of relevant matches with other organisms and, more specifically, with geminiviruses known to be present in Sicily.

3.3. *OEGV Real-Time LAMP Assay Optimization*

To evaluate the performances of the primer set designed for the real time LAMP assay in the identification of the presence of OEGV in olive DNA extracts, the LAMP assay was conducted using a subset of the samples listed in Table 1, selecting them among those that resulted positive in end point PCR. In the assay, a sample that tested negative was also included (i.e., cv. Giarraffa), together with an appropriate negative control (NC). The assay was conducted at 65 °C. As reported in Table 3 and Figure 2A, the positive samples showed exponential trends between 3 to 13 min. The melting curves of the positive LAMP reactions all had the same peak temperature of approximately 85 °C (Figure 2B). As expected, no signal was obtained with the negative control and, according to the end point PCR results, the samples from cv. Giarraffa could not be amplified by LAMP, even at late reaction times.

Table 3. Performance of the real time LAMP assay for the detection of OEGV in olive samples collected in Sicily.

Cultivar	No. of Different Samples Analyzed	ID Sample	Reaction Time (min)
Cavalieri Standard	2	1	10
		2	7
Cerasuola Standard	2	3	10
		4	7
Giarraffa	2	5	-
		6	-
Nocellara del Belice Giafalione	2	7	10
		8	13
Pizzutella	2	9	10
		10	9
Positive control	1	PC	3
Negative control	1	NC	-

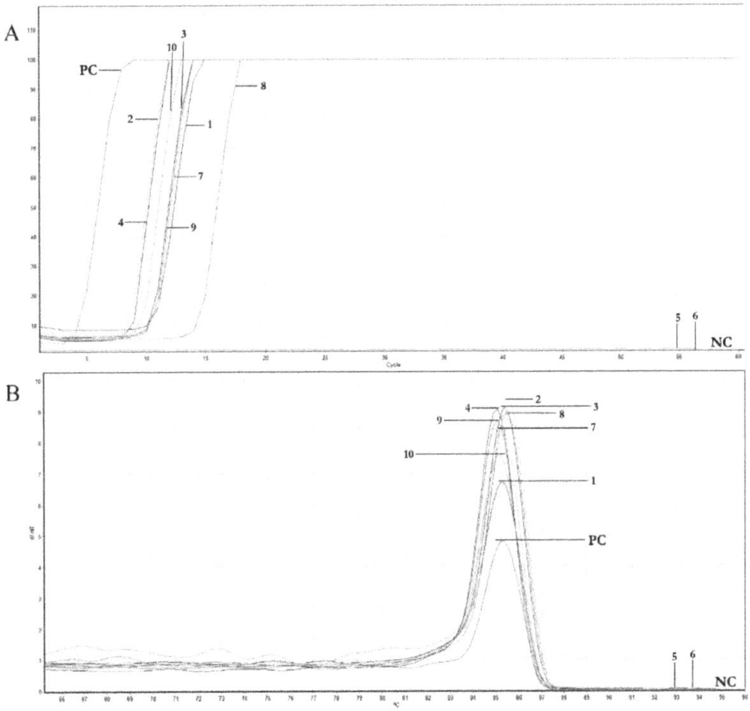

Figure 2. Results of the real time LAMP assay for the detection of OEGV. (**A**): Amplification curves of real-time LAMP assay; (**B**): Melting curves of the amplification curves previously obtained, including positive (PC) and negative control (NC).

3.4. Features of Real-Time LAMP Assay: Sensitivity and Comparison to Conventional PCR, Reaction Time and Specificity

To determine the sensitivity of the real-time LAMP assay compared to the end-point PCR and to evaluate the LAMP efficacy, a comparative experiment was conducted using as a template ten-fold serial dilutions of an amplicon obtained by end point PCR from an

OEGV-positive sample (pcr-DNA), starting from a concentration of 80.9 ng/µL. As can be observed in Table 4 and Figure 3, DNA concentration up to ~80.9×10^{-8} ng/µL was detected with the LAMP assay, while the end point PCR positive signals were obtained with DNA concentration up to ~80.9×10^{-7} ng/µL, indicating that real time LAMP was about ten times more sensitive than conventional PCR.

Table 4. Comparison of the sensitivity of the real time LAMP and end-point PCR.

	Starting DNA Concentration (80.9 ng/µL)										
Assay	10^{1}	10^{-1}	10^{-2}	10^{-3}	10^{-4}	10^{-5}	10^{-6}	10^{-7}	10^{-8}	10^{-9}	10^{-10}
End-point PCR	+	+	+	+	+	+	+	+	–	–	–
Real-time LAMP	+	+	+	+	+	+	+	+	+	–	–
Reaction Time (min)	3	4	5	6	7	8	9	10	10	–	–

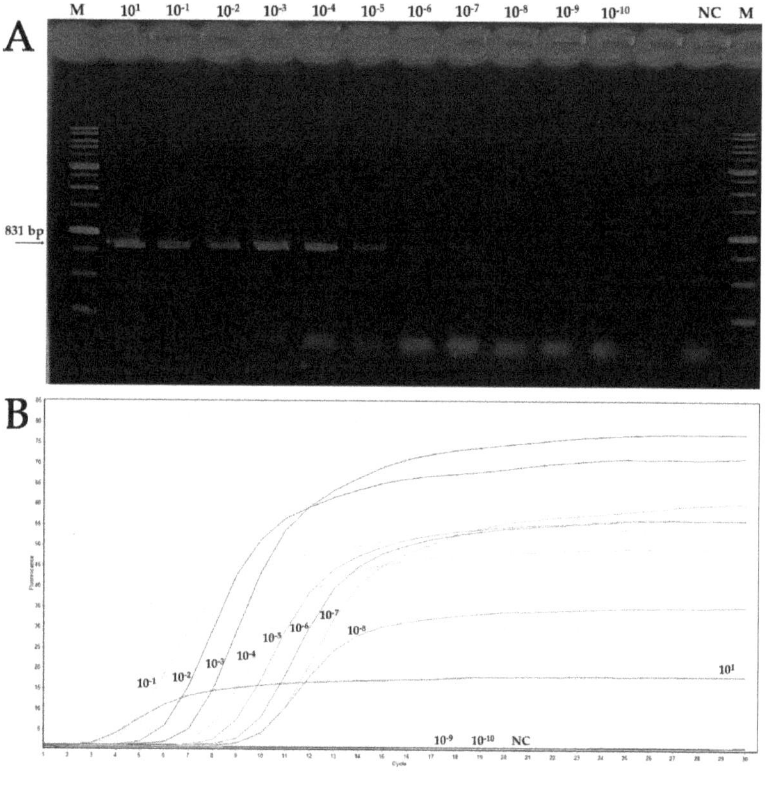

Figure 3. Sensitivity of the end point PCR (panel **A**) and real-time LAMP (panel **B**) for OEGV detection. The assay was conducted using 10-fold serial dilutions of pcr-DNA. Panel (**A**): Agarose gel electrophoresis of PCR products; M: 1 Kb ladder marker, NC: negative control. Panel (**B**). Fluorescence of the 10-fold serial dilutions analyzed. Fluorescence increased in positive sample curves (from 10^{-1} to 10^{-8}) after 3 to 10 min.

Moreover, even considering the lowest detectable concentration of the pcr-DNA sample in real time LAMP (~80.9×10^{-8} ng/µL), the results clearly showed that the time required to carry out the experiment was less than 30 min.

In addition, to evaluate the specificity of the LAMP assay and to assess potential non-specific cross reactions with other geminiviruses present in the agricultural areas where olive crop samples were collected, we conducted a LAMP assay using the geminiviruses reported in paragraph 2.6 as a template. Results showed that no signals were obtained with any of the geminiviruses used as the outgroup, while the two OEGV-positive olive DNA samples used as controls reacted in real time LAMP with a time value of 10 min and a single peak at 85 °C in the melting curve. This allowed us to confirm the specificity of the assay and to exclude the cross-reactivity with unrelated geminiviruses previously isolated in Sicily.

3.5. Set up of a Rapid Sample Preparation Method Suitable for the Real-Time LAMP Assay

With the purpose of identify a method that allows a simple and inexpensive sample preparation useful for real time LAMP, samples prepared with the two different procedures were tested. For this, the ten samples previously analyzed by end-point PCR and by real time LAMP assay (Table 3) were considered. As reported in Table 5, all samples tested positive in the LAMP assay when extracted with either procedure. Specifically, samples extracted with the commercial kit showed a fluorescence increase ranging between 3–14 min, while the same samples prepared with the "membrane spot crude extract" method could be detected in 10–24 min. This is worthy of note, as it indicates that the rapid method allows for the detection of the presence of OEGV with a delay of only a few minutes compared to the corresponding extract obtained with the commercial kit. As expected, even with this rapid procedure, no reaction was obtained with the samples from cv. Giaraffa.

Table 5. Comparison of two different sample preparation methods for the identification of the presence of OEGV in olive samples.

Cultivar	No. Samples Analyzed	Time Value		
		ID Sample	Total DNA Extraction by Commercial Kit (min)	Membrane Spot Crude Extract (min)
Cavalieri Standard	2	1	10	14
		2	7	12
Cerasuola Standard	2	3	10	16
		4	7	10
Giarraffa	2	5	-	-
		6	-	-
Nocellara del Belice Giafalione	2	7	10	15
		8	14	24
Pizzutella	2	9	10	16
		10	9	14
Positive control	1	PC	3	12
Negative control	1	NC	-	-

3.6. Spread of OEGV in Sicily

To investigate the spread of OEGV in different olive cultivars grown in Sicily, a new survey was conducted testing 70 samples, each consisting of eight different trees of the same cv. These samples were extracted with the rapid extraction protocol and tested in real-time LAMP, thus representing a total of 560 olive trees analyzed overall. This analysis showed that 30 out of the 70 cultivars (~43%) were positive for OEGV, indicating a relatively high incidence and prevalence of OEGV in the sampling locations and across cultivars (Table 6). When each of the eight plant samples present in the 30 positive batches were tested individually, the majority (235 out of 240 plants) resulted as being positive for OEGV, except the batch of cv. 'Calatina', where only three plants out of eight were positive (Table 6).

Table 6. Incidence of OEGV evaluated using a real-time LAMP assay on samples prepared with the "membrane spot crude extract" method.

Cultivar Analyzed	Real-Time LAMP	
	Cultivar Batch	Positive Plants/Tested Plants
Abunara	+	8/8
Aitana	−	NT
Arbequina	+	8/8
Bariddara	+	8/8
Biancolilla Caltabellotta	−	NT
Biancolilla Caltabellotta TA PC	+	8/8
Biancolilla Iacapa	−	NT
Biancolilla Napoletana	−	NT
Biancolilla Pantelleria	−	NT
Biancolilla Schimmenti	−	NT
Biancolilla Siracusana	−	NT
Biancuzza	−	NT
Bottone di Gallo Vassallo	−	NT
Brandofino	−	NT
Calamignara	−	NT
Calatina	+	3/8
Carasuola Cappuccia	+	8/8
Castricianella Rapparina	+	8/8
Cavalieri Standard	+	8/8
Cerasuola 1 Clone 2	+	8/8
Cerasuola Nilo Paceco	+	8/8
Cerasuola Standard	+	8/8
Conservolia	−	NT
Crastu Collesano	−	NT
Galatina	−	NT
Giarraffa	−	NT
Gordales	−	NT
Iacona	+	8/8
Indemoniata	−	NT
Koroneiki	+	8/8
Leucocarpa	−	NT
Lunga di Vassallo	+	8/8
Manzanilla	−	NT
Minna di Vacca	−	NT
Minuta	+	8/8
Monaca	+	8/8
Moresca	−	NT
Murtiddara Vassallo	+	8/8
Nasitana	+	8/8
Nocellara del Belice Giafalione	+	8/8
Nocellara del Belice Clone 1	−	NT
Nocellara del Belice Clone 7	−	NT
Nocellara del Belice Mazara del Vallo	−	NT
Nocellara del Belice Standard	−	NT
Nocellara Etnea	−	NT
Nocellara Messinese Ricciardi	−	NT
Nocellara Messinese Romana	−	NT
Ogliara Maltese	−	NT
Oliva Longa	−	NT
Olivo di Mandanici	+	8/8
Olivo di Monaci	+	8/8
Opera Pia	+	8/8
Passalunara di Lascari	−	NT

Table 6. *Cont.*

Cultivar Analyzed	Real-Time LAMP	
	Cultivar Batch	Positive Plants/Tested Plants
Picholine	−	NT
Piricuddara	+	8/8
Pizzo di Corvo	−	NT
Pizzuta d'Olio	+	8/8
Pizzutella	+	8/8
Salicina Vassallo	−	NT
Tonda Iblea	−	NT
Tortella Motticiana	−	NT
Tunnilidda	−	NT
Uovo di Piccione	−	NT
Vaddara	−	NT
Vaddarica	+	8/8
Verdella	+	8/8
Verdella Frutto Grosso	+	8/8
Verdello	+	8/8
Vetrana	+	8/8
Zaituna Floridia	−	NT

Note: +: positive sample; −: negative sample; NT: Not Tested.

4. Discussion

The olive tree is affected by many potential pathogens, including viruses. Some of them are reported to be transmitted by different vectors [9,27–29], but the use of infected propagating material might represent the major, though not the only, means of virus spread [29–33]. The first report on a probable viral disease of olive goes back to 1938 [34] and, since then, several virus-like diseases and viruses have been reported over the years in different areas where olive cultivation plays a prominent role [14,35–40]. Some of these are agents of recognized diseases, others cause latent infections with still undetermined effects on the host [29]. The discovery of OEGV adds to the list of unclassified members of the family *Geminiviridae*, whose genome sequences diverge significantly from those of classified members [16]. The identification of this new virus was facilitated by HTS, a technique that allows for the discovery of new plant viruses, especially when symptoms are not evident, as is the case of OEGV. Besides Apulia, OEGV was recently reported in different areas where the olive cultivation is widespread [15,17,18].

To our knowledge, this study represents the first report of OEGV in Sicily. Since PCR-based methods can be affected by several inhibitors [41,42], such as phenols and polysaccharides [43,44] and require nucleic acid extraction methods [45], we aimed to develop a rapid detection method for OEGV based on LAMP. Indeed, this detection technique showed optimal characteristics, providing rapid, sensitive, specific, and easy detection of several pathogens even in the field, showing a reduced sensitivity to inhibitors [42,46]. The AV1 (CP) gene of OEGV was selected as the target region for primer design and the set of the six LAMP primers showed good specificity and stability for OEGV detection. The specificity is crucial to obtain correct discrimination of OEGV from other viruses belonging to the large *Geminiviridae* family, and a high sensitivity is relevant to minimize false negatives. A LAMP assay optimization performed using DNA extracted from OEGV-infected olive samples revealed that the time required to carry out the experiment was 30 min. The LAMP assay could detect the virus presence from infected samples in as little as 3–14 min. Interestingly, the real-time LAMP developed here has proven to have a 10-fold higher sensitivity compared to the end-point PCR for detection of OEGV. Moreover, in this study, the conventional extraction method using a commercial kit was compared with a "membrane spot crude extract" method; the data obtained from this comparison suggests that the LAMP-based detection method could be suitable for direct use in the field, confirming that ease of sample preparation is a crucial requirement for future application for on-site detection. Specifically, we demonstrated that the rapid sample preparation method allowed for the avoidance of DNA extraction and could be applied for future epidemiological studies, drastically reducing the cost of the analysis. Furthermore, this real-time LAMP technique, associated with other rapid extraction methods developed in other works [47], could be fine-tuned for an efficient and rapid in-field diagnosis.

The rapid extraction method definitely simplified the surveys of the OEGV spread in different cultivars in Sicily. The effectiveness of the techniques developed is essential to understand its spread and to refine effective methods of crop protections, in order to quickly diagnose the presence of a new pathogen in different areas [48–50].

Our survey revealed a considerable presence of the virus in the olive crops in Sicily, probably due to the inadvertent movement of clonally propagated infected but asymptomatic germplasms. Related to this, the development of diagnostic protocols for plant virus detection [51,52] and the epidemiological studies [53] of viral diseases are among the most important and useful steps towards the containment of new epidemics [54–56].

In conclusion, the real-time LAMP assay described in this work is a rapid, simple, specific and sensitive technique for detecting the presence of the recently described OEGV, allowing for the processing of a great number of samples at the same time, especially if associated with the "membrane spot crude extract" method. For this reason, we propose to adopt this method for routine tests in the laboratory and field conditions for a timely detection of OEGV. In particular, this method represents a potential tool for rapidly screening olive plant material useful for large surveys of the spread and pathogenicity of this virus, which to date remain uncertain. Interestingly, as recently reported by Ruiz-García and co-workers [18], the high level of sequence conservation encountered among all OEGV accession so far isolated requires a prompt investigation of the evolutionary and biological significance of this geminivirus in olive, opening new scenarios about its mechanisms of spread in the major olive-growing areas of the world.

Author Contributions: Conceptualization, S.D. and S.P.; methodology, A.G.C., S.B., S.D. and S.P.; software, A.G.C., S.B. and S.D.; validation, A.G.C., S.B., S.M., E.N., S.D. and S.P.; formal analysis, A.G.C., S.B., D.T. and A.M.; investigation, A.G.C., S.B., S.D. and S.P.; resources, A.G., S.P. and S.D.; data curation, A.G.C., S.B., S.D. and S.P.; writing—original draft preparation, A.G.C., S.B., S.D. and S.P.; writing—review and editing, A.G.C., S.B., S.M., E.N., S.D. and S.P.; visualization S.M., E.N., A.M., D.T., A.G., M.I.F.S.A. and A.A.; supervision, S.D. and S.P.; project administration, A.G., S.D. and S.P.; funding acquisition, A.G., S.D. and S.P. All authors have read and agreed to the published version of the manuscript.

Funding: This research received no external funding.

Data Availability Statement: Not applicable.

Acknowledgments: The authors thank 'PSR SICILIA 2014-2020—Programma di Sviluppo Rurale—Misura 16—Cooperazione—Sottomisura 16.1—"Sostegno per la costituzione e la gestione dei gruppi operativi del PEI in materia di produttività e sostenibilità dell'agricoltura"—Gruppo Operativo: ATS ProOlivo—Titolo del progetto: Applicazione di tecnologie "smart" per il monitoraggio, prevenzione e diagnosi precoce delle malattie di interesse economico dell'olivo' for the technical support.

Conflicts of Interest: The authors declare that they have no conflict of interest.

References

1. Terral, J.F. Exploitation and management of the olive tree during prehistoric times in Mediterranean France and Spain. *J. Archaeol. Sci.* **2000**, *27*, 127–133. [CrossRef]
2. Goor, A. The place of the olive in the holy land and its history through the ages. *Econ. Bot.* **1966**, *20*, 223–243. [CrossRef]
3. de Graaff, J.; Eppink, L.A.A.J. Olive oil production and soil conservation in southern Spain, in relation to EU subsidy policies. *Land Use Policy* **1999**, *16*, 259–267. [CrossRef]
4. Food and Agriculture Organization of the United Nations (FAO). Available online: http://www.fao.org/faostat/en/#home (accessed on 8 January 2022).
5. ISTAT–Istituto Nazionale di Statistica–Agricoltura. Available online: https://www.istat.it/it/agricoltura/dati (accessed on 8 January 2022).

6. Caruso, T.; Marra, F.P.; Costa, F.; Campisi, G.; Macaluso, L.; Marchese, A. Genetic diversity and clonal variation within the main Sicilian olive cultivars based on morphological traits and microsatellite markers. *Sci. Hortic.* **2014**, *180*, 130–138. [CrossRef]
7. Las Casas, G.; Scollo, F.; Distefano, G.; Continella, A.; Gentile, A.; La Malfa, S. Molecular characterization of olive (*Olea europaea* L.) Sicilian cultivars using SSR markers. *Biochem. Syst. Ecol.* **2014**, *57*, 15–19. [CrossRef]
8. Marra, F.P.; Caruso, T.; Costa, F.; Di Vaio, C.; Mafrica, R.; Marchese, A. Genetic relationships, structure and parentage simulation among the olive tree (*Olea europaea* L. subsp. *europaea*) cultivated in Southern Italy revealed by SSR markers. *Tree Genet. Genomes* **2013**, *9*, 961–973. [CrossRef]
9. Faggioli, F.; Ferretti, L.; Albanese, G.; Sciarroni, R.; Pasquini, G.; Lumia, V.; Barba, M. Distribution of olive tree viruses in Italy as revealed by one-step RT-PCR. *J. Plant Pathol.* **2005**, *87*, 49–55.
10. Piccolo, S.L.; Mondello, V.; Giambra, S.; Conigliaro, G.; Torta, L.; Burruano, S. *Arthrinium phaeospermum*, *Phoma cladoniicola* and *Ulocladium consortiale*, New Olive Pathogens in Italy. *J. Phytopathol.* **2014**, *162*, 258–263. [CrossRef]
11. Saponari, M.; Boscia, D.; Nigro, F.; Martelli, G.P. Identification of DNA sequences related to Xylella fastidiosa in oleander, almond and olive trees exhibiting leaf scorch symptoms in Apulia (Southern Italy). *J. Plant Pathol.* **2013**, *95*. [CrossRef]
12. Martinelli, F.; Marchese, A.; Giovino, A.; Marra, F.P.; Della Noce, I.; Caruso, T.; Dandekar, A.M. In-field and early detection of Xylella fastidiosa infections in olive using a portable instrument. *Front. Plant Sci.* **2019**, *9*, 2007. [CrossRef]
13. Morelli, M.; García-Madero, J.M.; Jos, Á.; Saldarelli, P.; Dongiovanni, C.; Kovacova, M.; Saponari, M.; Baños Arjona, A.; Hackl, E.; Webb, S.; et al. *Xylella fastidiosa* in Olive: A Review of Control Attempts and Current Management. *Microorganisms* **2021**, *9*, 1771. [CrossRef]
14. Xylogianni, E.; Margaria, P.; Knierim, D.; Sareli, K.; Winter, S.; Chatzivassiliou, E.K. Virus Surveys in Olive Orchards in Greece Identify Olive Virus T, a Novel Member of the Genus Tepovirus. *Pathogens* **2021**, *10*, 574. [CrossRef] [PubMed]
15. Materatski, P.; Jones, S.; Patanita, M.; Campos, M.D.; Dias, A.B.; Félix, M.D.R.; Varanda, C.M. A Bipartite Geminivirus with a Highly Divergent Genomic Organization Identified in Olive Trees May Represent a Novel Evolutionary Direction in the Family Geminiviridae. *Viruses* **2021**, *13*, 2035. [CrossRef] [PubMed]
16. Chiumenti, M.; Greco, C.; De Stradis, A.; Loconsole, G.; Cavalieri, V.; Altamura, G.; Zicca, S.; Saldarelli, P.; Saponari, M. Olea Europaea Geminivirus: A Novel Bipartite Geminivirid Infecting Olive Trees. *Viruses* **2021**, *13*, 481. [CrossRef]
17. Alabi, O.J.; Diaz-Lara, A.; Erickson, T.M.; Al Rwahnih, M. Olea europaea geminivirus is present in a germplasm repository and in California and Texas olive (*Olea europaea* L.) groves. *Arch. Virol.* **2021**, *166*, 3399–3404. [CrossRef]
18. Ruiz-García, A.B.; Canales, C.; Morán, F.; Ruiz-Torres, M.; Herrera-Mármol, M.; Olmos, A. Characterization of Spanish Olive Virome by High Throughput Sequencing Opens New Insights and Uncertainties. *Viruses* **2021**, *13*, 2233. [CrossRef] [PubMed]
19. ICTV–International Committee on Taxonomy of Viruses. Available online: https://talk.ictvonline.org/taxonomy/ (accessed on 8 January 2022).
20. Ma, Y.; Navarro, B.; Zhang, Z.; Lu, M.; Zhou, X.; Chi, S.; Di Serio, F.; Li, S. Identification and molecular characterization of a novel monopartite geminivirus associated with mulberry mosaic dwarf disease. *J. Gen. Virol.* **2015**, *96*, 2421–2434. [CrossRef] [PubMed]
21. Gottwald, T.R.; Hughes, G. A new survey method for Citrus tristeza virus disease assessment. In Proceedings of the XIV International Organization of Citrus Virologists (IOCV), Sao Paulo, Brazil, 14–21 July 2000; pp. 77–87.
22. Davino, S.; Panno, S.; Arrigo, M.; La Rocca, M.; Caruso, A.G.; Bosco, G.L. Planthology: An application system for plant diseases management. *Chem. Eng. Trans.* **2017**, *58*, 619–624. [CrossRef]
23. Panno, S.; Caruso, A.G.; Troiano, E.; Luigi, M.; Manglli, A.; Vatrano, T.; Iacono, G.; Marchione, S.; Bertin, S.; Tomassoli, L.; et al. Emergence of tomato leaf curl New Delhi virus in Italy: Estimation of incidence and genetic diversity. *Plant Pathol.* **2019**, *68*, 601–608. [CrossRef]
24. Davino, S.; Napoli, C.; Dellacroce, C.; Miozzi, L.; Noris, E.; Davino, M.; Accotto, G.P. Two new natural begomovirus recombinants associated with the tomato yellow leaf curl disease co-exist with parental viruses in tomato epidemics in Italy. *Virus Res.* **2009**, *143*, 15–23. [CrossRef]
25. Davino, S.; Napoli, C.; Davino, M.; Accotto, G.P. Spread of Tomato yellow leaf curl virus in Sicily: Partial displacement of another geminivirus originally present. *Eur. J. Plant Pathol.* **2006**, *114*, 293–299. [CrossRef]
26. Panno, S.; Caruso, A.G.; Davino, S. The nucleotide sequence of a recombinant tomato yellow leaf curl virus strain frequently detected in Sicily isolated from tomato plants carrying the *Ty-1* resistance gene. *Arch. Virol.* **2018**, *163*, 795–797. [CrossRef] [PubMed]
27. Sabanadzovic, S.; Abou-Ghanem, N.; La Notte, P.; Savino, V.; Scarito, G.; Martelli, G.P. Partial molecular characterization and RT-PCR detection of a putative closterovirus associated with olive leaf yellowing. *J. Plant Pathol.* **1999**, *81*, 37–45.
28. Varanda, C.M.; Santos, S.; Clara, M.I.E.; do Rosário Félix, M. Olive mild mosaic virus transmission by *Olpidium virulentus*. *Eur. J. Plant Pathol.* **2015**, *142*, 197–201. [CrossRef]
29. Martelli, G. A brief outline of infectious diseases of olive. *Palest. Tech. Univ. Res. J.* **2013**, *1*, 10. [CrossRef]
30. Martelli, G.P. Infectious diseases and certification of olive: An overview. *EPPO Bull.* **1999**, *29*, 127–133. [CrossRef]
31. Loconsole, G.; Saponari, M.; Faggioli, F.; Albanese, G.; Bouyahia, H.; Elbeaino, T.; Materazzi, A.; Nuzzaci, M.; Prota, V.; Romanazzi, G.; et al. Inter-laboratory validation of PCR-based protocol for detection of olive viruses. *EPPO Bull.* **2010**, *40*, 423–428. [CrossRef]
32. Albanese, G.; Saponari, M.; Faggioli, F. Phytosanitary certification. In *Olive Germoplasm-The Olive Cultivation, Table Olive and Olive Oil Industry in Italy*, 1st ed.; Muzzalupo, I., Ed.; InTech: Rijeka, Croatia, 2012; pp. 107–132.

33. Roschetti, A.; Ferretti, L.; Muzzalupo, I.; Pellegrini, F.; Albanese, G.; Faggioli, F. Evaluation of the possibile effect of virus infections on olive propagation. *Petria* **2009**, *19*, 18–28.
34. Pesante, A. On a previously unknown disease of olive. *Boll. Della Stn. Patol. Veg. Roma* **1938**, *18*, 401–428.
35. Caglayan, K.; Fidan, U.; Tarla, G.; Gazel, M. First report of olive viruses in Turkey. *J. Plant Pathol.* **2004**, *86*, 91.
36. Barba, M. Viruses and virus-like diseases of olive 1. *EPPO Bull.* **1993**, *23*, 493–497. [CrossRef]
37. Savino, V.; Barba, M.; Galitelli, G.; Martelli, G.P. Two nepoviruses isolated from olive in Italy. *Phytopathol. Mediterr.* **1979**, *18*, 135–142.
38. Savino, V.; Gallitelli, D. Cherry leafroll virus in olive. *Phytopathol. Mediterr.* **1981**, *20*, 202–203.
39. Savino, V.; Gallitelli, D.; Barba, M. Olive latent ringspot virus, a newly recognised virus infecting olive in Italy. *Ann. Appl. Boil.* **1983**, *103*, 243–249. [CrossRef]
40. Martelli, G.P.; Sabanadzovic, S.; Savino, V.; Abu-Zurayk, A.R.; Masannat, M.C.F.A. Virus-like diseases and viruses of olive in Jordan. *Phytopathol. Mediterr.* **1995**, *34*, 133–136.
41. Ortega, S.F.; Tomlinson, J.; Hodgetts, J.; Spadaro, D.; Gullino, M.L.; Boonham, N. Development of loop-mediated isothermal amplification assays for the detection of seedborne fungal pathogens *Fusarium fujikuroi* and *Magnaporthe oryzae* in rice seed. *Plant Dis.* **2018**, *102*, 1549–1558. [CrossRef]
42. Panno, S.; Matić, S.; Tiberini, A.; Caruso, A.G.; Bella, P.; Torta, L.; Stassi, R.; Davino, S. Loop mediated isothermal amplification: Principles and applications in plant virology. *Plants* **2020**, *9*, 461. [CrossRef] [PubMed]
43. Tian, S.; Nakamura, K.; Kayahara, H. Analysis of phenolic compounds in white rice, brown rice, and germinated brown rice. *J. Agric. Food Chem.* **2004**, *543*, 4808–4813. [CrossRef] [PubMed]
44. Schrader, C.; Schielke, A.; Ellerbroek, L.; Johne, R. PCR inhibitors–occurrence, properties and removal. *J. Appl. Microbial.* **2012**, *113*, 1014–1026. [CrossRef]
45. Boonham, N.; Glover, R.; Tomlinson, J.; Mumford, R. Exploiting generic platform technologies for the detection and identification of plant pathogens. *Sustain. Dis. Manag. Eur. Context* **2008**, *121*, 355–363. [CrossRef]
46. Kaneko, H.; Kawana, T.; Fukushima, E.; Suzutani, T. Tolerance of loop-mediated isothermal amplification to a culture medium.and biological substances. *J. Biochem. Biophys. Methods* **2007**, *70*, 499–501. [CrossRef] [PubMed]
47. Panno, S.; Ruiz-Ruiz, S.; Caruso, A.G.; Alfaro-Fernandez, A.; San Ambrosio, M.I.F.; Davino, S. Real-time reverse transcription polymerase chain reaction development for rapid detection of Tomato brown rugose fruit virus and comparison with other techniques. *PeerJ* **2019**, *7*, e7928. [CrossRef] [PubMed]
48. Davino, S.; Panno, S.; Rangel, E.A.; Davino, M.; Bellardi, M.G.; Rubio, L. Population genetics of cucumber mosaic virus infecting medicinal, aromatic and ornamental plants from northern Italy. *Arch. Virol.* **2012**, *157*, 739–745. [CrossRef]
49. Davino, S.; Panno, S.; Iacono, G.; Sabatino, L.; D'Anna, F.; Iapichino, G.; Olmos, A.; Scuderi, G.; Rubio, L.; Tomassoli, L.; et al. Genetic variation and evolutionary analysis of Pepino mosaic virus in Sicily: Insights into the dispersion and epidemiology. *Plant Pathol.* **2017**, *66*, 368–375. [CrossRef]
50. Ferriol, I.; Rubio, L.; Pérez-Panadés, J.; Carbonell, E.A.; Davino, S.; Belliure, B. Transmissibility of Broad bean wilt virus 1 by aphids: Influence of virus accumulation in plants, virus genotype and aphid species. *Ann. Appl. Biol.* **2013**, *162*, 71–79. [CrossRef]
51. Tiberini, A.; Tomlinson, J.; Micali, G.; Fontana, A.; Albanese, G.; Tomassoli, L. Development of a reverse transcription-loop-mediated isothermal amplification (LAMP) assay for the rapid detection of Onion yellow dwarf virus. *J. Virol. Methods* **2019**, *271*, 113680. [CrossRef] [PubMed]
52. Waliullah, S.; Ling, K.S.; Cieniewicz, E.J.; Oliver, J.E.; Ji, P.; Ali, M.E. Development of loop-mediated isothermal amplification assay for rapid detection of Cucurbit leaf crumple virus. *Int. J. Mol. Sci.* **2020**, *21*, 1756. [CrossRef]
53. Jeger, M.J.; Holt, J.; Van Den Bosch, F.; Madden, L.V. Epidemiology of insect-transmitted plant viruses: Modelling disease dynamics and control interventions. *Physiol. Entomol.* **2004**, *29*, 291–304. [CrossRef]
54. Davino, S.; Calari, A.; Davino, M.; Tessitori, M.; Bertaccini, A.; Bellardi, M.G. Virescence of tenweeks stock associated to phytoplasma infection in Sicily. *Bull. Insectol.* **2007**, *60*, 279–280.
55. Panno, S.; Ferriol, I.; Rangel, E.A.; Olmos, A.; Han, C.; Martinelli, F.; Rubio, L.; Davino, S. Detection and identification of Fabavirus species by one-step RT-PCR and multiplex RT-PCR. *J. Virol. Methods* **2014**, *197*, 77–82. [CrossRef]
56. Panno, S.; Caruso, A.G.; Barone, S.; Lo Bosco, G.; Rangel, E.A.; Davino, S. Spread of tomato brown rugose fruit virus in Sicily and evaluation of the spatiotemporal dispersion in experimental conditions. *Agronomy* **2020**, *10*, 834. [CrossRef]

plants

MDPI

Article

Quantitative Real-Time PCR Assay for the Detection of *Pectobacterium parmentieri*, a Causal Agent of Potato Soft Rot

Anna A. Lukianova [1,2], Peter V. Evseev [1], Alexander A. Stakheev [1], Irina B. Kotova [2], Sergey K. Zavriev [1], Alexander N. Ignatov [3] and Konstantin A. Miroshnikov [1,*]

1 Shemyakin-Ovchinnikov Institute of Bioorganic Chemistry, Russian Academy of Sciences, Miklukho-Maklaya Str., 16/10, 117997 Moscow, Russia; a.al.lukianova@gmail.com (A.A.L.); petevseev@gmail.com (P.V.E.); stakheev.aa@gmail.com (A.A.S.); szavriev@ibch.ru (S.K.Z.)

2 Department of Biology, Lomonosov Moscow State University, Leninskie Gory, 1, Bldg. 12, 119234 Moscow, Russia; kira1959@gmail.com

3 AgroTechnical Institute, RUDN University, Miklukho-Maklaya Str., 6, 117198 Moscow, Russia; an.ignatov@gmail.com

* Correspondence: kmi@ibch.ru; Tel.: +7-(495)-335-5588

Abstract: *Pectobacterium parmentieri* is a plant-pathogenic bacterium, recently attributed as a separate species, which infects potatoes, causing soft rot in tubers. The distribution of *P. parmentieri* seems to be global, although the bacterium tends to be accommodated to moderate climates. Fast and accurate detection systems for this pathogen are needed to study its biology and to identify latent infection in potatoes and other plant hosts. The current paper reports on the development of a specific and sensitive detection protocol based on a real-time PCR with a TaqMan probe for *P. parmentieri*, and its evaluation. In sensitivity assays, the detection threshold of this protocol was 10^2 cfu/mL on pure bacterial cultures and 10^2–10^3 cfu/mL on plant material. The specificity of the protocol was evaluated against *P. parmentieri* and more than 100 strains of potato-associated species of *Pectobacterium* and *Dickeya*. No cross-reaction with the non-target bacterial species, or loss of sensitivity, was observed. This specific and sensitive diagnostic tool may reveal a wider distribution and host range for *P. parmentieri* and will expand knowledge of the life cycle and environmental preferences of this pathogen.

Keywords: *Pectobacterium parmentieri*; qPCR; bacterial taxonomy; bacterial identification; sensitivity; soft rot; pathogen detection

Citation: Lukianova, A.A.; Evseev, P.V.; Stakheev, A.A.; Kotova, I.B.; Zavriev, S.K.; Ignatov, A.N.; Miroshnikov, K.A. Quantitative Real-Time PCR Assay for the Detection of *Pectobacterium parmentieri*, a Causal Agent of Potato Soft Rot. *Plants* **2021**, *10*, 1880. https://doi.org/10.3390/plants10091880

Academic Editor: Alessandro Vitale

Received: 30 August 2021
Accepted: 8 September 2021
Published: 10 September 2021

1. Introduction

The potato (*Solanum tuberosum*) is one of the most important crops in the world. The world market for potato production exceeds 388 million tons per year (https://www.potatopro.com/world/potato-statistics (accessed on 7 April 2021)) and per capita consumption in Russia exceeds 110 kg (https://www.potatopro.com/russian-federation/potato-statistics (accessed on 7 April 2021)). Therefore, research related to optimising potato production, increasing yields and reducing losses associated with plant diseases and other factors is essential and urgent. Among the challenges faced by potato growers is potatoes' spoilage as a result of bacterial infections. In particular, the development of rot on tubers during storage and transportation can lead to severe losses—up to half of the harvest [1]. The leading cause of blackleg and soft rot in potatoes is the bacteria of the Pectobacteriaceae family, namely the group of Soft Rot Pectobacteriaceae (SRP), comprising phytopathogens of the genera *Pectobacterium* and *Dickeya* [2]. One of the representatives of this group is *P. parmentieri*.

P. parmentieri (Ppa) was first described by Khayi et al. in 2016. It is a species closely related to the previously known pathogen of Japanese horseradish, *P. wasabiae* (Pwa). Several Pwa strains, isolated from potatoes and which cause soft rot, have been scrutinised and finally reclassified as new species [3]. Later on, Ppa was identified among potato pathogens in circulation in Europe and Russia [4,5], Africa [6], Asia [7,8] and America [9,10]. Many strains isolated from potatoes earlier, and initially attributed as Pwa or *Pectobacterium carotovorum* subsp. *Carotovorum*, were proved to represent Ppa.

Thus, *P. parmentieri* can be considered as a worldwide pathogen (https://www.cabi.org/isc/datasheet/48069201 (accessed on 17 May 2021). Strains of Ppa studied are rather diverse [11,12], and two other species related to Ppa/Pwa, *P. polonicum* [13] and *P. punjabense* [14] have been established recently. A recent study of the distribution of *P. punjabense* in Europe [15] demonstrated the need for an appropriate method for discriminatory quantitative diagnostics for newly established SRP species.

Many PCR-based methods have already been developed for generalised and species-specific detection of SRP (reviewed in [16,17]). The accumulation of data on bacterial genomics and taxonomic redistributions has encouraged the design of an updated method for the specific diagnosis of newly established SRP species, particularly Ppa. Earlier, PCR diagnostic methods were proposed for Pwa detection, based on the amplification of the phytase/phosphatase (appA) gene [18] or tyrosine-aspartate (YD) repeat region [19]. Both of these assays enabled scientists to discriminate Ppa/Pwa from *P. carotovorum* and other SRP, but not between the former species. The analysis currently used in phytodiagnostics enables an assumption of the approximate specificity of the pathogen, taking into account the source of the isolation of the strain [4]. However, it still does not allow for species-specific detection and is somewhat outdated, due to the changed understanding of the taxonomy of the group. Recently, the authors developed a pipeline for searching unique sequences for genomic groups and tested it in the context of *P. atrosepticum*, a genetically distinct species of SRP [20]. This paper describes how this workflow can be used to design a quantitative real-time PCR assay to discriminate closely related species. The aim of the study was the development of a species-specific detection system for *P. parmentieri*.

2. Results

2.1. Phylogenetic Analysis

By early 2021, more than 200 bacterial genomes deposited in NCBI GenBank represented the family *Pectobacteriaceae*. *P. parmentieri* was represented by 30 complete and high coverage draft genomes (strains CFIA102, IFB5408, IFB5427, IFB5432, IFB5441, IFB5485, IFB5486, IFB5597, IFB5604, IFB5605, IFB5619, IFB5623, IFB5626, IPO1955, NY1532B, NY1533B, NY1540A, NY1548A, NY1584A, NY1585A, NY1587A, NY1588A, NY1712A, NY1722A, PB20, QK5, RNS-08-42-1A (type strain), SS90, WC19161 and WPP163).

Whole-genome comparisons made with orthoANI (Figure 1) demonstrate that all the strains assigned to *P. parmentieri* possess a close genome similarity, demonstrating high overall average nucleotide similarity (ANI) of 98.9% and above when compared to the type species, whereas the comparable ANI values of non-*parmentieri Pectobacterium* species lies in the range of 87–94%. The concatenated core genes phylogeny also places Ppa strains in a distinct clade (Figure 1).

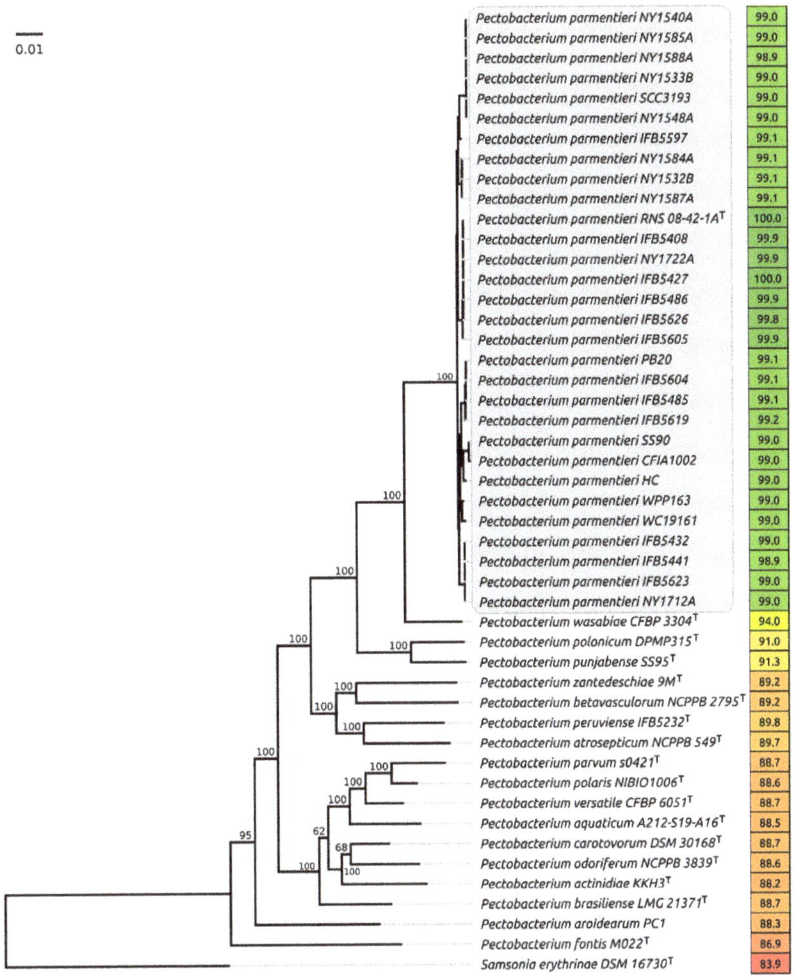

Figure 1. Phylogenetic tree based on the concatenated nucleotide sequences of 92 conservative genes, including the genes of ribosomal proteins and the proteins essential for the transcription and translation processes. Bootstrap support values are shown above their branch as a percentage of 1000 replicates. The scale bar shows 0.01 estimated substitutions per site, and the tree was rooted to *Samsonia erythrinae DSM 16730*. Average nucleotide identity (ANI) values compared to *P. parmentieri* RNA 08-42-1A type strain are shown to the right of the organism name and coloured according to a heat map scale, where a green colour corresponds to the highest value and a red colour corresponds to the lowest value.

2.2. Search for Species-Specific Primers

The search for species-specific sequences was carried out using the workflow described in a previous study [20]. Briefly, this workflow splits the genome of the type Ppa strain into short sections, then each section is compared with a negative database of "non-target" genomes and a positive database of "target genomes" and, as a result, regions are identified that occur in all Ppa genomes and are not found in genomes of other species.

Using this search, a set of unique Ppa species-specific sites was obtained. Regions belonging to the areas of the genome encoding no genes were manually rejected. Next, several potentially suitable sites within the housekeeping genes were selected for further preliminary testing in the conventional PCR mode (Section 2.3) and a further selection of the most appropriate sequence for qPCR analysis development was made (Section 2.4). Primers and probes were designed for these sites. Table 1 shows the sequences of primers, probe and amplicon for detection based on the ankyrin repeat domain-containing protein sequence that showed the best results and was therefore selected for further study.

Table 1. Primers for amplification of a species-specific region and *P. parmentieri* and the amplicon of ankyrin repeat domain-containing protein.

Type	Sequence
F primer	TAT CGC TGG CTC AGG CAA TT
R primer	TAC GCT GCG CAT ACT TGG AA
Probe	(6-FAM)-CGCCCGGG-(dT-BHQ-1)-GCCCAAGATATGACTT-(Pi)
Amplicon	TATCGCTGGCTCAGGCAATTGAA AAAAACGATGAATCAGAAGTCAA AAAACTTTCAGCTCATACGGACCTAAAT CGCCCGGGTGCCCAAGATATGACTTTAC TATTCTTCGCGATGCAAAATAGTTATGAC AAACAAGCCAAACATTTGTCGATAGTCT CATATTTGGTTAGTGCCGGAGCAAGTCC ATTACAGAAAGTTCCAAGTATGCGCAGCGTA

The selected species-specific sequence belongs to an ankyrin repeat domain-containing protein that is located adjacent to the components of a type VI secretion system. Interestingly, an avirulence factor was located several genes upstream of the locus shown in Figure 2. A type VI secretion system is important for plant-associated bacteria, including the *Pectobacterium* species. It contributes to virulence and grants fitness and colonisation advantages *in planta* [21]. It might be suggested that the gene containing the species-specific sequence is important for the bacterium. The sequence search conducted with BLAST using an nr/nt database confirmed that the chosen amplicon did not have close homologues in other organisms.

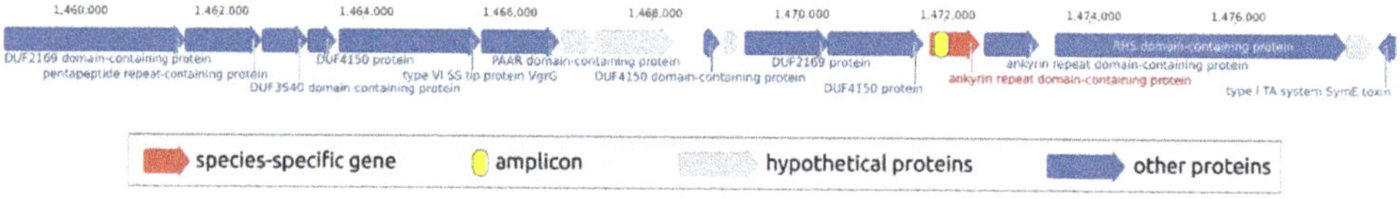

Figure 2. Region in the *P. parmentieri* RNS 08-42-1AT genome containing a species-specific sequence. The scheme was visualised using Geneious Prime 2021.2.2 (https://www.geneious.com, accessed on 20 January 2021).

2.3. Primary Analysis by Conventional PCR

For the initial assessment of the applicability of the primers obtained for the purpose of species-specific PCR detection, a conventional PCR test was carried out on a limited set of strains. The strains marked F ... are a part of the local collection of bacterial pathogens associated with potato soft rot. The collection includes comprehensively described type strains, strains with appropriate genomic characterisation and loosely characterised local isolates. The information on the strains used is provided in Supplementary Table S1. The primary testing strain set included several representatives of different Pectobacteriaceae species belonging to the genus *Pectobacterium* (F002, F004, F012, F016, F028, F041, F043, F048, F061, F109, F126, F131, F135, F152, F157, F160, F162, F164, F171, F182, F258), *Dickeya* (F012, F077, F082, F085, F097, F101, F102, F117, F155, F261) and an unrelated pectolytic isolate (F105). In the experiment described in this paper, amplification was expected only for Ppa (F034, F035, F127, F148, F149, F174), and with none of the others.

Figure 3 shows the results of such an analysis for the amplification of ankyrin repeat domain-containing protein, as a result of which significant amplification was demonstrated only with the target strains (marked in the boxes) and in the absence of false-positive results with all other strains. This enabled the assumption of this site's suitability for amplification in qPCR mode, and made it possible to proceed to the validation using an extended range of strains.

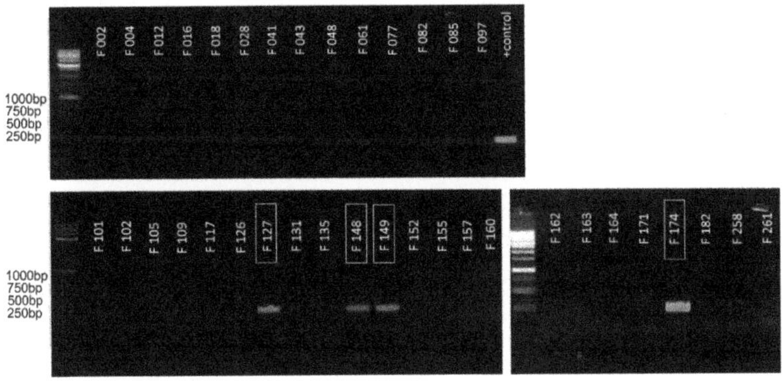

Figure 3. The results of conventional PCR visualised in 1.5% agarose gel. The numbers of the strains belonging to Ppa are marked with a frame. The remaining strains belonging to other species acted as negative controls. The lane designated as "+ control" contained PCR results with test plasmid, which served as a positive control. Evrogen 1 kb Ladder used for the evaluation of amplicons sizes.

2.4. *qPCR Analysis on an Extended Set of Strains*

This study involved seven strains previously attributed to being Ppa or Pwa on the basis of genomic sequencing or 16S rRNA gene sequencing. Two more strains were previously identified as Pwa using the diagnostic primer set PhF 5′-GGTTCAGTGCGTCAGGAGAG and PhR 5′-GCGGAGAGGAAGCGGTGAAG [18], which does not distinguish between Pwa and closely related Ppa (№ 1–9, Supplementary Table S1). A test was also conducted for 67 (№ 10–77) isolates of other *Pectobacteriaceae* species and 32 strains (№ 78–109) related to other species associated with crop rot. These strains were isolated from potato rots and passed through McConkey's medium to exclude *Salmonella* and Gram-positive isolates and SVP medium to ensure the presence of pectolytic activity.

As shown in Supplementary Table S1, all Ppa strains demonstrated a positive PCR signal. Among the strains with alternative Ppa/Pwa attribution (F035 and F178), F035 showed amplification and therefore can be more accurately classified as Ppa, while F178, revealing no positive signal, may be categorised as Pwa.

The historical strain Pwa F007 used in the study did not show any false positive amplification. No positive results were shown for other isolates with pectolytic activity, both Pectobacteriaceae and unrelated ones.

Additionally, in silico analysis using an nt-database did not presume any amplification of plant genomic DNA using the designed primers. No amplification was observed in the PCR reaction in vitro using potato DNA as a template. Thus, the authors are confident that the possibility of cross-amplification with potato DNA was excluded.

2.5. *Sensitivity*

Serially diluted plasmid and genomic DNA were used in qPCR reactions for a sensitivity test. Based on the threshold cycles (Cq) obtained for each concentration of copies in the sample (Table 2), standard curves were plotted. The resulting curves were linear (Figure 4). The correlation coefficient (R^2) was 0.99 for both curves, with a slope of −3.34 and −3.33 for plasmid and genomic DNA, respectively, corresponding to a PCR efficiency of 98.9% and 99.62%.

Table 2. Mean Cq values for qPCR carried out on serial dilutions of genomic DNA of the *P. parmentieri* F149 and corresponding plasmid. SD is standard deviation.

№	Plasmid			DNA		
	Copies Per Reaction	Cq	SD	Copies Per Reaction	Cq	SD
1	1.4×10^{10}	8.01	0.12	1.68×10^7	16.29	0.37
2	1.48×10^9	11.83	0.21	1.68×10^6	20.48	0.36
3	1.48×10^8	15.27	0.15	1.68×10^5	24.62	0.01
4	1.48×10^7	19.04	0.71	1.68×10^4	27.20	2.90
5	1.48×10^6	22.90	0.57	1.68×10^3	29.92	1.25
6	1.48×10^5	25.67	1.72	168	34.60	1.80
7	1.48×10^4	29.37	3.07	16.8	36.20	0.37
8	1.48×10^3	32.59	0.70	1.68	-	-
9	148	33.95	1.72	0.16	-	-

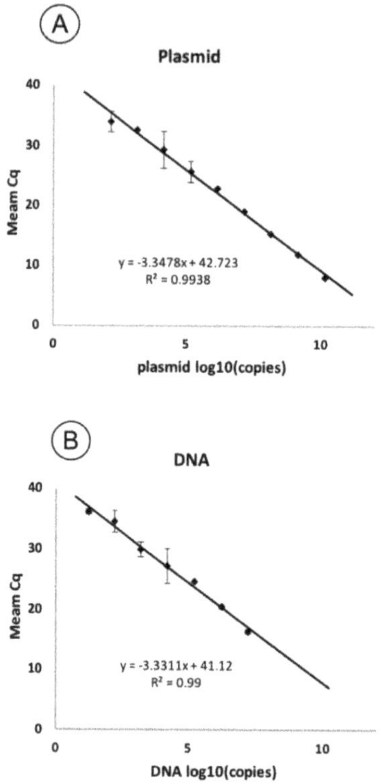

Figure 4. Standard curves showing the dependence of Cq on the concentration of pathogen DNA in the reaction. The curves are plotted based on the threshold cycles obtained for a series of ten-fold dilutions of the plasmid (**A**) and genomic DNA of the F149 strain (**B**). The standard deviation is shown as error bars.

The limit of detection (LoD) was nearly 16 copies per reaction, corresponding to 4 $\times 10^2$ copies/mL. Figure 5 shows the amplification curves for the sensitivity test and the good flare-up of the probe during the reaction, even at high dilutions.

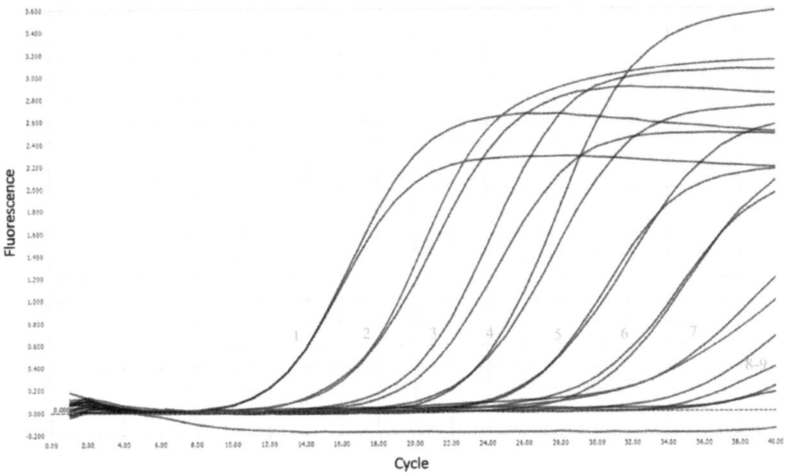

Figure 5. Amplification curves for a sensitivity test using the example of a series of dilutions of plasmid DNA. The numbers represent the dilution number.

2.6. Assays of Plant Samples

To conduct an experiment simulating a pathogen's detection in infected plants, the tubers of the "Gala" variety were used, one of the most widespread varieties in Russia, and one which is moderately resistant to bacterial diseases. The potatoes were soaked in a 10^6 cfu/mL suspension of the pathogen for infection and then incubated at 28 °C until the development of soft rot symptoms. On days 3, 4 and 5, a ~100 mg piece of peel was taken from the tubers and total DNA was isolated. Then, qPCR was performed from the DNA obtained, in the same way as in the previous experiments. Control tubers were soaked in a sterile LB medium.

As shown in Table 3, the pathogen was successfully detected in all cases, confirming the possibility of using the analysis to assess the contaminated material. With an increase in the duration of incubation, the titre of bacteria increased proportionally. Amplification was also recorded for the control tuber, indicating a trace presence of the pathogen, which did not lead to noticeable symptoms of rotting.

Table 3. Results of qPCR carried out on material obtained from artificially infected potatoes. APC permease gene of Ppa was detected using developed primers. SD is standard deviation.

Incubation, h	Cq	SD	Copies/mL
Control (120 h)	38.32	0.00	6.92
72	31.03	0.67	1.06×10^3
96	27.38	0.67	1.3×10^4
120	20.76	1.05	1.3×10^6

3. Discussion

According to the species definition, Ppa differs from Pwa by its ability to produce acid from melibiose, raffinose, lactose and D-galactose [3]. This feature was used to differentiate Ppa strains isolated from potato in Southern Europe [4]. However, the biochemical tests made the precise diagnostics more laborious and, thus, raised questions about the value of such fine analysis. Besides the obvious purpose of monitoring the causal agents of plant diseases, in order to develop adapted prevention actions in particular countries, regions or climate areas, some fundamental arguments exist.

Information on the role of Ppa in the bacterial pathogenesis of potatoes worldwide is contradictory [22]. According to national monitoring surveys, Ppa occurrence ranges from single, moderate cases [6] to severe breakouts [10]. While wet weather throughout the year is preferred for the development of the pathogen (https://www.cabi.org/isc/datasheet/48069201 (accessed on 17 May 2021)), a broad range of conditions is tolerated. The aggressiveness of Ppa is also debatable. As for other SRP, their pathogenesis relies on the production and secretion of plant cell wall-degrading enzymes, which cause the typical symptoms of soft rot. Enzyme synthesis depends on suitable environmental conditions [23]. Generally, the virulence of Ppa is considered to be moderate. However, a number of studies [24,25] have demonstrated that some strains of *P. parmentieri* can cause fast and severe maceration of tubers and plants comparable with *P. atrosepticum* and *P. brasiliense*, which are considered to be the most aggressive among *Pectobacterium*. It is worth noting that the bacterial community in rotting potato tissues is very complex [26] and may include several different pathogenic species. SRP pathogens may interact antagonistically [27] or synergistically [28] with respect to one another. Therefore, the study of the impact of a particular pathogen on the development of the disease requires quantitative differential identification of the SRP species, particularly with Ppa.

Currently, no effective control agents have been developed to prevent or to treat SRP infections [29,30]. A promising approach is the use of bacteriophages (phages), which are bacterial viruses that infect pathogenic bacteria. A number of successful applications of phage control of plant pathogens, including SRP, have been reported (reviewed in [31,32]). Some phages infecting Ppa have been isolated and investigated [33,34]. An important feature of phage therapy is to have a very selective host range of bacteriophages, usually limited to a bacterial species or even a group of strains within a species. This may be considered to be an advantage, because phage treatment does not affect commensal and endosymbiotic microflora of the plant attacking pathogenic bacteria only. However, scientifically sound use of therapeutic bacteriophages requires fine and precise diagnostics of the causative agent of the disease. Existing assays are often too general for efficient phage application, and more focused methods of discriminating SRP are needed.

Besides pectolytic enzymes, a number of other proteinaceous and carbohydrate factors and signal pathways have been found to participate in bacterial adhesion, the colonising of plant tissue and enhancement of the disease (reviewed in [23]). Essential intracellular effectors have been secreted into the plant cell via secretion systems type III (T3SS), type IV (T4SS) and type VI (T6SS) [35]. An important feature of Ppa/Pwa is the absence of a number of essential genes encoding T3SS in the genome [36,37]. This absence may explain the limited host range of *P. parmentieri*. In such conditions, the role of T6SS and other secretion systems becomes more important [38]. The genomic sequence unique to Ppa that was identified was located adjacent to the T6SS apparatus, and its conservation within a species may indicate a unique role in the functioning of the system. This sequence does not belong to any known mobile elements and, thus, may serve as a hallmark of Ppa genomes.

Another important area where qPCR detection of SRP is needed is the establishment of the threshold bacterial population necessary for the development of disease symptoms. While the occurrence of SRP-related blackleg, wilting and aerial rot of vegetating potato depends on numerous environmental factors (reviewed in [39]), the development of soft rot in stored ware and seed potato is a consequence of a latent infection of the tuber surface. The incidence of soft rot, as a minimum, correlates with the population of SRP as revealed by laboratory testing. Most in vitro experiments described in the literature use an application of 10^6–10^7 cfu/mL aliquots of SRP suspensions applied to unprotected potato tissue (tuber slices) to establish the stable development of soft rot symptoms. This work reports that, starting from almost negligible values, the population of Ppa grew fast at room temperature and reached ~10^6 cfu/mL, resulting in tissue rotting in a few days. On the other hand, undamaged potato tubers with a latent SRP population 10^4–10^6 cfu/mL on the skin revealed no signs of soft rot being stored in proper warehouse conditions (4–7 °C) [40]. Therefore, the monitoring of the bacterial insemination of the tubers may help to estimate the risk of soft rot development in the stored tubers and to reveal the dangerous threshold for each particular SRP species. The designed assay has been shown to be sensitive enough to detect Ppa within the range of natural latent infection level (10^2–10^5 cfu). Thus, this analysis is suitable for assessing the quality of potatoes and diagnosing the likely development of rot.

The reported protocol, based on the genomic analysis of an ample amount of recent GenBank data, was successfully tested and demonstrated high sensitivity and suitability for in vivo testing. The species-specific sequence revealed is not only unique to *Pectobacterium parmentieri,* but is also a part of a functional gene which can be important for pathogenic lifestyle of this economically important plant pathogen. The high specificity of the developed assay is particularly important for efficient phage application in the biocontrol of plant diseases caused by SRP bacteria.

4. Materials and Methods

4.1. Phylogenetic Analysis

Bacterial genomes were downloaded from the NCBI GenBank bacterial database (ftp://ftp.ncbi.nlm.nih.gov/genbank (accessed on 27 March 2021)). A phylogenetic tree was generated using an UBCG pipeline, based on 92 core genes including 43 ribosomal proteins, nine genes of aminoacyl-tRNA synthetases, DNA processing and translation proteins and other conservative genes. Bootstrap analysis phylogeny was conducted by aligned concatenated sequences of 92 core genes made by UBCG with MAFFT (FFT-NS-x1000, 200 PAM/k = 2). Then, bootstrap trees were constructed using the RAxML program (maximum likelihood method) (GTR Gamma I DNA substitution model). The robustness of the trees was assessed by fast bootstrapping (1000) [41].

Average nucleotide identity (ANI) was computed using orthoANI, with default settings [42].

4.2. Search for Species-Specific Sequences and Primer Design

To search for species-specific sequences, custom databases were constructed using BLAST (https://blast.ncbi.nlm.nih.gov/Blast.cgi (accessed on 25 February 2021)). The search for species-specific regions for amplification was carried out using the workflow presented in the previous study [20].

Primers and probes were generated with Primer3Plus (https://primer3.ut.ee/ (accessed on 15 March 2021)) and manually checked for the consistency of melting temperatures and for the absence of hairpins and dimers formation using the functions of Geneious Prime and Primer Biosoft (http://www.premierbiosoft.com/NetPrimer/AnalyzePrimerServlet (accessed on 20 March 2021)).

4.3. Bacterial Strains, Media and Culture Conditions

A complete list of bacterial strains engaged in this study, with an indication of their species, year and location of isolation, is shown in Supplementary Table S1. Strains were obtained from the Laboratory of Molecular Bioengineering, IBCh RAS. Pectolytic bacteria were cultivated at 28 °C on 1.5% LB agar. CVP medium was used to assess pectinolytic activity. *E. coli* NovaBlue strain was used for transformation during the preparation of a plasmid. *E. coli* was cultivated at 37 °C on LB agar medium with the addition of ampicillin.

4.4. Genomic DNA Isolation

Genomic DNA was isolated using overnight bacterial cultures, using a GeneJET Genomic DNA Purification Kit (ThermoScientific, Waltham, MA, USA), according to the manufacturer's protocol.

Potato DNA was extracted using a CTAB-based protocol. For this purpose, a piece of peel of 100 mg was mechanically homogenised with a 0.1% sodium pyrophosphate solution. The resulting homogenate was transferred into 1.5 mL tubes and centrifuged. 40 µL of lysozyme solution (100 µL/mL) and 60 µL of 10% SDS solution were added to the sediment, resuspended and incubated at 37 °C for 30 min. Then, 650 µL of 2% STAB was added to the mixture and incubated for another 30 min at 65 °C. Then, the mixture was cooled and 700 μL of chloroform was added, vortexed and precipitated at 12,000 rpm. The supernatant was mixed in a new tube with 600 μL of isopropanol. After subsequent centrifugation, the precipitate was washed twice with 75% ethanol and dried until the volatile solvents completely evaporated, and the resulting DNA was dissolved in water.

The concentration and quality of the extracted DNA was estimated using a NanoProteometer N60 (NanoProteometer, Munich, Germany). After extraction, DNA concentrations were diluted to a single value of 10 ng/μL.

4.5. PCR Conditions

The conventional PCR was carried out in a volume of 25 μL containing 5 μL of Evrogen ScreenMix (Evrogen, Moscow, Russia,), 0.35 μM of forward and reverse primers and 60 ng of template DNA. Amplification was performed using a T100 Thermal Cycler (Bio-Rad, Hercules, CA, USA) and in the following conditions: 94 °C for 300 s, then 45 cycles of 94 °C for 10 s, 62 °C for 10 s and 72 °C for 10 s. The resulting PCR products were separated by electrophoresis in 1.5% agarose/TA buffer gel and visualised by ethidium bromide staining. The size of the bands was eluted using a 1 kb DNA Ladder marker (Evrogen).

4.6. Plasmid Construction for Sensitivity Assay

For a precise evaluation of PCR sensitivity, we constructed a plasmid containing an insert of the target sequence amplified from the Ppa F149 strain. For this purpose, the product of PCR amplification was purified using ISOLATE II PCR and Gel Kit (Bioline, St. Petersburg, Russia) and cloned to pAL2-T vector using a QuickTA kit (Evrogen). Plasmid DNA used as standard was purified with a QIAprep Spin Miniprep Kit (Qiagen, Hilden, Germany), according to the manufacturer's instructions. Sanger sequencing of the corresponding region in the resulting plasmid confirmed the correctness of the insert.

4.7. qPCR

The qPCR was carried out in a LightCycler 96 (Roche, Basel, Switzerland). Each 35 μL reaction contained 200 μM of each dNTP, 0.2 μM of probe, 0.35 μM of forward and reverse primers and 60 ng of template DNA. The optimised amplification conditions were as listed in Section 4.5. Each reaction was carried out in four replicates. Water was used as a negative control. Plasmid-based internal control was used to exclude false-negative results, as described earlier [43].

The processing of the amplification curves obtained and the calculation of the threshold cycles were carried out using software supplied by Roche. A sensitivity analysis was carried out on serial three ten-fold dilutions of the test plasmid and genomic DNA of strain F149. The resulting samples were analysed by qPCR. For each defined threshold cycle, the mean and standard deviation were calculated using Roche software. To construct the standard curve, the threshold cycles' mean values were plotted against the concentration of copies of the target sequence in each reaction.

For all values, the standard deviation was calculated.

4.8. Testing the Detection System on Artificially Infected Tubers

For the experiment, potato tubers of the most widespread variety, "Gala", were obtained from a market. They were washed and soaked in a bacterial suspension to infect the tubers, following the same protocol as in a previous study [20]. Then, the tubers were incubated at 28 °C. On days three, four, five and six, DNA was extracted from 100 mg of the infected tuber's peel, as described in Section 4.4, and analysed by qPCR.

Supplementary Materials: The following are available online at https://www.mdpi.com/article/10.3390/plants10091880/s1, Table S1 Selectivity of the designed qPCR method.

Author Contributions: Conceptualisation, A.N.I. and K.A.M.; methodology, A.A.S. and S.K.Z.; investigation, A.A.L., P.V.E. and A.A.S.; software, P.V.E.; validation and formal analysis, A.A.L., A.A.S. and I.B.K.; data curation, A.A.L. and P.V.E.; writing—original draft preparation, A.A.L. and

P.V.E.; writing—review and editing, A.N.I. and K.A.M.; visualisation, A.A.L. and P.V.E.; project administration, K.A.M. All authors have read and agreed to the published version of the manuscript.

Funding: This research received no external funding.

Institutional Review Board Statement: Not applicable.

Informed Consent Statement: Not applicable.

Conflicts of Interest: The authors declare no conflict of interest.

References

1. Bhat, K.A.; Masood, S.; Bhat, N.; Bhat, M.A.; Razvi, S.; Mir, M.; Akhtar, S.; Wani, N.; Habib, M. Current Status of Post Harvest Soft Rot in Vegetables: A Review. *Asian J. Plant Sci.* **2010**, *9*, 200–208. [CrossRef]
2. Mansfield, J.; Genin, S.; Magori, S.; Citovsky, V.; Sriariyanum, M.; Ronald, P.; Dow, M.; Verdier, V.; Beer, S.V.; Machado, M.A.; et al. Top 10 plant pathogenic bacteria in molecular plant pathology. *Mol. Plant Pathol.* **2012**, *13*, 614–629. [CrossRef] [PubMed]
3. Khayi, S.; Cigna, J.; Chong, T.M.; Quêtu-Laurent, A.; Chan, K.-G.; Hélias, V.; Faure, D. Transfer of the potato plant isolates of *Pectobacterium wasabiae* to *Pectobacterium parmentieri* sp. nov. *Int. J. Syst. Evol. Microbiol.* **2016**, *66*, 5379–5383. [CrossRef] [PubMed]
4. Suarez, B.; Feria, F.J.; Martín-Robles, M.J.; Del Rey, F.J.; Palomo, J.L. *Pectobacterium parmentieri* Causing Soft Rot on Potato Tubers in Southern Europe. *Plant Dis.* **2017**, *101*, 1029. [CrossRef]
5. Ha, V.T.N.; Voronina, M.V.; Kabanova, A.P.; Shneider, M.M.; Korzhenkov, A.A.; Toschakov, S.V.; Miroshnikov, K.; Ignatov, A.N. First Report of *Pectobacterium parmentieri* Causing Stem Rot Disease of Potato in Russia. *Plant Dis.* **2019**, *103*, 144. [CrossRef]
6. Kamau, J.W.; Ngaira, J.; Kinyua, J.; Gachamba, S.; Ngundo, G.; Janse, J.; Macharia, I. Occurence of pectinolytic bacteria causing blackleg and soft rot of potato in Kenya. *J. Plant Pathol.* **2019**, *101*, 689–694. [CrossRef]
7. Cao, Y.; Sun, Q.; Feng, Z.; Handique, U.; Wu, J.; Li, M.W.; Zhang, R. First Report of *Pectobacterium parmentieri* Causing Blackleg on Potato in Inner Mongolia, China. *Plant Dis.* **2021**. [CrossRef] [PubMed]
8. Sarfraz, S.; Riaz, K.; Oulghazi, S.; Cigna, J.; Sahi, S.T.; Khan, S.; Hameed, A.; Alam, M.W.; Faure, D. First Report of *Pectobacterium parmentieri* and *Pectobacterium polaris* Causing Potato Blackleg Disease in Punjab, Pakistan. *Plant Dis.* **2019**, *103*, 1405. [CrossRef]
9. Ge, T.L.; Jiang, H.H.; Hao, J.J.; Johnson, S.B. First Report of *Pectobacterium parmentieri* Causing Bacterial Soft Rot and Blackleg on Potato in Maine. *Plant Dis.* **2018**, *102*, 437. [CrossRef]
10. McNally, R.R.; Curland, R.; Webster, B.T.; Robinson, A.P.; Ishimaru, C.A. First Report of Blackleg and Tuber Soft Rot of Potato Caused by *Pectobacterium parmentieri* in Minnesota and North Dakota. *Plant Dis.* **2017**, *101*, 2144. [CrossRef]
11. Zoledowska, S.; Motyka, A.; Zukowska, D.; Sledz, W.; Lojkowska, E. Population Structure and Biodiversity of *Pectobacterium parmentieri* Isolated from Potato Fields in Temperate Climate. *Plant Dis.* **2018**, *102*, 154–164. [CrossRef]
12. Zoledowska, S.; Motyka-Pomagruk, A.; Sledz, W.; Mengoni, A.; Lojkowska, E. High genomic variability in the plant pathogenic bacterium *Pectobacterium parmentieri* deciphered from de novo assembled complete genomes. *BMC Genom.* **2018**, *19*, 751. [CrossRef]
13. Waleron, M.; Misztak, A.; Waleron, M.; Jonca, J.; Furmaniak, M.; Waleron, K. *Pectobacterium polonicum* sp. nov. isolated from vegetable fields. *Int. J. Syst. Evol. Microbiol.* **2019**, *69*, 1751–1759. [CrossRef] [PubMed]
14. Sarfraz, S.; Riaz, K.; Oulghazi, S.; Cigna, J.; Sahi, S.T.; Khan, S.; Faure, D. *Pectobacterium punjabense* sp. nov., isolated from blackleg symptoms of potato plants in Pakistan. *Int. J. Syst. Evol. Microbiol.* **2018**, *68*, 3551–3556. [CrossRef] [PubMed]
15. Cigna, J.; Laurent, A.; Waleron, M.; Waleron, K.; Dewaegeneire, P.; van der Wolf, J.; Andrivon, D.; Faure, D.; Hélias, V. European Population of *Pectobacterium punjabense*: Genomic Diversity, Tuber Maceration Capacity and a Detection Tool for This Rarely Occurring Potato Pathogen. *Microorganisms* **2021**, *9*, 781. [CrossRef] [PubMed]
16. Humphris, S.N.; Cahill, G.; Elphinstone, J.G.; Kelly, R.; Parkinson, N.M.; Pritchard, L.; Toth, I.K.; Saddler, G.S. Detection of the Bacterial Potato Pathogens *Pectobacterium* and *Dickeya* spp. Using Conventional and Real-Time PCR. In *Plant Pathology*; Humana Press: New York, NY, USA, 2015; Volume 1302, pp. 1–16.
17. van der Wolf, J.M.; Cahill, G.; Van Gijsegem, F.; Helias, V.; Humphris, S.; Li, X.; Lojkowska, E.; Pritchard, L. Isolation, Detection and Characterization of *Pectobacterium* and *Dickeya* Species. In *Plant Diseases Caused by Dickeya and Pectobacterium Species*; Springer International Publishing: Cham, Switzerland, 2021; pp. 149–173.
18. De Boer, S.H.; Li, X.; Ward, L.J. *Pectobacterium* spp. Associated with Bacterial Stem Rot Syndrome of Potato in Canada. *Phytopathology* **2012**, *102*, 937–947. [CrossRef] [PubMed]
19. Kim, M.H.; Cho, M.S.; Kim, B.K.; Choi, H.J.; Hahn, J.H.; Kim, C.; Kang, M.J.; Kim, S.H.; Park, D.S. Quantitative Real-Time Polymerase Chain Reaction Assay for Detection of *Pectobacterium wasabiae* Using YD Repeat Protein Gene-Based Primers. *Plant Dis.* **2012**, *96*, 253–257. [CrossRef]
20. Lukianova, A.; Evseev, P.; Stakheev, A.; Kotova, I.; Zavriev, S.; Ignatov, A.; Miroshnikov, K. Development of qPCR Detection Assay for Potato Pathogen *Pectobacterium atrosepticum* Based on a Unique Target Sequence. *Plants* **2021**, *10*, 355. [CrossRef]
21. De Campos, S.B.; Lardi, M.; Gandolfi, A.; Eberl, L.; Pessi, G. Mutations in Two *Paraburkholderia phymatum* Type VI Secretion Systems Cause Reduced Fitness in Interbacterial Competition. *Front. Microbiol.* **2017**, *8*, 2473. [CrossRef]

22. Van Der Wolf, J.M.; Acuña, I.; De Boer, S.H.; Brurberg, M.B.; Cahill, G.; Charkowski, A.O.; Coutinho, T.; Davey, T.; Dees, M.W.; Degefu, Y.; et al. Diseases Caused by *Pectobacterium* and *Dickeya* Species around the World. In *Plant Diseases Caused by Dickeya and Pectobacterium Species*; Springer Nature: Cham, Switzerland, 2021; pp. 215–261.
23. Van Gijsegem, F.; Hugouvieux-Cotte-Pattat, N.; Kraepiel, Y.; Lojkowska, E.; Moleleki, L.N.; Gorshkov, V.; Yedidia, I. Molecular Interactions of *Pectobacterium* and *Dickeya* with Plants. In *Plant Diseases Caused by Dickeya and Pectobacterium Species*; Springer Nature: Cham, Switzerland, 2021; pp. 85–147.
24. Moleleki, L.N.; Onkendi, E.M.; Mongae, A.; Kubheka, G.C. Characterisation of *Pectobacterium wasabiae* causing blackleg and soft rot diseases in South Africa. *Eur. J. Plant Pathol.* **2013**, *135*, 279–288. [CrossRef]
25. van den Bosch, T.J.M.; Niemi, O.; Welte, C.U. Single gene enables plant pathogenic *Pectobacterium* to overcome host-specific chemical defence. *Mol. Plant Pathol.* **2020**, *21*, 349–359. [CrossRef] [PubMed]
26. Kõiv, V.; Roosaare, M.; Vedler, E.; Kivistik, P.A.; Toppi, K.; Schryer, D.W.; Remm, M.; Tenson, T.; Mäe, A. Microbial population dynamics in response to *Pectobacterium atrosepticum* infection in potato tubers. *Sci. Rep.* **2015**, *5*, 11606. [CrossRef] [PubMed]
27. Ge, T.; Ekbataniamiri, F.; Johnson, S.; Larkin, R.; Hao, J. Interaction between *Dickeya dianthicola* and *Pectobacterium parmentieri* in Potato Infection under Field Conditions. *Microorganisms* **2021**, *9*, 316. [CrossRef] [PubMed]
28. Leonard, S.; Hommais, F.; Nasser, H.W.; Reverchon, S. Plant-phytopathogen interactions: Bacterial responses to environmental and plant stimuli. *Environ. Microbiol.* **2017**, *19*, 1689–1716. [CrossRef] [PubMed]
29. Charkowski, A.O. The Changing Face of Bacterial Soft-Rot Diseases. *Annu. Rev. Phytopathol.* **2018**, *56*, 269–288. [CrossRef]
30. van der Wolf, J.M.; De Boer, S.H.; Czajkowski, R.; Cahill, G.; Van Gijsegem, F.; Davey, T.; Dupuis, B.; Ellicott, J.; Jafra, S.; Kooman, M.; et al. Management of Diseases Caused by *Pectobacterium* and *Dickeya* Species. In *Plant Diseases Caused by Dickeya and Pectobacterium Species*; Springer: Cham, Switzerland, 2021; pp. 175–214.
31. Holtappels, D.; Fortuna, K.; Lavigne, R.; Wagemans, J. The future of phage biocontrol in integrated plant protection for sustainable crop production. *Curr. Opin. Biotechnol.* **2021**, *68*, 60–71. [CrossRef]
32. Svircev, A.; Roach, D.; Castle, A. Framing the Future with Bacteriophages in Agriculture. *Viruses* **2018**, *10*, 218. [CrossRef]
33. Smolarska, A.; Rabalski, L.; Narajczyk, M.; Czajkowski, R. Isolation and phenotypic and morphological characterization of the first *Podoviridae* lytic bacteriophages ϕA38 and ϕA41 infecting *Pectobacterium parmentieri* (former *Pectobacterium wasabiae*). *Eur. J. Plant Pathol.* **2018**, *150*, 413–425. [CrossRef]
34. Kabanova, A.; Shneider, M.; Bugaeva, E.; Ha, V.T.N.; Miroshnikov, K.; Korzhenkov, A.; Kulikov, E.; Toschakov, S.; Ignatov, A.; Miroshnikov, K. Genomic characteristics of vB_PpaP_PP74, a T7-like *Autographivirinae* bacteriophage infecting a potato pathogen of the newly proposed species *Pectobacterium parmentieri*. *Arch. Virol.* **2018**, *163*, 1691–1694. [CrossRef]
35. Charkowski, A.; Blanco, C.; Condemine, G.; Expert, D.; Franza, T.; Hayes, C.; Cotte-Pattat, N.; Solanilla, E.A.L.; Low, D.; Moleleki, L.; et al. The Role of Secretion Systems and Small Molecules in Soft-Rot Enterobacteriaceae Pathogenicity. *Annu. Rev. Phytopathol.* **2012**, *50*, 425–449. [CrossRef]
36. Ma, B.; Hibbing, M.E.; Kim, H.-S.; Reedy, R.M.; Yedidia, I.; Breuer, J.; Breuer, J.; Glasner, J.D.; Perna, N.T.; Kelman, A.; et al. Host Range and Molecular Phylogenies of the Soft Rot Enterobacterial Genera *Pectobacterium* and *Dickeya*. *Phytopathology* **2007**, *97*, 1150–1163. [CrossRef] [PubMed]
37. Kim, H.-S.; Ma, B.; Perna, N.T.; Charkowski, A.O. Phylogeny and Virulence of Naturally Occurring Type III Secretion System-Deficient *Pectobacterium* Strains. *Appl. Environ. Microbiol.* **2009**, *75*, 4539–4549. [CrossRef]
38. Nykyri, J.; Mattinen, L.; Niemi, O.; Adhikari, S.; Kõiv, V.; Somervuo, P.; Fang, X.; Auvinen, P.; Mäe, A.; Palva, E.T.; et al. Role and Regulation of the Flp/Tad Pilus in the Virulence of *Pectobacterium atrosepticum* SCRI1043 and *Pectobacterium wasabiae* SCC3193. *PLoS ONE* **2013**, *8*, e73718. [CrossRef]
39. Toth, I.K.; Barny, M.-A.; Brurberg, M.B.; Condemine, G.; Czajkowski, R.; Elphinstone, J.G.; Helias, V.; Johnson, S.B.; Moleleki, L.N.; Pirhonen, M.; et al. Pectobacterium and *Dickeya*: Environment to Disease Development. In *Plant Diseases Caused by Dickeya and Pectobacterium Species*; Springer: Cham, Switzerland, 2021; pp. 39–84.
40. Bugaeva, E.; Voronina, M.; Vasiliev, D.; Lukianova, A.; Landyshev, N.; Ignatov, A.; Miroshnikov, K. Use of a Specific Phage Cocktail for Soft Rot Control on Ware Potatoes: A Case Study. *Viruses* **2021**, *13*, 1095. [CrossRef] [PubMed]
41. Na, S.-I.; Kim, Y.O.; Yoon, S.-H.; Ha, S.-M.; Baek, I.; Chun, J. UBCG: Up-to-date bacterial core gene set and pipeline for phylogenomic tree reconstruction. *J. Microbiol.* **2018**, *56*, 280–285. [CrossRef] [PubMed]
42. Lee, I.; Kim, Y.O.; Park, S.-C.; Chun, J. OrthoANI: An improved algorithm and software for calculating average nucleotide identity. *Int. J. Syst. Evol. Microbiol.* **2016**, *66*, 1100–1103. [CrossRef] [PubMed]
43. Stakheev, A.A.; Khairulina, D.R.; Zavriev, S.K. Four-locus phylogeny of *Fusarium avenaceum* and related species and their species-specific identification based on partial phosphate permease gene sequences. *Int. J. Food Microbiol.* **2016**, *225*, 27–37. [CrossRef] [PubMed]

plants

MDPI

Article

Paclobutrazol Improves the Quality of Tomato Seedlings to Be Resistant to *Alternaria solani* Blight Disease: Biochemical and Histological Perspectives

Tarek A. Shalaby [1,2,*], Naglaa A. Taha [3], Dalia I. Taher [4], Metwaly M. Metwaly [5], Hossam S. El-Beltagi [6,7,*], Adel A. Rezk [3,6], Sherif M. El-Ganainy [1,3], Wael F. Shehata [6,8], Hassan R. El-Ramady [9] and Yousry A. Bayoumi [2,10]

1 Department of Arid Land Agriculture, College of Agricultural and Food Science, King Faisal University, P.O. Box 400, Al-Ahsa 31982, Saudi Arabia; salganainy@kfu.edu.sa
2 Horticulture Department, Faculty of Agriculture, Kafrelsheikh University, Kafr El-Sheikh 33516, Egypt; yousry.bayoumi@agr.kfs.edu.eg
3 Agricultural Research Center, Plant Pathology Research Institute, Giza 12619, Egypt; naglaa_abdelbaset@yahoo.com (N.A.T.); arazk@kfu.edu.sa (A.A.R.)
4 Agricultural Research Center (ARC), Vegetable Crops Research Department, Horticulture Research Institute, Giza 12619, Egypt; daliataher1981@gmail.com
5 Agricultural Botany Department, Faculty of Agriculture, Kafrelsheikh University, Kafr El-Sheikh 33516, Egypt; metwalysalim@yahoo.com
6 Department of Agricultural Biotechnology, College of Agricultural and Food Science, King Faisal University, P.O. Box 400, Al-Ahsa 31982, Saudi Arabia; wshehata@kfu.edu.sa
7 Biochemistry Department, Faculty of Agriculture, Cairo University, Giza 12613, Egypt
8 Plant Production Department, College of Environmental Agricultural Science, El-Arish University, North Sinai 45511, Egypt
9 Soil and Water Department, Faculty of Agriculture, Kafrelsheikh University, Kafr El-Sheikh 33516, Egypt; hassan.elramady@agr.kfs.edu.eg
10 Physiology & Breeding of Horticultural Crops Laboratory, Horticulture Department, Faculty of Agriculture, Kafrelsheikh University, Kafr El-Sheikh 33516, Egypt
* Correspondence: tshalaby@kfu.edu.sa (T.A.S.); helbeltagi@kfu.edu.sa (H.S.E.-B.)

Citation: Shalaby, T.A.; Taha, N.A.; Taher, D.I.; Metwaly, M.M.; El-Beltagi, H.S.; Rezk, A.A.; El-Ganainy, S.M.; Shehata, W.F.; El-Ramady, H.R.; Bayoumi, Y.A. Paclobutrazol Improves the Quality of Tomato Seedlings to Be Resistant to *Alternaria solani* Blight Disease: Biochemical and Histological Perspectives. *Plants* **2022**, *11*, 425. https://doi.org/10.3390/plants11030425

Academic Editor: Alessandro Vitale

Received: 21 January 2022
Accepted: 1 February 2022
Published: 4 February 2022

Abstract: The production and quality of tomato seedlings needs many growth factors and production requirements besides controlling the phytopathogens. Paclobutrazol (PBZ) has benefit applications in improving crop productivity under biotic stress (*Alternaria solani*, the causal agent of early blight disease in tomatoes). In the current study, the foliar application of PBZ, at rates of 25, 50, and 100 mg L^{-1}, was evaluated against early blight disease in tomatoes under greenhouse conditions. The roles of PBZ to extend tomato seedling lives and handling in nurseries were also investigated by measuring different the biochemical (leaf enzymes, including catalase and peroxidase) and histological attributes of tomato seedlings. Disease assessment confirmed that PBZ enhanced the quality of tomato seedlings and induced resistance to early blight disease post inoculation, at 7, 14, and 21 days. Higher values in chlorophyll content, enzyme activities, and anatomical features of stem (cuticle thickness) and stomata (numbers and thickness) were recorded, due to applied PBZ. This may support the delay of the transplanting of tomato seedlings without damage. The reason for this extending tomato seedling life may be due to the role of PBZ treatment in producing seedlings to be greener, more compact, and have a better root system. The most obvious finding to emerge from this study is that PBZ has a distinguished impact in ameliorating biotic stress, especially of the early blight disease under greenhouse conditions. Further studies, which consider molecular variables, will be conducted to explore the role of PBZ in more detail.

Keywords: anatomy; biotic stress; catalase; early blight; paclobutrazol; peroxidase; phytopathogens; disease index; antifungal; chlorophyll content

1. Introduction

Tomato (*Solanum lycopersicum* L.) is one of the most valuable vegetable crops worldwide. The Egyptian cultivated area of tomato was 375,276 ha, with a productivity of 38.96 Mg ha^{-1} [1]. The highest Egyptian production of tomatoes during the last decade was 8.6 million metric tons in 2012 [2]. Globally, in 2019, the main producers of tomato included China, which produces alone about 63 million tons, ≈35%, of the total (181 million ton), followed by India, Turkey, the USA, and Egypt producing 19, 12.8, 10.9, and 6.9 million tons, respectively [1]. High seedling quality and their transplantation are mutual practices in the fruitful production of the tomato for fast, sustainable establishing, together with enhancement of earliness, uniform maturity and total yield, as well as quality [3]. The lack of a pre-contracting system for tomato seedlings between the nursery and farmers led the nursery to produce seedlings in a large quantity, sometimes causing a wait for sales [4]. These seedlings may be exposed to damage if they are not transplanted into the field at the appropriate time. Therefore, it is helpful to extend the seedlings life in the nursery, while maintaining high quality without losses. Several compounds can inhibit height growth, hence, extending the life of vegetable seedlings, such as chlormequat chloride (CCC) and daminozide [5], as well as paclobutrazol [6–8].

Paclobutrazol (PBZ), a triazole-type plant growth regulator or retardant, is well-known as anti-gibberellins. PBZ can block the conversion of *ent*-kaurene to *ent*-kaurenoic acid during biosynthesis pathway of gibberellin by inhibition of kaurene oxidase [9,10]. Foliar application of Paclobutrazol usually reduces shoot and root length by increasing the stiffness of the cell wall and decreasing cell wall expansion [11]. Several studies confirmed that applied PBZ improved various kinds of compatible solutes and osmo-protectants, such as proline, which increases the plant's tolerance to water deficits [9,12]. Many benefits of PBZ application have been intensively reported, including improving crop productivity, plant stress tolerance, fruit/grain quality, plant water relation, and membrane stability index [13,14]. In addition, PBZ prevents sucker re-growth in bananas [15], promoting fruit sets in many crops (such as olives) [16], as well as inhibiting the biosynthesis of gibberellin, early fruit set, and reduced stem growth [17]. Concerning the toxicity of PBZ, for living organisms, showed low toxicity via the dermal route in animals, whereas it caused moderate toxicity via human oral and inhalation routes. Based on the available researches, PBZ is considered unlikely to be genotoxic or carcinogenic to humans [18,19].

Fungal diseases are considered one of the core problems facing and affecting tomato seedlings in the nurseries. Among the pathogens that affect tomato seedlings are soil-borne (causing root decay or damping-off) and foliar diseases, including *Alternaria solani* and *Phytophthora infestans*, which reduce yield quality [20]. Among foliar pathogens, *A. solani*, which caused an early blight disease in tomatoes, is a highly destructive pathogen on both open field and greenhouse tomatoes [21,22]. *A. solani* causes infections on foliage, basal stems of transplants, stems of mature plants (stem lesions), and fruits (fruit rot) of tomato [23]. Early blight disease may cause crop losses of up to 78% to solanaceous crops [24].

Regardless of the promising results of chemical treatments in controlling fungal pathogens, phytotoxicity, and chemical residues are major problems that lead to environmental pollution and human health hazards.

Keeping the roles of PBZ under stress in view, the main aim of the present study is (1) to find out whether the foliar application of paclobutrazol has any growth regulatory outcomes on tomato seedlings under both normal and early blight stress disease, (2) to document the extending impact of the applied paclobutrazol handling of tomato seedlings in nurseries, and (3) to observe the biochemical and histological responses of tomato seedlings to paclobutrazol foliar application, under both control and biotic stress conditions.

2. Results

2.1. *Pathogenic Ability of the Four A. solani Isolates on Tomato Plants*

An experiment was conducted to assess virulence of four *A. solani* isolates by using a susceptible tomato hybrid (Alissa F_1) under greenhouse conditions. While all of the four obtained isolates were pathogenic to tomato seedlings, causing identical early blight disease symptoms, isolate number 1 (I_1) was the greatest virulence isolate in the experiment (Table 1), compared with the other isolates. Isolate I_1 had the highest disease index percent on tomato plants (23.07, 45.92, and 80.5% after 7, 14, and 21 dpi, respectively). However, the other isolates were varied in their degrees of pathogenicity. Therefore, isolate I_1 of the pathogen was chosen for the following studies. Differences in the pathogenicity of the tested pathogenic isolates may be due to their physiological and biochemical components. It may also relate to the genetic makeup of host variety and pathogen, as far as their interactions are concerned [25].

Table 1. Pathogenic ability of the isolates of *Alternaria solani* on tomato seedlings in pots under greenhouse conditions.

Isolate No.	Disease Index (%)		
	7 Days	14 Days	21 Days
Isolate no. I1	23.05 ± 0.89 a	45.92 ± 2.36 a	80.50 ± 2.39 a
Isolate no. I2	17.07 ± 1.63 b	38.07 ± 1.62 b	54.00 ± 3.26 b
Isolate no. I3	9.80 ± 0.73 d	29.44 ± 1.62 d	40.50 ± 1.24 d
Isolate no. I4	15.60 ± 0.77 c	32.50 ± 1.63 c	45.80 ± 1.67 c
F. test	**	*	**

Where: I1, I2, I3m and I4 are the four isolates of *Alternaria solani*, which were used in the study, Mean values in each column, followed by the same letter, are not significant ($p < 0.05$), * and ** indicating significant and highly significant, respectively.

2.2. *Impact of Paclobutrazol on the Linear Growth of A. solani*

The in vitro antifungal activity of paclobutrazol showed that all evaluated concentrations of PBZ indicated antifungal activity and significantly inhibited mycelial growth percentage of *A. solani* (Figure 1 and Table 2). All concentrations (25, 50, and 100 mg L^{-1}) showed the highest reduction of mycelial radial growth (2, 2.7, and 3 cm), without significant differences in between, suggesting similar potency, compared to control treatment (without PBZ), which resulted no inhibition of mycelia growth (9 cm). The results showed that PBZ had the highest antagonistic effect against of *A. solani* (Figure 1 and Table 2). The reduction percentage of *A. solani*, due to PBZ applications, were the highest values, especially when using the rate of 100 mg L^{-1} (77.8%), without significant differences at 25 and 50 mg L^{-1} rates, 66.7 and 70%, respectively.

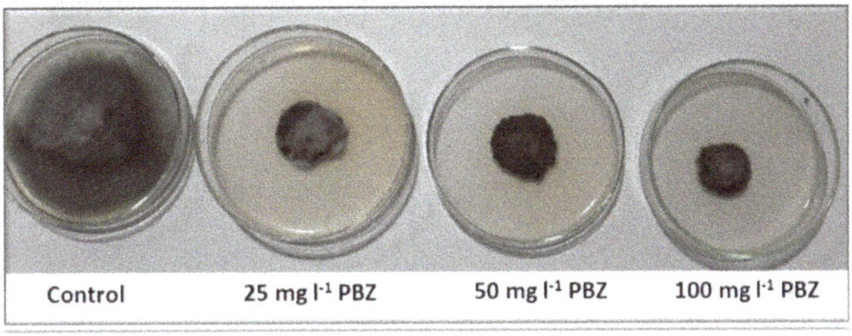

Figure 1. In vitro antifungal activity of paclobutrazol (PBZ) against *Alternaria solani*. The highest applied dose of PBZ (100 mg L^{-1}) recorded the highest control of this disease, compared to the control.

Table 2. The inhibition of *Alternaria solani* mycelial growth by PBZ application in different concentrations on PDA medium.

Treatments	Mycelial Growth (cm)	Reduction Rate (%)
Control	9.0 ± 0.0 a	0.01 ± 0.004 b
PBZ 25 mg L^{-1}	3.0 ± 0.082 b	66.7 ± 0.735 a
PBZ 50 mg L^{-1}	2.7 ± 0.082 b	70.0 ± 2.450 a
PBZ 100 mg L^{-1}	2.0 ± 0.082 b	77.8 ± 1.563 a
F. test	**	**

Mean values in each column, followed by the same letter, are not significant ($p < 0.05$). PDA: potato dextrose agar medium. ** indicating highly significant.

2.3. Development of Early Blight Disease Due to PBZ Applications

In general, in vitro experiment exogenous application with PBZ obviously reduced the of early blight intensity on tomato leaves, in comparison to the control seedlings, after 7, 14, and 21 days post-inoculation (dpi) (Figure 2 and Table 3). Although, an advanced rise was observed in both the disease incidence and disease severity on control tomato seedlings during the experiment. All PBZ applications significantly decreased the disease incidence (DI) and its severity (DS) percent, at 7, 14, and 21 dpt, until finishing the experiment. Applied PBZ at 25 mg L^{-1} was the most efficient treatment and had the lowermost DI and DS (%) after the previously mentioned periods. It is worth mentioning that PBZ, at all doses, significantly reduced both the DI and DS percent at the three studied stages, and compared with the control treatment. In the same manner, all PBZ applications (25, 50, and 100 mg L^{-1}) showed the highest efficacy (87.6, 87.2, and 84.3%, respectively), compared to the untreated plants (Table 3).

Figure 2. Effect of Paclobutrazol (PBZ) on progress symptoms of early blight disease on tomato seedlings under greenhouse conditions at 7, 14, and 21 dpi (day post inoculation).

Table 3. Effect of PBZ applications on both disease incidence (DI) and disease severity (DS); % of tomato early blight pathogen (*A. solani*) under greenhouse conditions at 7, 14, and 21 dpi and an efficacy % at 21 dpi.

Treatments	After 7 Days		After 14 Days		After 21 Days		
	DI (%)	DS (%)	DI (%)	DS (%)	DI (%)	DS (%)	Efficacy (%)
Control	50.2 ± 1.95 a	35.4 ± 3.17 a	60.7 ± 5.10 a	50.6 ± 5.47 a	87.9 ± 7.05 a	79.2 ± 4.59 a	00.0 ± 0.0
PBZ 25 mg L^{-1}	9.4 ± 3.57 c	3.6 ± 0.95 b	10.2 ± 1.65 c	5.5 ± 1.25 b	18.4 ± 3.18 b	9.8 ± 1.58 b	87.6 ± 5.19
PBZ 50 mg L^{-1}	10.3 ± 2.07 c	3.6 ± 0.89 b	18.5 ± 2.57 b	8.1 ± 1.98 b	25.2 ± 1.95 b	10.1 ± 1.99 b	87.2 ± 4.66
PBZ 100 mg L^{-1}	15.8 ± 1.85 b	3.6 ± 1.05 b	25.1 ± 3.07 b	10.3 ± 2.05 b	30.1 ± 3.07 b	12.4 ± 2.17 b	84.3 ± 5.05
F. test	**	**	**	**	**	**	-

Mean values in each column, followed by the same letter, are not significant ($p < 0.05$). ** indicating highly significant.

2.4. Response of Vegetative Growth and Chlorophyll Content to Applied PBZ

To evaluate the response of tomato seedlings to applied doses of PBZ, different vegetative growth parameters, besides chlorophyll content, were measured after 10, 20, and 30 days from PBZ applications (during 2021) (Table 4). The seedlings treated with PBZ, after 10 days, represent the study of the role of applying different doses of PBZ, without infected seedlings and with *A solani* as a control, whereas after 20 and 30 days, as infected seedlings. The commonly known impact of PBZ as a plant growth retarder is clear, due to its decreased seedling height. From Table 4, it can be seen that the seedling height was deceased by increasing the applied doses of PBZ, whereas, in general, seedling height values was increased from 10 to 30 days after foliar-applied PBZ. Although, the increasing rate of seedling height was the highest in control seedlings, as compared with the PBZ application in all stages (10, 20, and 30 days after applications). The applied dose of 50 mg L^{-1} PBZ recorded, in general, the highest values of chlorophyll content, as well as after 10 or 20 or 30 days after foliar applied PBZ; then, the differences were not significant with the control treatment after only 10 days. Stem diameter values increased by an increasing period after PBZ application from 10 to 30, recording the highest value at dose of 100 mg L^{-1} (0.336 mm), compared to control seedlings, which resulted the lowest diameter in all stages. For seedling fresh weight, it was significantly influenced by treatments at all growth stages. After 10 days, the highest values were shown from control seedlings, compared to all PBZ applications. Nonetheless, all doses from PBZ produced the highest values of seedling fresh weight at 20 and 30 days after application, especially at a dose of 100 mg L^{-1}. The most obvious observation to appear from the statistics comparison was the dry biomass per seedling, which was significantly differed after 10, 20, and 30 days and the highest values were recorded from applied PBZ at dose of 100 mg L^{-1} (0.37 and 0.62 g, respectively), after 20 and 30 days from application. Root fresh and dry weights were significantly affected by PBZ applications at the three growth stages. PBZ, at a rate of 100 mg L^{-1}, resulted highest value of root fresh weight; however, a PBZ dose of 50 mg L^{-1} resulted in the highest value of root dry weight in most cases, compared to the other doses and control.

2.5. Response of Enzyme Activities to Applied PBZ

Two plant enzymes (catalase and peroxidase) were evaluated as bioindicators for survival tomato seedlings under biotic stress (Table 5). There was a highly significant relation between applied doses of PBZ and values of the studied enzymes, where the highest applied dose of PBZ (100 mg L^{-1}) recorded the highest value of CAT (102.88) and POD (0.057) as mM H_2O_2 g^{-1} FW min^{-1}, which did not significantly differ with PBZ at a dose of 50 mg L^{-1} in enzyme activities. These results confirmed that the foliar application of PBZ enhanced the cultivated tomato seedlings quality under biotic stress, through promoting and producing higher plant enzymes, which support cultivated seedlings under biotic and abiotic stresses.

Table 4. Response of some vegetative growth parameters and chlorophyll content to PBZ doses after 10, 20, and 30 days from PBZ foliar application (during April 2021) (with/without infection by *Alternaria solani*).

Treatments	Seedling Height (cm)	Stem Diameter (mm)	Seedling FW (g)	Seedling DW (g)	Root FW (g)	Root DW (g)	Chlorophyll Content (SPAD)
After 10 days (Not infected seedlings by *Alternaria solani*)							
Control	13.24 ± 1.50 a	0.207 ± 0.021 a	2.65 ± 0.501 a	0.30 ± 0.006 a	0.86 ± 0.14 a	0.085 ± 0.005 a	28.4 ± 2.35 a
PBZ 25 mg L^{-1}	8.83 ± 1.68 ab	0.215 ± 0.015 a	1.62 ± 0.452 b	0.25 ± 0.005 ab	0.47 ± 0.05 b	0.071 ± 0.005 a	26.2 ± 2.17 b
PBZ 50 mg L^{-1}	6.83 ± 1.29 b	0.238 ± 0.011 a	1.48 ± 0.363 b	0.18 ± 0.003 c	0.44 ± 0.07 b	0.054 ± 0.003 a	27.6 ± 3.02 ab
PBZ 100 mg L^{-1}	9.56 ± 1.55 ab	0.266 ± 0.03 a	1.54 ± 0.295 b	0.22 ± 0.005 bc	0.36 ± 0.07 b	0.063 ± 0.003 a	24.3 ± 1.95 c
F. test	**	NS	*	**	*	NS	**
After 20 days (Infected seedlings by *Alternaria solani*)							
Control	14.75 ± 2.17 a	0.275 ± 0.007 a	3.05 ± 0.524 ab	0.33 ± 0.005 b	0.94 ± 0.18 c	0.082 ± 0.005 b	28.8 ± 3.05 c
PBZ 25 mg L^{-1}	11.94 ± 1.66 b	0.296 ± 0.025 a	3.12 ± 0.354 ab	0.34 ± 0.007 b	1.02 ± 0.25 b	0.088 ± 0.007 b	31.3 ± 2.84 b
PBZ 50 mg L^{-1}	9.94 ± 1.59 b	0.299 ± 0.034 a	2.88 ± 0.441 b	0.29 ± 0.006 c	0.89 ± 0.18 c	0.120 ± 0.008 a	38.6 ± 3.10 a
PBZ 100 mg L^{-1}	11.22 ± 2.09 b	0.310 ± 0.033 a	3.23 ± 0.455 a	0.37 ± 0.007 a	1.05 ± 0.20 a	0.089 ± 0.007 b	33.9 ± 1.99 b
F. test	**	NS	*	**	**	**	**
After 30 days (Infected seedlings by *Alternaria solani*)							
Control	27.33 ± 3.17 a	0.254 ± 0.028 b	3.15 ± 0.625 ab	0.42 ± 0.008 b	0.93 ± 0.19 b	0.091 ± 0.009 c	24.9 ± 2.25 c
PBZ 25 mg L^{-1}	11.79 ± 1.29 b	0.325 ± 0.033 a	3.42 ± 0.605 ab	0.57 ± 0.009 ab	1.14 ± 0.24 a	0.098 ± 0.009 b	32.1 ± 2.19 b
PBZ 50 mg L^{-1}	9.86 ± 1.45 b	0.318 ± 0.029 a	2.96 ± 0.385 b	0.44 ± 0.008 b	0.95 ± 0.23 b	0.137 ± 0.012 a	38.9 ± 3.55 a
PBZ 100 mg L^{-1}	11.48 ± 1.88 b	0.336 ± 0.017 a	3.55 ± 0.550 a	0.62 ± 0.009 a	1.22 ± 0.026a	0.132 ± 0.015 a	32.7 ± 3.06 b
F. test	**	**	**	*	**	**	**

Mean values in each column, followed by the same letter, are not significant ($p < 0.05$). Root fresh or dry weights were per one seedling, FW (fresh weight) and DW (dry weight), * and ** indicating significant and highly significant, respectively.

Table 5. Impact of applied treatments on enzyme activities (catalase, CAT, peroxidase, and POD) after 10 days form PBZ foliar application (after 10 days from PBZ application and without infection by *Alternaria solani*).

Treatments	POD (mM H_2O_2 g^{-1} FW min^{-1})	CAT Activity (mM H_2O_2 g^{-1} FW min^{-1})
Control	0.012 ± 0.011 ^{c}	2.82 ± 0.429 ^{c}
PBZ 25 mg L^{-1}	0.032 ± 0.020 ^{b}	64.87 ± 4. 298 ^{b}
PBZ 50 mg L^{-1}	0.049 ± 0.029 ^{ab}	90.30 ± 5.556 ^{a}
PBZ 100 mg L^{-1}	0.057 ± 0.025 ^{a}	102.88 ± 5.939 ^{a}
F. test	**	**

Mean values in each column, followed by the same letter, are not significant ($p < 0.05$). ** indicating highly significant.

2.6. Response of Anatomical Features to Applied PBZ

The internal structure of the tomato seedling stem is similar to the other dicotyledonous plants and built-up, essentially, of parenchyma ground tissue, including cortex tissue and pith tissue, regular vascular bundles, and medullary rays, which connect between the cortex and pith tissues. It is clear that, from the present data in Table 6 and Figure 3, the application of paclobutrazol has a positive impact on stem anatomical features, which led to enhancing most of investigated the anatomical measurements of tomato stem, especially the second dose (50 mg L^{-1}), which increased the thickness values of the cuticle layer and tissues of epidermis, cortex, xylem, and phloem, as well as the diameter of stem cross-sections and xylem vessels. These obtained results were compared with the control and other concentrations of PBZ used.

Stomata measurements have included stomata density and stomata dimensions (length and width). Data presented in Table 7 and Figure 4 showed that there was an increase in density of stomata, due to the application of PBZ, with various concentrations. These obtained results were compared to the control treatment. This increase of stomata density is due to the negative effect of PBZ on the leaf area, as well as the inhibition of it [26]. The highest values of stomata density were recorded in the application of the second concentration of PBZ, compared with other concentrations. Besides that, using PBZ with investigated concentrations enhanced the values of the stomata dimension (length and width), compared to the control treatment, due to the PBZ application stacking of the stomata, per area unit of leaves.

Table 6. Anatomical measurements of tomato stems, as affected by the application of various concentrations of PBZ substance (after 10 days from PBZ application and without infection by *Alternaria solani*).

Anatomical Measurements		Applied PBZ (mg L^{-1}) Doses at the Second True Leaf Stage				
		Control	25	50	100	LSD 0.05
Thickness (µm)	Cuticle	4.37 ± 0.69 b	7.24 ± 0.72 a	7.55 ± 0.81 a	5.73 ± 0.78 b	1.42
	Epidermis	33.70 ± 2.09 b	37.84 ± 1.19 a	26.93 ± 0.82 c	27.22 ± 0.48 c	3.46
	Cortex	444.65 ± 14.36 a	304.87 ± 11.20 c	435.64 ± 8.66 a	396.43 ± 34.04 b	37.25
	Xylem	387.53 ± 8.58 c	476.97 ± 30.32 a	444.30 ± 15.09 ab	415.46 ± 10.36 bc	34.31
	Phloem	95.00 ± 4.80 c	88.76 ± 2.50 c	155.48 ± 6.80 a	109.03 ± 6.11 b	10.07
Diameter (µm)	Xylem vessels	60.87 ± 2.00 a	46.47 ± 1.46 a	58.42 ± 2.01 a	53.58 ± 2.25 a	11.70
	Stem	2194.72 ± 162.23 b	2291.66 ± 167.19 b	2477.26 ± 149.38 a	2581.89 ± 138.94 a	70.40

Mean values in each column, followed by the same letter, are not significant ($p < 0.05$).

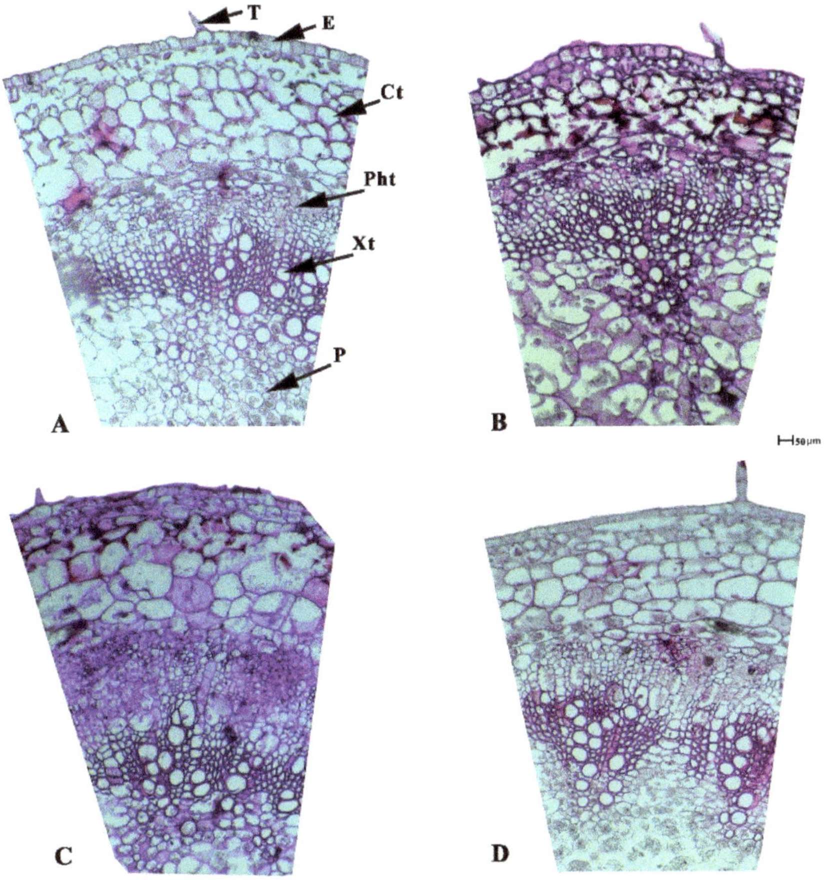

Figure 3. Transverse sections through the tomato stems, as affected by the application of various concentrations of PBZ substance, 10 days after applications, where (**A–D**) represent the treatments of the control, applied PBZ at 25, 50, and 100 mg L^{-1}. The abbreviations: T (cuticle), E (epidermis), Ct (cortex tissue), Pht (phloem tissue), Xt (xylem tissue), P (pith), and MR (medullary rays).

Table 7. Stomata measurements of tomato leaves, as affected by PBZ treatments (after 10 days from PBZ application and without infection by *Alternaria solani*).

Treatments	Stomata Measurements		
	Thickness (μm)	Width (μm)	Numbers
Control	42.09 ± 1.83 b	83.66 ± 2.07 c	23.33 ± 1.15 c
PBZ 25 mg L^{-1}	38.16 ± 1.60 c	88.89 ± 1.80 b	24.67 ± 0.57 c
PBZ 50 mg L^{-1}	46.21 ± 0.65 a	93.01 ± 1.47 a	36.00 ± 1.00 b
PBZ 100 mg L^{-1}	41.61 ± 1.68 b	86.14 ± 1.58 bc	38.33 ± 0.57 a
LSD 0.05	2.86	3.29	1.63

Mean values in each column, followed by the same letter, are not significant ($p < 0.05$).

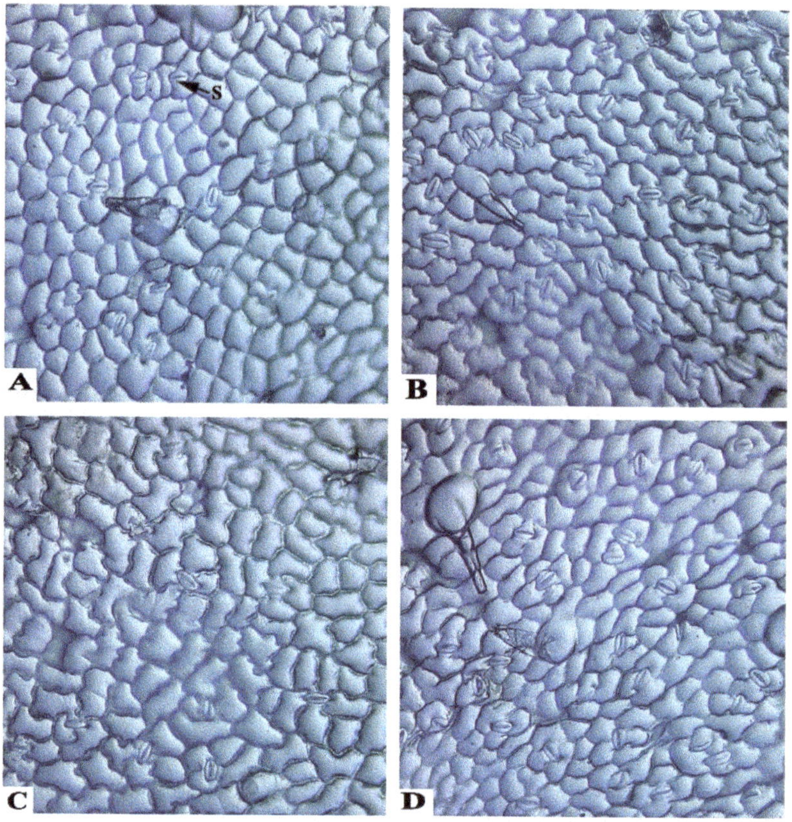

Figure 4. Photographs of stomata on adaxial (upper) surfaces of the tomato leaves, as affected by the application of various concentrations of PBZ substance, 10 days after applications, where (**A–D**) represent the treatments of the control, applied PBZ at 25, 50, and 100 mg L^{-1}. The abbreviation: S, stomata.

3. Discussion

High-quality tomato seedling production needs to save the required growth factors, including the environmental and practical issues. These issues may include controlling the pests and diseases, as well as proper agricultural practices (e.g., fertilization, irrigation, lighting, etc.). It is very important to produce vigorous tomato seedlings with a long shelf life, especially under intensive work in greenhouses, which sometimes needs a delay in transferring and transplanting seedlings into farming field [27]. This target will be more complicated under conditions of plant diseases, particularly the early blight. To extend the life and juvenility of tomato seedlings, without losses, many substances have been applied, such as chlormequat chloride (CCC), daminozide, and paclobutrazol. The role of applying different doses of PBZ to control the early blight, resulting from *Alternaria solani*, is investigated in the current study. After selecting the most aggressive, isolated inoculant of *Alternaria solani*, which had the highest disease index after 7, 14, and 21 days from the inoculation, different doses of applied PBZ were investigated on disease incidence and its severity of early blight pathogen (*A. solani*) on tomato seedlings under greenhouse conditions (Figures 1 and 2 and Tables 1–3).

Many studies published on the role of paclobutrazol in growing tomato seedlings, including applied different doses of PBZ (i.e., 50 and 100 mg L^{-1}) through seed treatment or watering seedlings [28], production of tomato seedlings by applying PBZ (50, 100, and 150 mg L^{-1}) using two tomato hybrids [29], and accelerating growth of tomato seedlings by applying 25, 50, and 100 mg L^{-1} PBZ [30]; however, to our knowledge, there are no studies on the crucial role of PBZ against early blight (*Alternaria solani*). On the other hand, many studies reported on the reduction of vegetative growth via applied PBZ on many cultivated crops, such as potato [31], mango [32], *Leonotis leonurus* L. [33], and olive [16], as well as its potential under stress, i.e., drought [13,14,34] and salinity [9,35].

In the current study, in response to the question: what is the impact of paclobutrazol on the vegetative growth of tomato seedlings, a range of responses were elicited. The overall response to this question was very positive. The vegetative growth parameters were tested after 10, 20, and 30 days after the applied PBZ doses, where parameters after 10 days were without infected seedlings, but after 20 and 30 days, as infected seedlings. The direct cause of the increase in leaf darkness and greenness of treated seedlings with PBZ may be due to the increase in chlorophyll content. Surprisingly, PBZ was found to decrease the dry weight and chlorophyll content of seedlings after 10 days, but increased by increasing PBZ levels after 20 and 30 days. The increased chlorophyll content could be due to an increase in the activity of oxidative enzymes, which changed in the levels of carotenoids, ascorbate, and ascorbate peroxidase [31]. Plant enzymes, including CAT and POD, were increased by increasing applied doses of PBZ up to 100 mg L^{-1}, recording 0.057 and 102.88, respectively (Table 5). This study produced results that corroborate the findings of a great deal of previous work in promoting growth of plants under stress by applying PBZ. PBZ improves plant tolerance against different stresses by increasing proline content and enzymatic antioxidants [14], increases fruit yield (due to the relatively stouter canopy of PBZ-treated plants), improving rooting system (which may increase the uptake of water and nutrients) [36], regulating photosynthetic capacity and delaying leaf senescence [37], improving the resistance against many plant pathogens [38], and acting as a systemic fungicide against several economically fungal diseases [39]. The mode of action of paclobutrazol may include the inhibition of gibberellic acid synthesis in plants, which reduces gibberellins level, slows cell division and elongation (without causing toxicity to cells), and increases cytokinin content, as well as the root activity and C: N ratio. Therefore, PBZ can delay senescence and extend the juvenility of seedlings, which increased the seedlings life without losses; additionally, it increased the resistance against most of pathogens in the nursery. Interestingly, few studies have reported the potential of paclobutrazol in improving the levels of chlorophyll, antioxidants, and proline contents under various biotic and abiotic stresses, as well as extending the plant growth cycle by delaying physiological maturity [40–43].

All anatomical features of the seedlings were influenced by the different doses of PBZ applied, particularly the stomata measures, including thickness, width, and numbers, which increased by increasing the applied doses of PBZ, up to 100 mg L^{-1} (Table 7). This increase in tomato stem diameter is achieved via the application of paclobutrazol treatment, due to its role to induce an increase in the vascular bundles' thickness, thicker cortex tissue, and wider pith tissue diameter, which is associated with larger medullary cells [44]. The highest value in the thickness of the stem cuticle (7.55 μm) was achieved by applying 50 mg L^{-1} PBZ, which may support the resistance of tomato seedlings to early blight (Table 6). The highest numbers or values of stomata of numbers (38.33), thickness (46.21 μm), and width (93.01 μm) may lead to an increase in the efficiency of photosynthesis. Similarly, Tekalign and Hammes [31] stated that applying PBZ on potato leaves increased the anatomical parameters, i.e., thickness cortex, pith diameter, and size of the vascular bundles, and resulted in the thickest stems. This might be due to the radial enlargement of cells because of the decreased endogenous gibberellin activities in response to the treatment. In addition, using PBZ with investigated concentrations led to enhanced values of stomata dimension (length and width), compared to the control treatment, due to the fact that PBZ application stacked the stomata per area unit of leaves. This increase of stomata density is due to the negative effect of PBZ on the leaf area, as well as the inhibition of it [26].

4. Materials and Methods

4.1. Isolation, Purification, and Identification of Causal Organism

Four pathogenic fungi of *A. solani* were isolated from different tomato fields in Kafr El-Sheikh governorate, Egypt. Briefly, tomato plants, showing typical symptoms of early blight disease, were collected from different locations. Infected leaves and stems were washed using tap water and cut into small parts (5 mm), sterilized using solution of 2% sodium hypo chloride for 2–3 min, then washed three times by sterilized distilled water (SDW). They were then dried between two layers of sterilized filter papers to remove excess water and plated onto petri dishes, 9 cm (in diameter), containing 15 mL potato dextrose agar (PDA) medium, amended with 100 mg L^{-1} streptomycin sulphate at 28 ± 2 °C for 7 days [45]. The hyphal tip technique was used to purify the developed fungal cultures. Four isolates were characterized as *A. solani*, based on the morphological characters counting conidia size, number of longitudinal and transverse septa, and length of a beak [46]. Then, the four isolates were assured as *A. solani*, based on their pathogenicity and typical early blight disease symptoms on tomato plants.

4.2. Pathogenicity Test

Pathogenicity test of four *A. solani* isolates were confirmed on a highly susceptible hybrid of tomato (Alisa F_1 hybrid) in pot trials under greenhouse conditions. Shortly, mycelial mats were harvested from seven-day old cultures of *A. solani*, then milled in 100 mL sterilized distilled water using sterilized mortar; it was filtered and put in a test tube, according to [47]. Thereafter, spore suspensions (10^6 spores mL^{-1}) were prepared for each of the four isolates in sterilized water. At 40 days old, tomato transplants were sprayed with tested inoculum of *A. solani* isolates, as spore suspension (30 mL $plant^{-1}$), while control plants were treated with same amount of distilled water. Inoculated plants were kept under polythene bags for 48 h to raise humidity and then incubated under greenhouse conditions. Disease index (DI%) was assessed and results were recorded three times frequently (7, 14, and 21 days post-inoculation (dpi)) to detect development of early blight disease. In this trial, six replicates were used; each replicate contains five pots (20 cm diameter) with two plants in each pot.

4.3. Antifungal Activity

In vitro antifungal activities of PBZ were evaluated by the agar diffusion technique [48]. Briefly, three concentrations of PBZ (25, 50, and 100 mg L^{-1}), besides the control, were mixed in proper volumes, individually concentrate with 100 mL of the PDA medium, in sterilized Petri dishes to find required concentration. The negative control was sterilized with PDA medium. Then, the pre-prepared Petri dishes were inoculated with 5 mm diameter mycelial mass of freshly prepared culture of the pathogen (I_1 isolate), incubated at 27 ± 1 °C, and fungal growth was recorded for 7 days post-inoculation (dpi). The whole experiment used six replications for all treatments. The inhibition percentage of the mycelial growth has been determined using this equation:

$$Inhibition\ (\%) = \frac{C - T}{C} \times 100$$

where "C" shows the mycelial growth in negative control dish, and "T" show the mycelial growth in different treatments.

4.4. Plant Materials and Growth Conditions

This study was carried out in a nursery of the Faculty of Agriculture, April 2021, Kafrelsheikh University in Kafr El-Sheikh governorate, to examine the extending of tomato seedlings life using different doses of PBZ. Paclobutrazol was obtained from Shoura Company for chemicals, Cairo, Egypt, as super coltar 25% PBZ. Paclobutrazol was dissolved in water to make solutions of four concentrations that have been used in the present study (i.e., 0, 25, 50, and 100 mg L^{-1}). Throughout the current study, tomato genotype (*Solanum lycopericum* L.—Alissa F_1 hybrid), which was more susceptible to early blight disease, was used as an experiential plant. Tomato seeds were obtained from the Nunhems Netherlands BV Company, Nunhem, Netherlands and sown in seedling trays in the nursery of a protected cultivation center, Faculty of Agriculture, Kafrelsheikh University, Egypt. Styrofoam trays, with 209 compartments, were filled with a mixture of coco peat: vermiculite (1:1 as v/v). Treatments were arranged in six replicates; each replication was one tray per treatment (209 cells). All trays were planted manually, with 209 seeds per tray, and covered with the above-mentioned media. After sowing, the trays were put in a plastic house with temperatures ranging from 20 to 30 °C. Trays were watered every 2–3 days using a sprinkler system to maintain substrate at field capacity. During the growth of the seedlings, they were fertilized one time in each trial, after over emergence by a soluble compound fertilizer.

At the second true leaf growth stage, seedlings were sprayed with four treatments of paclobutrazol (0, 25, 50, and 100 mg L^{-1} PBZ. After 10 days from PBZ applications, tomato seedlings were sprayed with a spore suspension of *A. solani.* inoculum (10^6 spores mL^{-1}), as 250 mL seedling $tray^{-1}$, whereas the similar amount of distilled water was sprayed on the control seedlings. The most aggressive isolate of *A. solani* (isolate number 1, I_1) was used in our study. Tomato seedlings were sprayed using a manual pump sprayer, with an appropriate flow rate, until runoff.

Disease incidence (DI) of early blight was evaluated three times after inoculation, at 7, 14, and 21 days post inoculation (dpi). At the 7th, 14th, and 21st dpt, disease severity (DS) of the typical symptoms of early blight disease was assessed [47]. For all treatments, six replications were investigated.

4.5. Assessment of Vegetative Growth and Chlorophyll Content

Vegetative growth parameters of tomato seedlings were assessed for all treatments after 10, 20, and 30 days from PBZ application. The growth parameters included seedling height (cm), stem diameter (mm), and fresh and dry weights of seedling and roots (g) per one seedling. Dry mass was measured after drying at 65 °C for 48 h. Total chlorophyll content was recorded in the fully expanded seedling leaf, via the SPAD-501 chlorophyll meter (SPAD-501, Konica Minolta, Tokoyo, Japan), according to [49].

4.6. Enzyme Activities

For enzyme analysis, after 10 days from foliar application of PBZ, samples from fresh leaves tissues were used to measure the total soluble enzymes activity of Catalase (CAT) activity, according to [50], and peroxidases (POD), according to [51].

4.7. Anatomical Measurements

For anatomical investigation, transverse sections were taken from the tomato seedlings ten days after PBZ application. The selected treatments samples used the second internode of the stem from the apex. The chosen samples were killed and fixed for 48 h in (FAA) solution (10 mL formalin, 5 mL glacial acetic acid, and 85 mL ethyl alcohol 70%), then washed in ethyl alcohol 70% twice. The dehydration of the samples was performed by passing it in a series of concentrations of ethyl alcohol, followed by embedding it in paraffin wax of 54 °C melting point. Sectioning, at a thickness of 12 (μm), was done with a rotary microtome (model Leica RM 2125, Leica company, Wetzler, Germany)), followed by staining with safranin and light green. The samples were cleared in xylene and mounted in Canada balsam, prepared for microscopic examination [52]. Five reading from each slide were examined with electric microscope (Leica DM LS, Wetzler, Germany) and digital camera (Leica DC300, Wetzler, Germany), then photographed. The histological manifestations were calculated using Leica IM 1000 image management software. Leica software was calibrated utilizing a 1 cm stage micrometer, scaled at 100 μm increments (604364 Leitz Wetzler, Germany) at 10× magnification. The chosen sections were examined microscopically to detect histological features to follow the changes occurring in the stems of tomato plants, as affected by the application of three different concentrations of PBZ, i.e., 25, 50, and 100 mg L^{-1}. The histological features in the stem sections are the vascular bundle dimensions (thickness and width), xylem vessels diameter, and thickness of xylem and phloem tissuc. One developed mature leaf was randomly chosen after 10 days from PBZ application. Upper epidermis imprints were formed from the middle of each leaflet blade using Cyanoacrylate adhesive (Amir Alpha, www.amazon.eg, accessed on 21 January 2022). A drop of the adhesive was placed on a microscopic slide and quickly pressed on the desired spot of a leaflet, baked by hand. After hardening, the adhesive forms replica of the leaf surface; it was gently peeled off, and the slides were kept for microscopic measurements [53]. Each imprint was examined and photographed with an electric microscope with a digital camera; from each photograph, the number of contained stomata were counted in square microns (μm^2) using the Leica IM 1000 image management software, Wetzler, Germany).

4.8. *Statistical Analyses*

All the obtained results of the experiments were tabularized and statistically analyzed using analysis of variance method, by means of Co-STAT computer software package, IBM, Armonk, NY, USA, and Duncan's multiple range test was used to compare between means of treatments [54].

5. Conclusions

Healthy and vigorous tomato seedlings are necessary for tomato production, where strong seedlings can support plant productivity in the farming field. The extending of seedling life is also considered an important agro-practice in several nurseries, especially under intensive work to avoiding seedlings losses or damages. The current study was carried out to evaluate whether applying different doses of PBZ can enhance the growth and quality of tomato seedlings, as well as suppress the early blight under greenhouse conditions. The results of this research support the idea that PBZ is not only a plant retardant or plant growth regulator but also a stress ameliorant. Applying PBZ enhanced tomato seedlings quality, as the vegetative growth, through the inhibition of stem cell elongation, reduced the length of internodes of the stem, as well as the size and volume of leaves, and increased chlorophyll production. This is the first study on PBZ that examines the associations between applied doses of PBZ on tomato seedling resistance to the early blight pathogen (*A. solani*). Taken together, these findings suggest the role of PBZ in promoting the life of tomato seedlings, with high quality and without losses. The findings of this investigation complement those of earlier studies in the field of tomato seedling production, particularly under different stresses in particular biotic ones. These findings raised important theoretical issues that have a bearing on the environmental dimension of PBZ: are there any ecotoxicological impacts of applied PBZ on the agroecosystem?

Author Contributions: Conceptualization and visualization, T.A.S., H.S.E.-B., N.A.T., and D.I.T.; resources, M.M.M., A.A.R., W.F.S., S.M.E.-G., H.R.E.-R., and Y.A.B.; methodology, M.M.M., A.A.R., W.F.S., S.M.E.-G., H.R.E.-R., and Y.A.B.; software, T.A.S., H.S.E.-B., N.A.T., and D.I.T.; validation, M.M.M., A.A.R., W.F.S., S.M.E.-G., and H.R.E.-R.; investigation, T.A.S., H.S.E.-B., N.A.T., and D.I.T.; data curation, M.M.M., A.A.R., W.F.S., S.M.E.-G., H.R.E.-R., and Y.A.B.; writing—original draft preparation, T.A.S., H.S.E.-B., N.A.T., and D.I.T.; writing—review and editing, T.A.S., H.S.E.-B., N.A.T., D.I.T., M.M.M., A.A.R., W.F.S., S.M.E.-G., H.R.E.-R., and Y.A.B.; funding acquisition, H.S.E.-B., T.A.S., A.A.R., and W.F.S. All authors have read and agreed to the published version of the manuscript.

Funding: Deanship of Scientific Research, Vice Presidency for Graduate Studies, and Scientific Research, King Faisal University, Saudi Arabia: grant no. AN00050.

Institutional Review Board Statement: Not applicable.

Informed Consent Statement: Not applicable.

Data Availability Statement: Not applicable.

Acknowledgments: Authors acknowledge the Deanship of Scientific Research Vice Presidency for Graduate Studies and Scientific Research, at King Faisal University, for the financial support, under annual research project (grant no.AN00050).

Conflicts of Interest: There is no conflict of interest among the authors.

References

1. FAOSTAT. *Tomato Production in 2019, Crops/Regions/World List/Production Quantity (Pick Lists)*; Corporate Statistical Database (FAOSTAT); UN Food and Agriculture Organization: Rome, Italy, 2020.
2. Statista. Tomatoes Production Volume in Egypt from 2010 to 2019 (In Million Metric Tons). Available online: https://www.statista.com/statistics/1066432/egypt-tomatoes-production-volume/ (accessed on 12 December 2021).
3. Hoffmann, A.B.; Poorter, H. Avoiding bias in calculations of relative growth rate. *Ann. Bot.* **2002**, *80*, 37–42. [CrossRef] [PubMed]
4. Kelley, W.T.; Boyhan, G. *Commercial Tomato Production Handbook*; UGA Cooperative Extension Bulletin 1312; University of Georgia: Athens, GA, USA, 2017.
5. Wanderley, C.D.; de Faria, R.T.; Ventura, M.U.; Vendrame, W. The effect of plant growth regulators on height control in potted *Arundina graminifolia* orchids (Growth regulators in *Arundina graminifolia*). *Acta Sci. Agron.* **2014**, *36*, 489–494. [CrossRef]
6. Koukourikou-Petridou, M.A. Paclobutrazol affects the extension growth and the levels of endogenous IAA of almond seedlings. *Plant Growth Regul.* **1996**, *18*, 187–190. [CrossRef]
7. Aphalo, P.; Rikala, R.; Sanchez, R.A. Effect of CCC on the morphology and growth potential of containerised silver birch seedlings. *New For.* **1997**, *14*, 167–177. [CrossRef]
8. Ellis, G.D.; Knowles, L.O.; Knowles, N.R. Increasing the Production Efficiency of Potato with Plant Growth Retardants. *Am. J. Potato Res.* **2020**, *97*, 88–101. [CrossRef]
9. Detpitthayanan, S.; Romyanon, K.; Songnuan, W.; Metam, M.; Pichakum, A. Paclobutrazol Application Improves Grain 2AP Content of Thai Jasmine Rice KDML105 under Low-Salinity Conditions. *J. Crop Sci. Biotechnol.* **2019**, *22*, 275–282. [CrossRef]
10. Abdalla, N.; Taha, N.; Bayoumi, Y.; El-Ramady, H.; Shalaby, T.A. Paclobutrazol applications in agriculture, plant tissue cultures and its potential as stress ameliorant, A Mini-Review. *Environ. Biodivers. Soil Secur.* **2021**, *5*, 245–257. [CrossRef]
11. Yang, T.; Davies, P.J.; Reid, J.B. Genetic dissection of the relative roles of auxin and gibberellin in the regulation of stem elongation in intact light-grown peas. *Plant Physiol.* **1996**, *110*, 1029–1034. [CrossRef]
12. Tesfahun, W. A review on, Response of crops to paclobutrazol application. *Cogent Food Agric.* **2018**, *4*, 1525169. [CrossRef]
13. Fan, Z.X.; Li, S.C.; Sun, H.L. Paclobutrazol Modulates Physiological and Hormonal Changes in *Amorpha fruticosa* under Drought Stress. *Russ. J. Plant Physiol.* **2020**, *67*, 122–130. [CrossRef]
14. Iqbal, S.; Parveen, N.; Bahadur, S.; Ahmad, T.; Shuaib, M.; Nizamani, M.M.; Urooj, Z.; Rubab, S. Paclobutrazol mediated changes in growth and physio-biochemical traits of okra (*Abelmoschus esculentus* L.) grown under drought stress. *Gene Rep.* **2020**, *21*, 100908. [CrossRef]
15. Luo, L.-N.; Cui-Ling, L.; Wu, F.; Li, S.-P.; Han, S.-Q.; Li, M.-F. Desuckering effect of KH_2PO_4 mixed with paclobutrazol and its influence on banana (*Musa paradisiaca* AA) mother plant growth. *Sci. Hortic.* **2018**, *240*, 484–491. [CrossRef]
16. Ajmi, A.; Larbi, A.; Morales, M.; Fenollosa, E.; Chaari, A.; Munné-Bosch, S. Foliar paclobutrazol application suppresses olive tree growth while promoting fruit set. *J. Plant Growth Regul.* **2020**, *39*, 1638–1646. [CrossRef]
17. Lucho, S.R.; do Amaral, M.N.; Milech, C.; Bianchi, V.J.; Almagro, L.; Ferrer, M.A.; Calderón, A.A.; Braga, E.J.B. Gibberellin reverses the negative effect of paclobutrazol but not of chlorocholine chloride on the expression of SGs/GAs biosynthesis-related genes and increases the levels of relevant metabolites in *Stevia rebaudiana*. *Plant Cell Tissue Organ Cult.* **2021**, *146*, 171–184. [CrossRef]
18. Wang, W.D.; Wu, C.Y.; Lonameo, B.K. Toxic effects of Paclobutrazol on developing organs at different exposure times in zebrafish. *Toxics* **2019**, *7*, 62. [CrossRef] [PubMed]

19. Kumar, G.; Lal, S.; Bhatt, P.; Ram, R.A.; Bhattacherjee, A.K.; Dikshit, A.; Rajan, S. Mechanisms and kinetics for the degradation of paclobutrazol and biocontrol action of a novel *Pseudomonas putida* strain T7. *Pestic. Biochem. Physiol.* **2021**, *175*, 104846. [CrossRef] [PubMed]
20. Terna, T.P.; Akomolafe, G.F.; Ubhenin, A.; Abok, J. Disease responses of different tomato (*Solanum lycopersicum* L.) cultivars inoculated with culture filtrates of selected fungal pathogens. *Vegetos* **2020**, *33*, 166–171. [CrossRef]
21. Adhikari, P.; Oh, Y.; Panthee, D.R. Current status of early blight resistance in Tomato, An Update. *Int. J. Mol. Sci.* **2017**, *18*, 2019. [CrossRef]
22. El-Nagar, A.; Elzaawely, A.; Taha, N.; Nehela, Y. The antifungal activity of gallic acid and its derivatives against *Alternaria solani*, the causal agent of tomato early blight. *Agronomy* **2020**, *10*, 1402. [CrossRef]
23. Panno, S.; Davino, S.; Caruso, A.G.; Bertacca, S.; Crnogorac, A.; Mandic, A.; Noris, E.; Matic, S. A Review of the most common and economically important diseases that undermine the cultivation of tomato crop in the Mediterranean basin. *Agronomy* **2021**, *11*, 2188. [CrossRef]
24. El-Ganainy, S.M.; El-Abeid, S.E.; Ahmed, Y.; Iqbal, Z. Morphological and molecular characterization of large-spored Alternaria species associated with potato and tomato early blight in Egypt. *Int. J. Agric. Biol.* **2021**, *25*, 1101–1110. [CrossRef]
25. Henning, R.G.; Alexander, L.J. Evidence of existence of physiologic races of *Alternaia solani*. *Plant Dis. Rep.* **1959**, *43*, 298–308.
26. Berova, M.; Zlatev, Z. Physiological response and yield of paclobutrazol treated tomato plants (*Lycopersicon esculentum* Mill.). *Plant Growth Regul.* **2000**, *30*, 117–123. [CrossRef]
27. Misu, H.; Mori, M.; Okumura, S.; Kanazawa, S.-I.; Ikeguchi, N.; Nakai, R. High-quality tomato seedling production system using artificial light. *SEI Tech. Rev.* **2018**, *86*, 119–124.
28. Seleguini, A.; Júnior, M.J.A.F.; Benett, K.S.S.; Lemos, O.L.; Seno, S. Strategies for tomato seedlings production using paclobutrazol. *Semin. Agric. Sci. Londrina* **2013**, *34*, 539–548. (In Portuguese) [CrossRef]
29. Bene, K.S.S.; De Araújo Faria, M.J.; Bene, C.G.S.; Seleguini, A.; Lemos, O.L. Use of paclobutrazol in the production of tomato seedlings. *Comun. Sci.* **2014**, *5*, 164–169.
30. Uçan, U.; Uğur, A. Acceleration of growth in tomato seedlings grown with growth retardant. *Turk. J. Agric. For.* **2021**, *45*, 669–679. [CrossRef]
31. Tekalign, T.; Hammes, P. Growth and biomass production in potato grown in the hot tropics as influenced by paclobutrazol. *Plant Growth Regul.* **2005**, *45*, 37–46. [CrossRef]
32. Upreti, K.K.; Reddy, Y.T.N.; Shivu Prasad, S.R.; Bindu, G.V.; Jayaram, H.L.; Rajan, S. Hormonal changes in response to paclobutrazol induced early flowering in mango cv Totapuri. *Sci. Hortic.* **2013**, *150*, 414–418. [CrossRef]
33. Teto, A.A.; Laubscher, C.P.; Ndakidemi, P.A.; Matimati, I. Paclobutrazol retards vegetative growth in hydroponically-cultured *Leonotis leonurus* (L.) R. Br. Lamiaceae for a multipurpose flowering potted plant. *S. Afr. J. Bot.* **2016**, *106*, 67–70. [CrossRef]
34. Mohammadi, M.H.S.; Etemadi, N.; Arab, M.M.; Aalifar, M.; Arab, M.; Pessarakli, M. Molecular and physiological responses of Iranian Perennial ryegrass as affected by Trinexapac ethyl, Paclobutrazol and Abscisic acid under drought stress. *Plant Physiol. Biochem.* **2017**, *111*, 129–143. [CrossRef] [PubMed]
35. Forghani, A.H.; Almodares, A.; Ehsanpou, R.A.A. The Role of Gibberellic Acid and Paclobutrazol on Oxidative Stress Responses Induced by In Vitro Salt Stress in Sweet Sorghum. *Russ. J. Plant Physiol.* **2020**, *67*, 555–563. [CrossRef]
36. Mehmood, M.Z.; Qadir, G.; Afzal, O.; Ud Din, A.M.; Raza, M.A.; Khan, I.; Hassan, M.J.; Awan, S.A.; Ahmad, S.; Ansar, M.; et al. Paclobutrazol Improves Sesame Yield by Increasing Dry Matter Accumulation and Reducing Seed Shattering Under Rainfed Conditions. *Int. J. Plant Prod.* **2021**, *15*, 337–349. [CrossRef]
37. Kamran, M.; Ahmad, S.; Ahmad, I.; Hussain, I.; Meng, X.; Zhang, X.; Javed, T.; Ullah, M.; Ding, R.; Xu, P.; et al. Paclobutrazol application favors yield improvement of maize under semiarid regions by delaying leaf senescence and regulating photosynthetic capacity and antioxidant system during grain-filling stage. *Agronomy* **2020**, *10*, 187. [CrossRef]
38. Roseli, A.N.M.; Ahmad, M.F. In Vitro Evaluation of Paclobutrazol against selected pathogenic soil fungi. *J. Trop. Plant Physiol.* **2019**, *11*, 13–21.
39. Desta, B.; Amare, G. Paclobutrazol as a plant growth regulator. *Chem. Biol. Technol. Agric.* **2021**, *8*, 1. [CrossRef]
40. Waqas, M.; Yaning, C.; Iqbal, H.; Shareef, M.; Rehman, H.; Yang, Y. Paclobutrazol improves salt tolerance in quinoa, Beyond the stomatal and biochemical interventions. *J. Agron. Crop Sci.* **2017**, *203*, 315–322. [CrossRef]
41. Jaleel, C.A.; Gopi, R.; Manivannan, P.; Panneerselvam, R. Responses of antioxidant defense system of *Catharanthus roseus* (L.) G. Don. to paclobutrazol treatment under salinity. *Acta Physiol. Plant.* **2007**, *29*, 205–209. [CrossRef]
42. El-Beltagi, H.S.; Ahmad, I.; Basit, A.; Shehata, W.F.; Hassan, U.; Shah, S.T.; Haleema, B.; Jalal, A.; Amin, R.; Khalid, M.A.; et al. Ascorbic acid enhances growth and yield of sweet peppers (*Capsicum annum*) by mitigating salinity stress. *Gesunde Pflanz.* **2022**, *74*, 1–11. [CrossRef]
43. El-Beltagi, H.S.; Ahmad, I.; Basit, A.; Abd El-Lateef, H.M.; Yasir, M.; Shah, S.T.; Ullah, I.; Mohamed, M.E.M.; Ali, I.; Ali, F.; et al. Effect of azospirillum and azotobacter species on the performance of cherry tomato under different salinity levels. *Gesunde Pflanz.* **2022**, *74*, 1–13. [CrossRef]
44. Tsegaw, T.; Hammes, S.; Robbertse, J. Paclobutrazol-induced leaf, stem and root anatomical modification in potato. *Hortscience* **2005**, *40*, 1343–1346. [CrossRef]
45. Benhamou, N.; Bélanger, R.R. Benzothiadiazole-mediated induced resistance to *Fusarium oxysporum* f. sp. *radicis-lycopersici* in tomato. *Plant Physiol.* **1998**, *118*, 1203–1212. [CrossRef] [PubMed]

46. Siciliano, I.; Gilardi, G.; Ortu, G.; Gisi, U.; Gullino, M.L.; Garibaldi, A. Identification and characterization of Alternaria species causing leaf spot on cabbage, cauliflower, wild and cultivated rocket by using molecular and morphological features and mycotoxin production. *Eur. J. Plant Pathol.* **2017**, *149*, 401–413. [CrossRef]
47. Pandey, K.K.; Pandey, P.K.; Kalloo, G.; Banerjee, M.K. Resistance to early blight of tomato with respect to various parameters of disease epidemics. *J. Gen. Plant Pathol.* **2003**, *69*, 364–371. [CrossRef]
48. Grover, R.; Moore, J.D. Toximetric studies of fungicides against the brown root organisms, *Sclerotinia fructicola* and *S. laxa*. *Phytopathology* **1962**, *52*, 876–880.
49. Yadawa, U.L. A rapid and nondestructive method to determine chlorophyll in intact leaves. *HortScience* **1986**, *21*, 1449–1450.
50. Aebi, H. Catalase in vitro. *Methods Enzymol.* **1984**, *105*, 121–126.
51. Rathmell, W.G.; Sequeira, L. Soluble Peroxidase in Fluid from the Intercellular Spaces of Tobacco Leaves. *Plant Physiol.* **1974**, *53*, 317–318. [CrossRef]
52. Ruzin, S.E. *Plant Microtechnique and Microscopy*, 1st ed.; Oxford University Press: New York, NY, USA, 1999.
53. El-Yamany, W.Z. Genetic, Physiological and Anatomical Response of Some Wheat Crosses to Water Stress. Ph.D. Thesis, Kafrelsheikh University, Kafr El Sheikh, Egypt, 2009.
54. Snedecor, G.W.; Cochran, W.G. *Statistical Methods*, 8th ed.; Iowa State University Press: Ames, IA, USA, 1989.

plants

MDPI

Article

Identification and Pathogenicity of *Paramyrothecium* Species Associated with Leaf Spot Disease in Northern Thailand

Patchareeya Withee [1], Sukanya Haituk [1], Chanokned Senwanna [1], Anuruddha Karunarathna [1], Nisachon Tamakaew [1], Parichad Pakdeeniti [1], Nakarin Suwannarach [2,3], Jaturong Kumla [2,3], Piyawan Suttiprapan [1,4], Paul W. J. Taylor [5], Milan C. Samarakoon [1,*] and Ratchadawan Cheewangkoon [1,2,4,*]

1 Department of Entomology and Plant Pathology, Faculty of Agriculture, Chiang Mai University, Chiang Mai 50200, Thailand; patchareeya_withee@cmu.ac.th (P.W.); sukanya_hai@cmu.ac.th (S.H.); chanokned.swn@gmail.com (C.S.); anumandrack@yahoo.com (A.K.); nisachon_t@cmu.ac.th (N.T.); parichad_pakdee@cmu.ac.th (P.P.); piyawan.s@cmu.ac.th (P.S.)
2 Research Center of Microbial Diversity and Sustainable Utilization, Chiang Mai University, Chiang Mai 50200, Thailand; suwan.462@gmail.com (N.S.); jaturong_yai@hotmail.com (J.K.)
3 Department of Biology, Faculty of Science, Chiang Mai University, Chiang Mai 50200, Thailand
4 Innovative Agriculture Research Centre, Faculty of Agriculture, Chiang Mai University, Chiang Mai 50200, Thailand
5 Faculty of Veterinary and Agricultural Sciences, The University of Melbourne, Parkville, VIC 3010, Australia; paulwjt@unimelb.edu.au
* Correspondence: milanchameerasamar.s@cmu.ac.th (M.C.S.); ratchadawan.c@cmu.ac.th (R.C.)

Abstract: Species of *Paramyrothecium* that are reported as plant pathogens and cause leaf spot or leaf blight have been reported on many commercial crops worldwide. In 2019, during a survey of fungi causing leaf spots on plants in Chiang Mai and Mae Hong Son provinces, northern Thailand, 16 isolates from 14 host species across nine plant families were collected. A new species *Paramyrothecium vignicola* sp. nov. was identified based on morphology and concatenated (ITS, *cmdA*, *rpb2*, and *tub2*) phylogeny. Further, *P. breviseta* and *P. foliicola* represented novel geographic records to Thailand, while *P. eichhorniae* represented a novel host record (*Psophocarpus* sp., *Centrosema* sp., *Aristolochia* sp.). These species were confirmed to be the causal agents of the leaf spot disease through pathogenicity assay. Furthermore, cross pathogenicity tests on *Coffea arabica* L., *Commelina benghalensis* L., *Glycine max* (L.) Merr., and *Dieffenbachia seguine* (Jacq.) Schott revealed multiple host ranges for these pathogens. Further research is required into the host–pathogen relationship of *Paramyrothecium* species that cause leaf spot and their management. Biotic and abiotic stresses caused by climate change may affect plant health and disease susceptibility. Hence, proper identification and monitoring of fungal communities in the environment are important to understand emerging diseases and for implementation of disease management strategies.

Keywords: climate change; diversity; food security; multi-gene phylogeny; new species; plant pathology; taxonomy

Citation: Withee, P.; Haituk, S.; Senwanna, C.; Karunarathna, A.; Tamakaew, N.; Pakdeeniti, P.; Suwannarach, N.; Kumla, J.; Suttiprapan, P.; Taylor, P.W.J.; et al. Identification and Pathogenicity of *Paramyrothecium* Species Associated with Leaf Spot Disease in Northern Thailand. *Plants* **2022**, *11*, 1445. https://doi.org/10.3390/plants11111445

Academic Editor: Alessandro Vitale

Received: 17 May 2022
Accepted: 26 May 2022
Published: 29 May 2022

1. Introduction

Plant diseases have a high impact on food security [1] and fungi play a major role in plant diseases [2]. Foliar fungal pathogens severely affect the yield and health of commercial crops [3]. Leaf spots are an early indicator of foliar diseases and may initially occur on the adaxial leaf surfaces and then appear on the abaxial leaf surface.

Paramyrothecium species have been frequently identified to cause leaf spot and blight disease on a wide range of vegetables, ornamental plants, and economic crops [4–7]. Disease symptoms caused by *Paramyrothecium* may also include stem and crown canker and fruit rot [8–10]. Lombard et al. [4] designated an epitype for the generic type *Paramyrothecium roridum* (≡*M. roridum*). *Paramyrothecium* species are distinguished from related *Myrothecium* sensu stricto and other myrothecium-like genera by the presence of 1–3 septate, thin-walled setae surrounding the sporodochia. Currently, there are 19 species listed in Index Fungorum (http://www.indexfungorum.org/; accessed on 14 April 2022).

Paramyrothecium roridum and *P. foliicola* are well-known pathogens that cause leaf spot or leaf blight and have been reported on many commercial crops and a wide range of hosts, such as soybean, strawberry, and muskmelon [5,8,9]. Rennberger and Keinath [11] isolated *P. foliicola* and *P. humicola* from watermelon and two other cucurbits and confirmed their pathogenicity on watermelons, tomatoes, and southern peas. Aumentado and Balendres [12] reported *P. foliicola* causing crater rot in eggplant and 45 plant species from 21 plant families and were tested for the pathogenicity on detached fruit or leaf assays. Furthermore, *P. foliicola* is pathogenic to cucumber seedlings and watermelon, causing stem canker [13]. Due to the lignicolous nature of the *Paramyrothecium*, they are being used as bio-pesticides for the control of weeds and insects [14,15]. Interestingly, several important secondary metabolites or toxins found in *Paramyrothecium* include trichothecenes macrolides such as roridin, verrucarin, and mytoxin B, which are important for some medicinal and biotechnological applications [16–18].

In Thailand, only *P. eichhorniae* has been reported and this was identified as the cause of the leaf blight disease of water hyacinth [19]. The diversity of *Paramyrothecium* species in Thailand is unknown. As a result, surveys and additional research on the distribution of *Paramyrothecium* in Thailand is required. The objective of this study was to identify and describe *Paramyrothecium* spp. from northern Thailand and assess their pathogenicity across a broad range of potential host plant species.

2. Results

2.1. Symptoms

Leaf spots varying in size and shape, depending on the host, were most visible on the upper surface. The leaf spots consisted of small brown spots or necrotic lesions with a dark border, while in older lesions, small sporodochia were visible (Figure 1f,n,o). Necrotic lesions appeared dark gray or black on *Centrosema* sp., *Coccinia grandis, Oroxylum indicum, Solanum virginianum, Tectona grandis, Vigna mungo, Vigna* sp., and *V. unguiculata* (Figure 1a,d,e,g,h–k,m), and surrounded by a prominent yellow halo on *Lablab purpureus, Psophocarpus* sp., and *Spilanthes* sp. (Figure 1b,c,l). Lesions on *Aristolochia* sp., *Coffea arabica*, and *Commelina benghalensis* consisted of light to dark brown concentric rings with a target-like appearance, and small sporodochia that appeared on lower and upper surfaces (Figure 1f,n,o).

Figure 1. Symptoms on different hosts caused by *Paramyrothecium* (**left**) and sporodochia on the host surface (**right**); (**a**) *Solanum virginianum*; (**b**) *Lablab purpureus*; (**c**) *Psophocarpus* sp.; (**d**,**i**) *Vigna* sp.; (**e**) *Coccinia grandis*; (**f**) *Commelina benghalensis*; (**g**) *Tectona grandis*; (**h**) *Vigna mungo*; (**j**) *Vigna unguiculata*; (**k**) *Oroxylum indicum*; (**l**) *Spilanthes* sp.; (**m**) *Centrosema* sp.; (**n**) *Aristolochia* sp.; (**o**) *Coffea arabica*. Scale bars: (**c**,**g**,**j**) = 1 mm; (**b**,**d**–**f**,**l**) = 2 mm; (**a**,**i**,**k**) = 4 mm; (**h**,**m**) = 5 mm; (**o**) = 6 mm; (**n**) = 1 cm.

2.2. Culture Morphology

Diverse culture characters were observed on PDA at room temperature (25–30 °C) (Figure 2). Eleven isolates of *Paramyrothecium* sp. (SDBR-CMU374, SDBR-CMU375, SDBR-CMU376, SDBR-CMU377, SDBR-CMU378, SDBR-CMU379, SDBR-CMU380, SDBR-CMU382, SDBR-CMU387, SDBR-CMU388, and SDBR-CMU389) (Figure 2a–j,l) formed whitish colonies with entire to slightly undulated margins, radial or in concentric rings with sporodochia, covered with slimy olivaceous green to black conidial masses, while the other four isolates (SDBR-CMU383, SDBR-CMU384, SDBR-CMU385, and SDBR-CMU386) (Figure 2m–p) formed abundant white aerial mycelium with sporodochia forming on the stroma and surface of the medium, covered by slimy olivaceous green to black conidial masses. Isolate SDBR-CMU381 (Figure 2k) produced exudates with brown pigment into the medium.

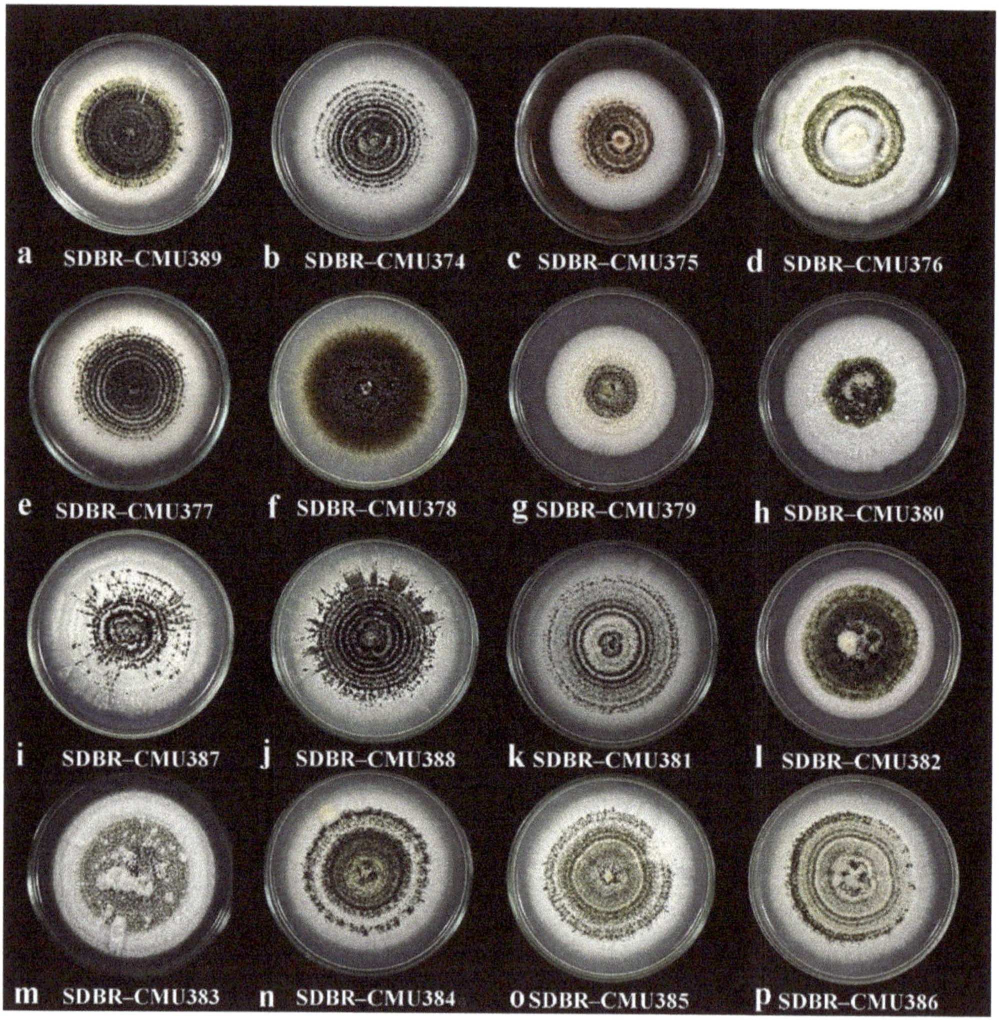

Figure 2. Colonies of *Paramyrothecium* species on PDA after 15 days at 25–30 °C.

2.3. Phylogenetic Analysis

The phylogenetic tree topologies of the ML and BI analyses for concatenated ITS, *cmdA*, *rpb2*, and *tub2* were similar. Hence, a phylogenetic tree from ML analyses is used to represent the results of both ML and BI analyses. The dataset comprised 53 taxa with 1760 characters (ITS: 1–542; *cmdA*: 543–824; *rpb2*: 825–1548; *tub2*: 1549–1760), including gaps. The GTR+G+I model was the best-fit model for all loci. The best scoring likelihood tree was selected on the basis of the ML analysis, with a final ML optimization likelihood value of −8176.4871, as shown in Figure 3. Sixteen new isolates were clustered into four distinct clades in *Paramyrothecium* (see the notes).

2.4. Taxonomy

Isolates from symptomatic living leaves of different hosts were recognized under *Paramyrothecium* based on taxonomy (Table 1) and multi-gene phylogeny (Figure 3). The morphologies of the *Paramyrothecium* species are described herein.

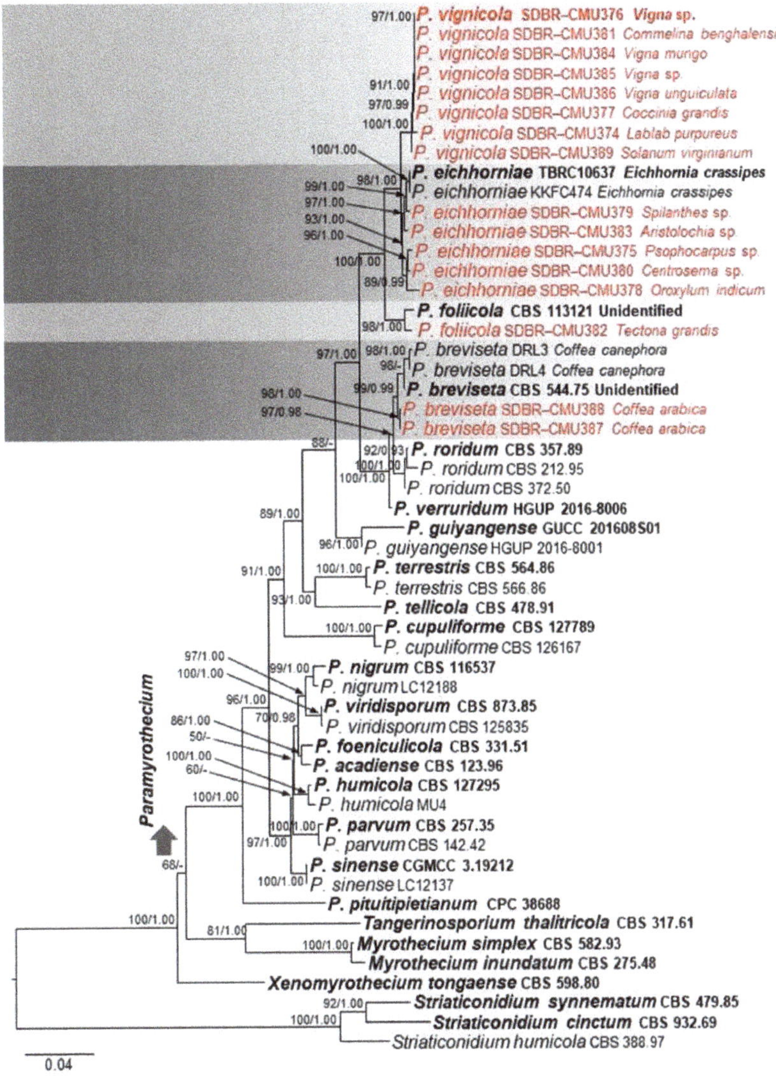

Figure 3. Phylogram generated from maximum likelihood analysis based on combined ITS, *cmdA*, *rpb2*, and *tub2* sequenced data. Fifty-three strains are included in the combined sequence analyses, which comprise 1760 characters with gaps. Single gene analyses were also performed, and topology and clade stability were compared from combined gene analyses. *Striaticonidium cinctum* (CBS 932.69), *S. humicola* (CBS 388.97), and *S. synnematum* (CBS 479.85) are used as the outgroup taxa. The best scoring RAxML tree with a final likelihood value of −8176.4871 is presented. The matrix had 524 distinct alignment patterns. Estimated base frequencies were as follows; A = 0.2266, C = 0.2915, G = 0.2681, T = 0.2138; substitution rates AC = 1.1215, AG = 5.1556, AT = 1.0792, CG = 1.2292, CT = 11.1203, GT = 1.0000; gamma distribution shape parameter α = 0.3855. The bootstrap support ($\geq$50%) of ML and the posterior probability values ($\geq$0.9) of BI analyses are indicated above or below the respective branches. The fungal isolates from this study are indicated in red. The type species are indicated in bold.

Table 1. Synopsis of *Paramyrothecium* type species.

Species	Host	Location	Conidiophores (μm)	Conidiogenous Cells (μm)	Conidia (μm)	Setae (μm)	References
Paramyrothecium acadiense	*Tussilago farfara*	Canada	9–14 × 2–2.5	–	0–1-septate, 5.5–16.5 × 1.5–2.5	–	[20]
P. breviseta	unknown	India	6–9 × 2–4	6–11 × 1–2	aseptate, 4–5 × 1–2	1–3-septate, 25–40 × 2–3	[4]
P. cupuliforme	Soil	Namibia	15–25 × 2–4	4–11 × 1–3	aseptate, 6–8 × 1–2	1–3-septate, 45–90 × 2–3	[4]
P. eichhorniae	*Eichhornia crassipes*	Thailand	15–40 × 2–3	(8–)11–17(–20) × 2–3	aseptate,5– 6.5 × 1.5–2.5	1–3-septate, 40–120 × 2–3	[19]
P. foeniculicola	*Foeniculum vulgare*	Netherlands	7–17 × 2–3	6–16 × 1–2	aseptate, 5–7 × 1–2	–	[4]
P. foliicola	Decaying leaf	Brazil	15–25 × 2–3	8–14 × 1–2	aseptate, 5–6 × 1–2	1–3-septate, 60–100 × 2–3	[4]
P. guiyangense	Soil	China	10−60 × 1−3	8−18 × 1.6−2.7	aseptate, 6.6−9.0 × 2−3	1−3-septate, 60−120 × 1−3	[21]
P. humicola	Soil	USA	12–22 × 2–3	8–13 × 1–3	aseptate, 6–7 × 1–2	1–2-septate, 55–65 × 2–3	[4]
P. lathyri	*Lathyrus tuberosus*	Russia	5–10 × 2–3.5	5–10 × 2–3	aseptate, (8–)9 (–10) × 2(–2.5)	3–10-septate, up to 300 × 3–4	[22]
P. nigrum	Soil	Spain	25–45 × 2–4	8–13 × 1–2	aseptate, 5–6 × 1–2	1–3-septate, 60–100 × 2–3	[4]
P. parvum	*Viola* sp.	UK	12–26 × 2–4	7–23 × 1–2	aseptate, 4–5 × 1–2	–	[4]
P. pituitipietianum	*Grielum humifusum*	South Africa	20–35 × 3–4	20–35 × 3–4	aseptate, (7–)9–10(–12) × (2–)2.5	7–10-septate, 100–300 × 4–5	[23]
P. roridum	*Gardenia* sp.	Italy	15–40 × 2–4	7–33 × 2–3	aseptate, (5–)6.5–7.5(–8) × 2	1–3(–4)-septate, 60–100 × 2–6	[4]
P. salvadorae	*Salvadora persica*	Namibia	20–40 × 3–4	8–15 × 2–2.5	aseptate, (8–)10–12 (–13) × 2–2.5	5–10- septate, 100–200 × 2.5–3	[24]
P. sinense	Rhizosphere soils of *Poa* sp.	China	20–30 × 2–3	7–16 × 1–3	aseptate, 6–7 × 2–3	1–3-septate, 45–90 × 1–3	[25]
P. tellicola	Soil	Turkey	15–30 × 2–4	7–17 × 1–3	aseptate, (7–)7.5–8.5 (–9) × 1–3	1–3-septate, 45–80 × 2–3 μm	[4]
P. terrestris	Soil	Turkey	15–30 × 2–3	7–12 × 2–3	aseptate, (7–)7.5–8.5 (–10) × 1–3	1–3-septate, 35–70 × 2–3	[4]
P. verruridum	Soil	China	20−40 × 1.5−2.5	12−20 × 1.7−2.7	aseptate, 6.8−7.8 × 2−2.7	1−3-septate, 40−120 × 2−3	[21]
P. vignicola	*Vigna* sp.	Thailand	40–60 × 2–3	11–16 × 1–3	aseptate, 5–7 × 1–3 μm	3–8-septate, 80–155 × 2–3	This study
P. viridisporum	Soil	Turkey	15–35 × 2–3	6–12 × 3–5	aseptate, 3–5 × 2 μm	1–3-septate, 60–140 × 2–3	[4]

Paramyrothecium vignicola Withee & Cheew., sp. nov. (Figure 4).

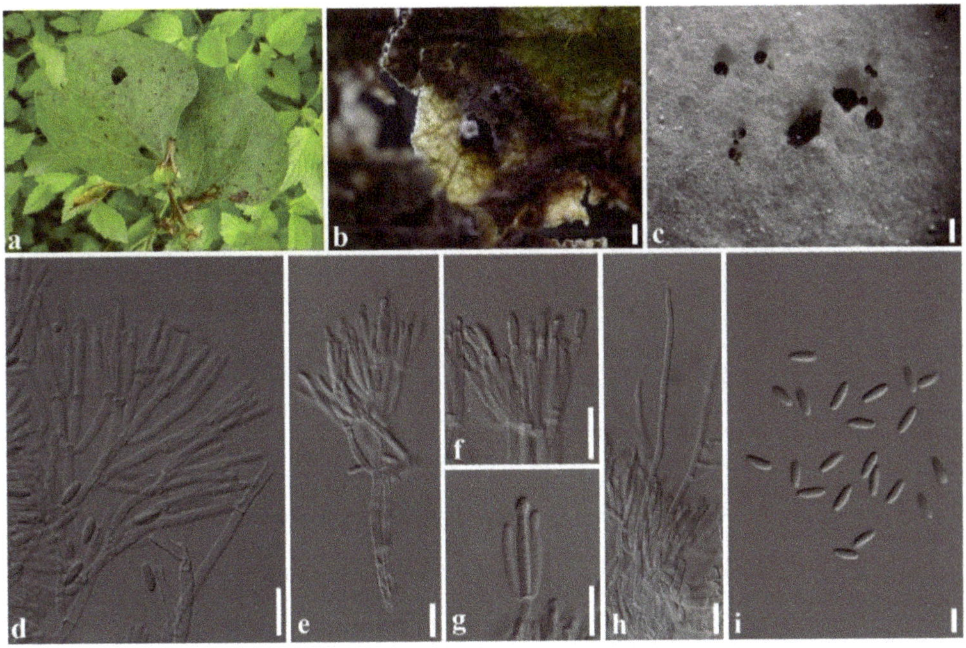

Figure 4. *Paramyrothecium vignicola* (CRC4-H, holotype); (**a**) leaf spot of *Vigna* sp.; (**b**) sporodochia on leaf; (**c**) sporodochial conidiomata on PDA; (**d**,**e**) conidiophores and conidiogenous cells; (**f**,**g**) conidiogenous cells; (**h**) setae; (**i**) conidia. Scale bars: (**b**,**c**) = 1 mm; (**d**–**h**) = 10 μm; (**i**) = 5 μm.

Mycobank: MB 843763.

Etymology: Name reflects the host genus Vigna, from which the species was collected.

Holotype: SDBR–CMU376.

Description: Sexual morph: unknown. Asexual morph: *Conidiomata* sporodochial, stromatic, superficial, cupulate, scattered or gregarious, oval or irregular in outline, (60–)90–300(–385) μm diam, (70–)140–180(–200) μm deep, with a white to creamy setose fringe surrounding an olivaceous green agglutinated slimy mass of conidia. *Stroma* poorly developed, hyaline. *Setae* arising from the stroma thin-walled, hyaline, 3–8-septate, straight becoming sinuous above the apical septum, 80–155 μm long, 2–3 μm wide, tapering to an acutely rounded apex. *Conidiophores* arising from the basal stroma, consisting of a stipe and a penicillately branched conidiogenous apparatus; stipes unbranched, hyaline sometimes covered by a green mucoid layer, septate, smooth, 40–60 × 2–3 μm; primary branches aseptate, unbranched, smooth, 10–26 × 2–3 μm ($\overline{x}$ = 18 × 3 μm, n = 20); secondary branches aseptate, unbranched, smooth, 10–17 × 2–3 μm ($\overline{x}$ = 13 × 3 μm, n = 20); terminating in a whorl of 3–6 conidiogenous cells; conidiogenous cells phialidic, cylindrical to subcylindrical, hyaline, smooth, straight to slightly curved, 11–16 × 1–3 μm ($\overline{x}$ = 13 × 2 μm, n = 20), with conspicuous collarettes and periclinal thickenings. *Conidia* aseptate, hyaline, smooth, cylindrical to ellipsoidal, 5–7 × 1–3 μm ($\overline{x}$ = 6 × 2 μm, n = 20), rounded at both ends.

Culture characteristics: Colonies on PDA, dense, circular, flattened, slightly raised, floccose, white aerial mycelium, radiating with concentric ring of sporodochia forming, covered by slimy olivaceous green to black conidial masses.

Material examined: Thailand, Mae Hong Son Province, on living leaf of *Vigna* sp. (*Fabaceae*), 11 September 2019, N. Tamakaew, CRC4-H (holotype), ex-type living culture SDBR-CMU376; *ibid.*, on living leaf of *Solanum virginianum* (*Solanaceae*), 11 September 2019,

N. Tamakaew, CRC1-H, living culture SDBR-CMU389; *ibid.*, on living leaf of *Lablab purpureus* (*Fabaceae*), 11 September 2019, N. Tamakaew, CRC2-H, living culture SDBR-CMU374; *ibid.*, on living leaf of *Coccinia grandis* (*Cucurbitaceae*), CRC6-H, living culture SDBR-CMU377; Chiang Mai province, on living leaf of *Commelina benghalensis* (*Commelinaceae*), 20 November 2019, P. Withee, CRC14-H, living culture SDBR-CMU381; *ibid.*, on living leaf of *Vigna mungo* (*Fabaceae*), 5 December 2019, N. Tamakaew, CRC144-H, living culture SDBR-CMU384; *ibid.*, on living leaf of *Vigna* sp. (*Fabaceae*), CRC145-H, living culture SDBR-CMU385; *ibid.*, on living leaf of *Vigna unguiculata* (*Fabaceae*), 10 February 2020, P. Withee, CRC146-H, living culture SDBR–CMU386.

Notes: Based on ITS, *cmdA*, *rpb2* and *tub2* phylogeny (Figure 3) and *cmdA* and *tub2* (data not shown), *Paramyrothecium foliicola* formed two distinct clades. The clade with *Paramyrothecium foliicola* type (CBS 113121) was treated as the *Paramyrothecium* sensu stricto. Eight of the new strains clustered with eight previously described *Paramyrothecium* strains (as *P. foliicola*) and formed a well-supported clade (100% BS/1.00 PP) (*Paramyrothecium* sensu lato) closely related to *P. eichhorniae* and *P. foliicola* (Figure 3). Based on morphology and phylogeny, we introduce a new species to accommodate taxa in *P. foliicola* sensu lato. *Paramyrothecium vignicola* differs from *P. eichhorniae* and *P. foliicola* with longer setae (up to 155 μm vs. up to 120 μm and up to 100 μm). The conidia of *P. vignicola* (5–7 × 1–3 μm) are slightly larger than those of *P. eichhorniae* (5–6.5 × 1.5–2.5 μm) [20] and *P. foliicola* (5–6 × 1–2 μm) [4]. *Paramyrothecium vignicola* differs from other *Paramyrothecium* species by its 3–8-septate, thin-walled setae surrounding the sporodochia. In BLAST searches of NCBI GenBank, the closest matches of the sequences are *Paramyrothecium*: *P. foliicola* (CBS 11321) with 98.98% similarity in ITS sequence, 93.89% similarity in *cmdA*. *P. vignicola*, 96.32% in *tub2* with *P. foliicola* (CBS 11321). Based on phylogenetic evidence and morphological differences, *P. vignicola* is a new species.

Paramyrothecium breviseta L. Lombard & Crous, in Lombard et al., Persoonia 36: 207 (2016) (Figure 5).

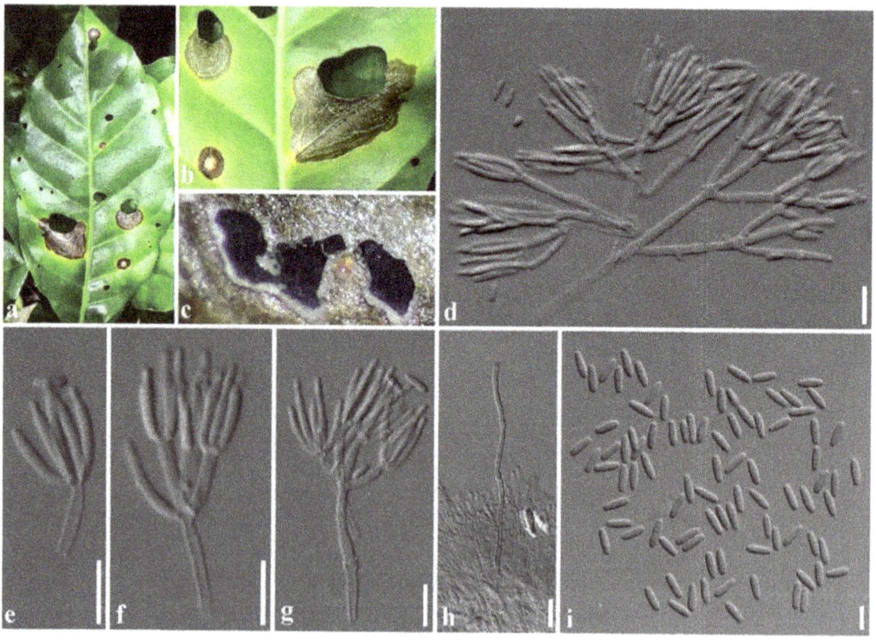

Figure 5. *Paramyrothecium breviseta* (CRC13-H); (**a**,**b**) leaf spot of *Coffea arabica*; (**c**) sporodochia on leaf; (**d**) conidiophores and conidiogenous cells; (**e**–**g**) conidiogenous cells; (**h**) setae; (**i**) conidia. Scale bars: (**d**) = 20 μm; (**e**–**h**) = 10 μm; (**i**) = 5 μm.

Description: Sexual morph: unknown. Asexual morph: *Conidiomata* sporodochial, stromatic, cupulate, superficial, scattered or rarely gregarious, oval or irregular in outline, 135–790 µm diam, 9–15 µm deep, with a white setose fringe surrounding an olivaceous green to black agglutinated slimy mass of conidia. *Setae* arising from the stroma thin-walled, hyaline, 1–5-septate, straight to flexuous, 25–120 µm long, 2–3 µm wide, tapering to an acutely rounded apex. *Conidiophores* arising from the basal stroma, consisting of a stipe and a penicillately branched conidiogenous apparatus; stipes unbranched, hyaline, septate, smooth, 6–9 × 2–4 µm; primary branches aseptate, unbranched, smooth, 12–24 × 3–4 µm ($\bar{x}$ = 18 × 3 µm, n = 20); secondary branches aseptate, unbranched, smooth, 10–17 × 2–4 µm ($\bar{x}$ = 12 × 3 µm, n = 20); terminating in a whorl of 3–6 conidiogenous cells; conidiogenous cells phialidic, cylindrical to subcylindrical, hyaline, smooth, straight to slightly curved, 6–11 × 1–2 µm ($\bar{x}$ = 9 × 2 µm, n = 20), with conspicuous collarettes and periclinal thickenings. *Conidia* aseptate, hyaline, smooth, cylindrical to ellipsoidal, 5–7 × 1–2 µm ($\bar{x}$ = 6 × 2 µm, n = 20), rounded at both ends.

Culture characteristics: Colonies on PDA, dense, circular, flattened, slightly raised, floccose, white aerial mycelium, radiating with concentric ring of sporodochia forming, covered by slimy olivaceous green to black conidial masses.

Material examined: Thailand, Chiang Mai, on living leaf of *Coffea arabica* (*Rubiaceae*), 20 November 2019, R. Cheewangkoon and P. Withee, CRC13-H, living culture SDBR-CMU387; *ibid.*, CRC12-H, living culture SDBR-CMU388.

Notes: Phylogenetically, SDBR-CMU387 and SDBR-CMU388 formed a well-supported clade closely related to *Paramyrothecium breviseta* L. Lombard & Crous (Figure 2). *Paramyrothecium breviseta* was collected on an unknown substrate in India [4] and in this study, we collected *P. breviseta* from *Coffea arabica* (*Rubiaceae*) in Chiang Mai Province. The morphology of the fresh specimen is similar to that described by Lombard et al. [4], but the conidia (5–7 × 1–2 vs. 4–5 × 1–2 µm) and setae (25–120 × 2–3 vs. 25–40 × 2–3 µm) are longer. However, this is the first host report of leaf spot causing *P. breviseta* on *C. arabica* in Thailand.

Paramyrothecium eichhorniae J. Unartngam, A. Unartngam & U. Pinruan, in Pinruan et al., Mycobiology 50: 17 (2022) (Figure 6).

Description: Sexual morph: unknown. Asexual morph: *Conidiomata* sporodochial, stromatic, superficial, cupulate, scattered or gregarious, oval or irregular in outline, (60–)70–250(–500) µm diam, (60–)70–270(−370) µm deep, with a white setose fringe surrounding an olivaceous green to dark green slimy mass of conidia. *Setae* arising from the stroma thin-walled, hyaline, 1–5-septate, straight to flexuous, 60–120 µm long, 2–3 µm wide, tapering to an acutely rounded apex. *Conidiophores* arising from the basal stroma, consisting of a stipe and a penicillately branched conidiogenous apparatus; stipes unbranched, hyaline, septate, smooth, 15–40 × 2–3 µm; primary branches aseptate, unbranched, smooth, 10–17 × 2–3 µm ($\bar{x}$ = 12 × 3 µm, n = 20); secondary branches aseptate, unbranched, smooth, 7–14 × 2–3 µm ($\bar{x}$ = 10 × 3 µm, n = 20); terminating in a whorl of 3–6 conidiogenous cells; conidiogenous cells phialidic, cylindrical to subcylindrical, hyaline, smooth, straight to slightly curved, 11–17 × 2–3 µm ($\bar{x}$ = 14 × 2 µm, n = 20), with conspicuous collarettes and periclinal thickenings. *Conidia* aseptate, hyaline, smooth, cylindrical to ellipsoidal, 5–7 × 1–2 µm ($\bar{x}$ = 6 × 2 µm, n = 20), rounded at both ends.

Culture characteristics: Colonies on PDA, entire to slightly undulated margins, with sporodochia forming on the surface of the medium, covered by slimy olivaceous green to black conidial masses.

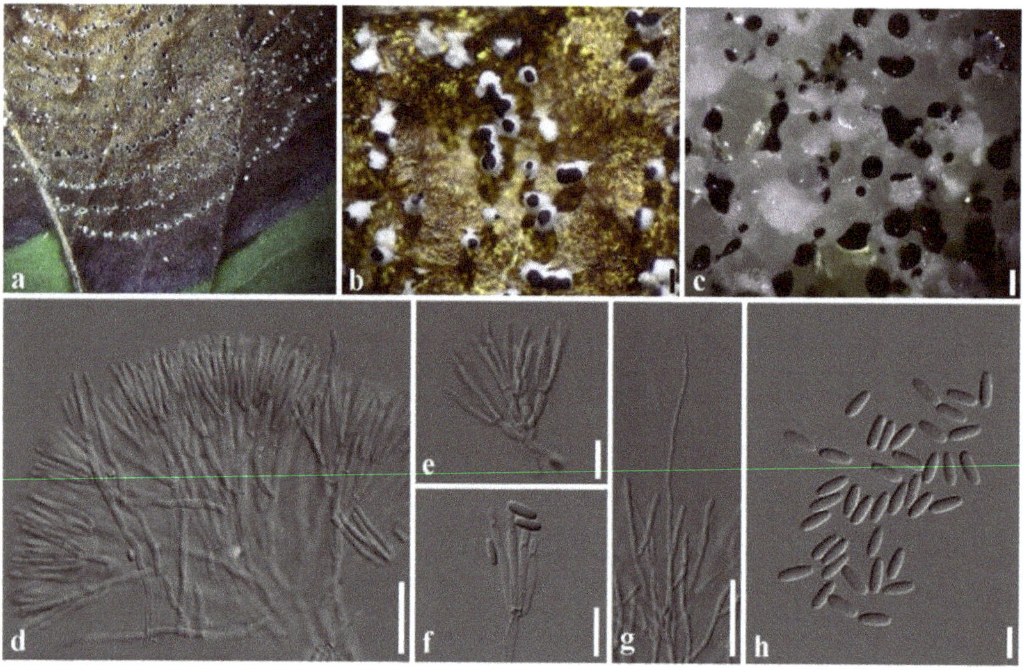

Figure 6. *Paramyrothecium eichhorniae* (CRC143); (**a**) leaf spot of *Aristolochia* sp.; (**b**) sporodochia on leaf; (**c**) sporodochial conidiomata on PDA; (**d**) sporodochia; (**e**,**f**) conidiogenous cells; (**g**) setae; (**h**) conidia. Scale bars: (**b**,**c**) = 1 mm; (**d**,**g**) = 20 μm; (**e**,**f**) = 10 μm; (**h**) = 5 μm.

Material examined: Thailand, Mae Hong Son Province, on living leaf of *Psophocarpus* sp. (*Fabaceae*), 11 September 2019, N. Tamakaew, CRC3-H, living culture SDBR-CMU375; *ibid.*, on living leaf of *Oroxylum indicum* (*Bignoniaceae*), 11 September 2019, N. Tamakaew, CRC8-H, living culture SDBR-CMU378; *ibid.*, on living leaf of *Spilanthes* sp. (*Asteraceae*), 11 September 2019, N. Tamakaew, CRC148-H, living culture SDBR-CMU379; *ibid.*, on living leaf of *Centrosema* sp. (*Fabaceae*), 11 September 2019, N. Tamakaew, CRC11-H, living culture SDBR-CMU380; Chiang Mai province, on living leaf of *Aristolochia* sp. (*Aristolochiaceae*), January 2020, P. Suttiprapan, CRC143-H, living culture SDBR-CMU383.

Note: Based on multigene phylogeny, five isolates in this study clustered with *Paramyrothecium eichhorniae*, which was associated with water hyacinth (*Eichhornia crassipes*) and recently described from Thailand [10]. Morphologically, the conidiogenous cells of our collections are similar to those of the holotype of *P. eichhorniae*. However, the conidia of *P. eichhorniae* in this study are thinner than reported by Pinruan et al. [19] (5–7 × 1–2 μm vs. 5–6.5 × 1.5–2.5 μm) and have more septa in setae than the holotype (1–5 vs. 1–3 septate). This is the first report of *P. eichhorniae* on *Psophocarpus* sp., *Centrosema* sp., and *Aristolochia* sp. from Thailand.

Paramyrothecium foliicola L. Lombard & Crous, in Lombard et al., Persoonia 36: 209 (2016) (Figure 7).

Description: Sexual morph: unknown. Asexual morph: *Conidiomata* sporodochial, stromatic, superficial, cupulate, scattered or gregarious, oval or irregular in outline, (60–)100–170(–245) μm diam, (70–)140–165(–200) μm deep, with a white to creamy setose fringe surrounding an olivaceous green agglutinated slimy mass of conidia. *Stroma* poorly developed, hyaline. *Setae* arising from the stroma thin-walled, hyaline, 1–4(–8)-septate, straight becoming sinuous above the apical septum, 35–175 μm long, 2–3 μm wide, tapering to an acutely rounded apex. *Conidiophores* arising from the basal stroma, consisting of a stipe and a penicillately branched conidiogenous apparatus; stipes unbranched, hyaline sometimes covered by a green mucoid layer, septate, smooth, 20–75 × 2–4 µm; primary branches aseptate, unbranched, smooth, (10–)17–21(–26) × 2–3(–4) µm ($\overline{x}$ = 15 × 3 µm, n = 20); secondary branches aseptate, unbranched, smooth, (7–)9–17(–19) × 2–3(–4) µm ($\overline{x}$ = 14 × 3 µm, n = 20); terminating in a whorl of 3–6 conidiogenous cells; conidiogenous cells phialidic, cylindrical to subcylindrical, hyaline, smooth, straight to slightly curved, 10–17 × 1–3 µm ($\overline{x}$ = 13 × 2 µm, n = 20), with conspicuous collarettes and periclinal thickenings. Conidia aseptate, hyaline, smooth, cylindrical to ellipsoidal, 5–8 × 1–3 µm ($\overline{x}$ = 7 × 2 µm, n = 20), rounded at both ends.

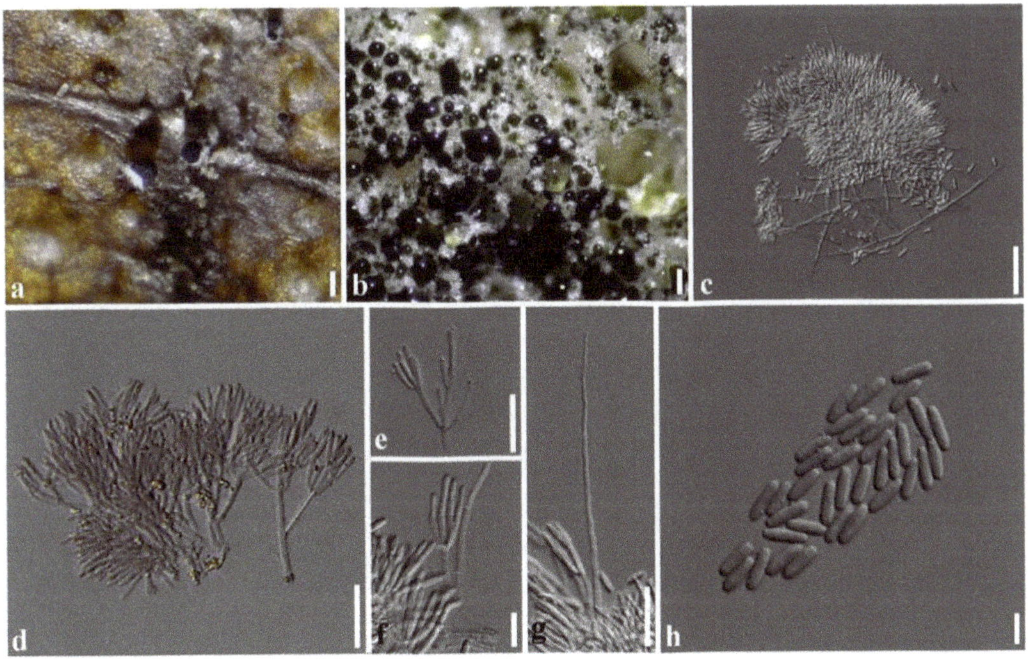

Figure 7. *Paramyrothecium foliicola* (CRC15); (**a**) sporodochia on leaves of *Tectona grandis* (**b**) sporodochial conidiomata on PDA; (**c**,**d**) sporodochia (**e**,**f**) conidiogenous cells; (**g**) setae; (**h**) condia. Scale bars: (**a**) = 500 µm; (**b**) = 1 mm; (**c**) = 30 µm; (**d**,**g**) = 20 µm; (**h**) = 5 µm.

Culture characteristics: Colonies on PDA, abundant white aerial mycelium with sporodochia forming on the aerial mycelium and surface of the medium, covered by slimy olivaceous green to black conidial masses.

Materials examined: Thailand, Chiang Mai, on living leaf of *Tectona grandis* (*Lamiaceae*), 20 November 2019, P. Withee, CRC15-H, living culture SDBR-CMU382.

Notes: Based on our phylogenetic analysis (Figure 3), SDBR-CMU382 isolates were clustered with *Paramyrothecium foliicola*. The morphology of our collection (CRC15-H) is similar to that of *P. foliicola* described by Lombard et al. [4]. However, our collection has longer conidiophores (20–75 × 2–4 vs. 15–25 × 2–3 µm) and more septa in setae (1–4(–8) vs. 1–3 septate), conidiogenous cells (10–17 × 1–3 vs. 8–14 × 1–2 µm) and conidia (5–8 × 1–3 vs. 5–6 × 1–2 µm). This may be due to distribution, environment, and morphological variability within the species. This is the first report of *P. foliicola* from *Tectona grandis* in Thailand.

2.5. Pathogenicity Test and Cross Pathogenicity

Koch's postulates confirmed that all the fungal isolates were able to cause disease in unwounded leaves of *Commelina benghalensis* and *Glycine max* (Figure 8b,c). The SDBR–CMU383 isolate infected all inoculated plants and was highly aggressive on most, except for *C. benghalensis*. No infection was observed in the unwounded inoculation of *Coffea arabica* and *Dieffenbachia seguine* (Figure 8a,d). Leaves receiving sterilized distilled water remained healthy. The fungi were re-isolated from the diseased leaf tissues in each experiment, and each isolated fungus was identical to the inoculated fungus. Further, Koch's postulates confirmed that all isolates of *Paramyrothecium vignicola*, *P. breviseta*, *P. eichhorniae*, and *P. foliicola* were pathogenic to their original host plants. Cross pathogenicity tests showed that all isolates infected inoculated (wounded) *C. arabica*, *C. benghalensis*, *G. max*, and *D. seguine* leaves (Table 2). The symptoms showed light to dark brown and irregular to round lesions, which had scattered olive-colured sporodochia and dark exudates of spore masses (Figure 8).

Figure 8. Pathogenicity test (**a**,**b**) and cross pathogenicity (**c**,**d**); Control (left); (**a**) *Paramyrothecium brevista* on *Coffea arabica;* (**b**) *P. vignicola* on *Commelina benghalensis;* (**c**) *P. vignicola* on *Glycine max;* (**d**) *P. vignicola* on *Dieffenbachia seguine;* (w) wound and (uw) unwound. Scale bars: (**a–c**) = 1 cm; (**d**) = 6 cm.

Table 2. Pathogenicity test and cross pathogenicity of *Paramyrothecium* species on original hosts and other plant species.

Species	Original Host	Isolates	Plant Hosts							
			Coffea arabica		*Commelina benghalensis*		*Glycine max*		*Dieffenbachia seguine*	
			w	uw	w	uw	w	uw	w	uw
P. vignicola	*Vigna* sp.	SDBR-CMU376 ^{T}	+	-	+	+	+	+	+	-
	Lablab purpureus	SDBR-CMU374	+	-	+	+	+	+	+	-
	Coccinia grandis	SDBR-CMU377	+	-	+	+	+	+	+	-
	Commelina benghalensis	SDBR-CMU381	+	-	+	+	+	+	+	-
	Vigna mungo	SDBR-CMU384	+	-	+	+	+	+	+	-
	Vigna sp.	SDBR-CMU385	+	-	+	+	+	+	+	-
	Vigna unguiculata	SDBR-CMU386	+	-	+	+	+	+	+	-
	Solanum virginianum	SDBR-CMU389	+	-	+	+	+	+	+	-
P. brevista	*Coffea arabica*	SDBR-CMU387	+	-	+	+	+	+	+	-
	Coffea arabica	SDBR-CMU388	+	-	+	+	+	+	+	-
P. eichhorniae	*Psophocarpus* sp.	SDBR-CMU375	+	-	+	+	+	+	+	-
	Oroxylum indicum	SDBR-CMU378	+	-	+	+	+	+	+	-
	Spilanthes sp.	SDBR-CMU379	+	-	+	+	+	+	+	-
	Centrosema sp.	SDBR-CMU380	+	-	+	+	+	+	+	-
	Aristolochia sp.	SDBR-CMU383	+	-	-	-	+	+	+	-
P. foliicola	*Tectona grandis*	SDBR-CMU382	+	-	+	+	+	+	+	-

Note: (-) No symptoms (+) Symptoms (w) wound and (uw) unwound.; Superscript "T" indicates type species.

3. Discussion

The new species *Paramyrothecium vignicola* was described using morphology and multi-gene phylogeny and the host range included *Solanum virginianum* (*Solanaceae*), *Lablab purpureus* (*Fabaceae*), *Coccinia grandis* (*Cucurbitaceae*), *Commelina benghalensis* (*Commelinaceae*), *Vigna* sp., *V. mungo*, and *V. unguiculata* (*Fabaceae*). Multi-gene phylogeny using ITS, *cmdA*, *rpb2*, and *tub2* sequence data clearly identified *P. eichhorniae*, *P. vignicola*, *P. breviseta*, and *P. foliicola* as distinct species within *Paramyrothecium*. Further, multi-gene phylogeny precisely demonstrated the species delineation of *Paramyrothecium*.

The pathogenicity assays showed that *P. vignicola*, *P. breviseta*, *P. eichhorniae*, and *P. foliicola* isolated from different hosts from different locations in northern Thailand can all cause leaf spot disease on different host families, including *Rubiaceae*, *Fabaceae*, *Commelinaceae*, and *Araceae*. However, the disease severity was related to the plant species and inoculation methods, where *Paramyrothecium* spp. could not cause disease in *Coffea arabica* and *Dieffenbachia seguine* without wounding. Wounding involves the breakage of the plant's first barrier of defense; cuticle and epidermal cells. The tissue then becomes more susceptible to the pathogens. Some species cannot infect non-wounded leaves, hence they are weakly aggressive on these hosts [15]. On the other hand, *Commelina benghalensis* and *Glycine max* were susceptible to all isolates. These results are similar to those of Rennberger and Keinath [11] and Aumentado and Balendres [12], in which *Parammyrothecium* species were able to infect original and non-original hosts within the same family (host shift ability) and different families (host jump ability).

For species diversity and distribution, more gene studies and more reference sequences are needed to resolve the species boundaries of *Paramyrothecium*. Field inspections are needed to confirm the importance of this pathogen and prove that diseases associated with *Paramyrothecium* species are threats to economic crops in Thailand. The information on the spread of related species to new areas is necessary as climate change may enable saprotrophic fungi to switch their nutritional mode across a wider host range, even if an area is predicted to be at risk from an introduced pathogen. It may be the case that few of the susceptible host species are present in this predicted area [26], so for the risk to be realized, climate change should also favor the migration of susceptible species or increase the susceptibility of the resident hosts.

Paramyrothecium leaf spot occurs in commercially important plants (*Coffea arabica*, *Tectona grandis*, *Vigna mungo*, and *V. unguiculata*) as well as on non-commercial plants (*Aristolochia* sp., *Centrosema* sp., *Coccinia grandis*, *Commelina benghalensis*, *Lablab purpureus*, *Oroxylum indicum*, *Psophocarpus* sp., *Solanum virginianum*, *Spilanthes* sp., and *Vigna* sp.). In cross pathogenicity assays, all the isolates from host plants could induce the disease on non-original hosts. *Paramyrothecium* species can stay in non-commercial plants, and they can infect commercially important crops. Hence, *Paramyrothecium* leaf spot disease has the potential to be an emerging fungal disease in Thailand. Thus, more research on *Paramyrothecium* is required for epidemiology studies and management strategies in agriculture, horticulture, and plantation forestry.

4. Materials and Methods

4.1. Sample Collection

Symptomatic plant leaves were collected from fields or forests in different locations in northern Thailand. The name of the host, location, and collection dates were recorded. Specimens were taken to the lab, and infected leaves were examined directly using the stereo microscope (Zeiss Stemi 305) to observe the fungal structures (sporodochia). Symptomatic leaves without fungal structures were also incubated in moist chambers (Petri dishes containing moist filter paper). Leaves were inspected daily for Paramyrothecium-like fungi.

4.2. Fungal Isolation and Taxonomic Description

Fungal structures on leaf samples were mounted in lactic acid and photographed under a light microscope (Axiovision Zeiss Scope-A1). Measurements were made with the Tarosoft (R) Image Frame Work program (Tarosoft, Bangkok, Thailand). The fungi were isolated using the single spore isolation technique [27]. Cultures were plated onto fresh PDA and incubated at 25–30 °C in daylight to promote sporulation. Cultural characteristics were observed after 14 days. The specimens were deposited in the fungal collection library at the Department of Entomology and Plant Pathology (CRC), Faculty of Agriculture, Chiang Mai University, Chiang Mai, Thailand. Pure fungal isolates were deposited in the Culture Collection of the Sustainable Development of Biological Resources Laboratory (SDBR), Faculty of Science, Chiang Mai University, Chiang Mai, Thailand.

4.3. DNA Extraction, Amplification, and Analyses

Fungal mycelia were grown on PDA at 25–30°C for 7 days and DNA was extracted by using the DNA Extraction Mini Kit (FAVORGEN, Ping-Tung, Taiwan) following the manufacturer's instructions. DNA amplifications were performed by polymerase chain reaction (PCR). The relevant primer pairs used in this study are listed in Table 3.

Table 3. Gene regions and primer sequences used in this study.

Gene Regions	Primers	Sequence (5′→3′)	Length (bp)	References
ITS	ITS5 ITS4	GGA AGT AAA AGT CGT AAC AAG G TCC GCT TAT TGA TAT GC	ca. 600	[28]
cmdA	CAL-228F CAL-737R CAL2Rd	GAG TTC AAG GAG GCC TTC TCC C CAT CTT TCT GGC CAT GG TGR TCN GCC TCD CGG ATC ATC TC	CAL-228F–CAL-737R: 470–570 CAL-228F–CAL2Rd: 680–745	[29,30]
rpb2	RPB2-5F RPB2-7cR	GAY GAY MGW GAT CAY TTY GG CCC ATR GCT TGY TTR CCC AT	ca. 1000	[31]
tub2	Bt2a Bt2b	GGT AAC CAA ATC GGT GCT TTC ACC CTC AGT GTA GTG ACC CTT GGC	ca. 320	[32]

The quality of PCR amplification was confirmed on 1% agarose gel electrophoresis and viewed under ultraviolet light, and the sizes of amplicons were determined against a HyperLadderTM I molecular marker (BIOLINE). Further purification of PCR products was performed using the PCR Clean-up Gel Extraction NucleoSpin ® Gel and PCR Clean-Up Kit (Macherey-Nagel, Düren, Germany). The purified PCR fragments were sent to the 1st Base Company (Kembangan, Selangor, Malaysia). The obtained nucleotide sequences were deposited in GenBank.

Sequences were assembled using SeqMan 5.00 and the closely related taxa for newly generated sequences were selected from GenBank® based on BLAST searches of the NCBI nucleotide database (http://blast.ncbi.nlm.nih.gov/; accessed on 4 March 2022). The reference nucleotide sequences of representative genera in *Stachybotriaceae* are in Table 4. The individual gene sequences were initially aligned by MAFFT version 7 [33] (http://mafft.cbrc.jp/align-ment/server/; accessed on 4 March 2022) and improved manually where necessary in BioEdit v.7.0.9.1 [34]. The final alignment of the combined multigene dataset was analyzed and inferred the phylogenetic trees based on maximum likelihood (ML) and Bayesian inference (BI) analyses. The ML analyses were carried out on RAxML-HPC2 on XSEDE (v. 8.2.8) [35,36] via the CIPRES Science Gateway platform [37]. Maximum likelihood bootstrap values (BS) equal or greater than 50% are defined above each node. The BI analyses were performed by MrBayes on XSEDE, MrBayes 3.2.6 [38] via the CIPRES Science Gateway. Bayesian posterior probabilities (PP) [39,40] were determined by Markov Chain Monte Carlo Sampling (BMCMC). Six simultaneous Markov chains were run from random trees for 2,000,000 generations, and trees were sampled every 100th generation. The run was stopped when the standard deviation of split frequencies was reached at less than 0.01. The first 20% of generated trees representing the burn-in phase of the analysis were discarded, and the remaining trees were used for calculating PP in the majority rule consensus tree. The Bayesian posterior probabilities (BYPP) equal to or greater than 0.9 are defined above the nodes. The phylogenetic tree was visualized in FigTree v.1.4.3 [41] and edited in Adobe Illustrator CC 2021 version 23.0.3.585 and Adobe Photoshop CS6 version 13.0. (Adobe Systems, New York, USA).

4.4. Pathogenicity Tests and Cross Pathogenicity

Koch's postulates were used to confirm the pathogenicity of all the isolates on their original hosts. Cross pathogenicity of all the isolates was performed in healthy leaves of selected economically important plants in northern Thailand, including *Coffea arabica* (*Rubiaceae*) and *Glycine max* (*Fabaceae*) and widespread herbaceous plants including *Commelina benghalensis* (*Commelinaceae*) and *Dieffenbachia seguine* (*Araceae*). Healthy leaves were surface disinfected with 70% ethanol, washed two times with sterile distilled water, and air-dried under laminar flow. Conidial suspensions (10^6 conidia/mL) were prepared for all fungal isolates in sterile distilled water. The conidia (10 µL of spore suspension) were placed on the upper surface of the leaves. In addition, the leaves were also wounded before inoculation. The upper epidermis was wounded approximately 2 cm from the mid-vein by pricking with a sterile needle to about 1 mm depth. Three wounds were made for each leaf, vertically on each side of the mid-vein. Control leaves received drops of sterile distilled water. All inoculated leaves were placed in a moist chamber at 25–30 °C under daylight condition. After 7 days, symptoms were recorded, compared, and confirmed with the original morphology and molecular relationships.

Table 4. Taxa used in the phylogenetic analyses and their corresponding GenBank numbers.

Species	Isolate No.	Substrate	Location	GenBank Accession Numbers			
				ITS	*cmdA*	*rpb2*	*tub2*
Myrothecium inundatum	CBS 275.48 ^{T}	On decaying pileus of *Russula nigricans*	England	KU846452	KU846435	–	KU846533
M. simplex	CBS 582.93 ^{T}	On decaying agaric	Japan	KU846456	KU846439	–	KU846537
Paramyrothecium acadiense	CBS 123.96 ^{T} = DAOMC 221473 = UAMH 7653	On leaves of *Tussilago farfara*	Canada	KU846288	–	KU846350	KU846405
P. breviseta	CBS 544.75 ^{T}	unknown	India	KU846289	KU846262	KU846351	KU846406
	DRL3	On leaves of *Coffea canephora*	China	MT853067	MT897897	–	MT897899
	DRL4	On leaves of *C. canephora*	China	MT853068	MT897898	–	MT897900
	SDBR-CMU387	**On living leaf of *C. arabica***	**Thailand**	**MZ373251**	**OM810407**	**ON033773**	**OM982450**
	SDBR-CMU388	**On living leaf of *C. arabica***	**Thailand**	**MZ373252**	**OM810408**	**ON033774**	**OM982451**
P. cupuliforme	CBS 127789 ^{T}	On surface soil in desert	Namibia	KU846291	KU846264	KU846353	KU846408
	CBS 126167	On surface soil in desert	Namibia	KU846290	KU846263	KU846352	KU846407
P. eichhorniae	TBRC 10637 ^{T}	On leaf of *Eichhornia crassipes*	Thailand	MT973996	MT975319	MT975317	MT977540
	KKFC 474	On leaf of *E.crassipes*	Thailand	MT973995	MT975318	MT977541	MT975316
	SDBR-CMU375	**On living leaf of *Psophocarpus* sp.**	**Thailand**	**MZ373241**	**OM810411**	**ON033781**	**ON033770**
	SDBR-CMU378	**On living leaf of unidentified plant**	**Thailand**	**MZ373246**	**OM810414**	**ON033782**	**ON033772**
	SDBR-CMU379	**On living leaf of unidentified plant**	**Thailand**	**MZ373247**	**OM810415**	**ON033783**	**ON033768**
	SDBR-CMU380	**On living leaf of *Centrosema* sp.**	**Thailand**	**MZ373250**	**OM810416**	**ON033784**	**ON033771**
	SDBR-CMU383	**On living leaf of *Aristolochia* sp.**	**Thailand**	**MZ373255**	**OM810418**	**ON033785**	**ON033769**
P. foeniculicola	CBS 331.51 ^{T} = IMI 140051	On leaf sheath *Foeniculum vulgare*	The Netherlands	KU846292	–	KU846354	KU846409
P. foliicola	CBS 113121 ^{T} = INIFAT C02/104 ^{T}	On rotten leaf of unknown host	Brazil	KU846294	KU846266	–	KU846411
	SDBR-CMU382	**On decaying leaf of *Tectona grandis***	**Thailand**	**MZ373254**	–	**ON033775**	**OM982452**

Table 4. *Cont.*

Species	Isolate No.	Substrate	Location	GenBank Accession Numbers			
				ITS	*cmdA*	*rpb2*	*tub2*
P. guiyangense	GUCC 201608S01 T	From soil	China	KY126418	KY196193	–	KY196201
	HGUP 2016–8001	From soil	China	KY126417	KY196192	–	KY196200
P. humicola	CBS 127295 T	from tallgrass prairie soil	USA	KU846295	–	KU846356	KU846412
	MU4	On leaf of *Citrullus lanatus*	USA	MN227389	MN593629	MN397959	MN398054
P. nigrum	CBS 116537 T = AR 3783	From soil	Spain	KU846296	KU846267	KU846357	KU846413
	LC12188	Rhizosphere soils of *Poa* sp.	China	MK478871	MK500252	MK500261	MK500269
P. parvum	CBS 257.35 T = IMI 140049	On *Viola* sp.	UK	KU846298	–	KU846359	KU846415
	CBS 142.42 = IMI 155923 = MUCL 7582	From dune sand	France	KU846297	KU846268	KU846358	KU846414
P. pituitipietianum	CPC38688 T	On stems of *Grielum humifusum*	South Africa	MW175358	MW173100	–	MW173139
P. roridum	CBS 357.89 T	On *Gardenia* sp.	Italy	KU846300	KU846270	KU846361	KU846417
	CBS 212.95	From water	The Netherlands	KU846299	KU846269	KU846360	KU846416
	CBS 372.50 = IMI 140050	On twig of *Coffea* sp.	Colombia	KU846301	KU846271	KU846362	KU846418
P. sinense	CGMCC3.19212 T = LC12136	Rhizosphere soils of *Poa* sp.	China	MH793296	MH885437	MH818824	MH793313
	LC12137	Rhizosphere soils of *Poa* sp.	China	MH793295	MH885436	MH818822	MH793312
P. tellicola	CBS 478.91 T	From soil	Turkey	KU846302	KU846272	KU846363	KU846419
P. terrestris	CBS 564.86 T	From soil under *Lycopersicon esculentum*	Turkey	KU846303	KU846273	KU846364	KU846420
	CBS 566.86	From soil beneath *Helianthus annuus*	Turkey	KU846305	KU846275	KU846366	KU846422
P. verruridum	HGUP 2016–8006 T	From soil	China	KY126422	KY196197	–	KY196205

Table 4. *Cont.*

Species	Isolate No.	Substrate	Location	GenBank Accession Numbers			
				ITS	*cmdA*	*rpb2*	*tub2*
P. vignicola	**SDBR-CMU389**	**On living leaf of *Solanum virginianum***	**Thailand**	**MZ373239**	**OM810409**	**ON033776**	**ON009013**
	SDBR-CMU374	**On living leaf of *Lablab purpureus***	**Thailand**	**MZ373240**	**OM810410**	**ON033777**	**ON009014**
	SDBR-CMU376[T]	**On living leaf of *Vigna* sp.**	**Thailand**	**MZ373242**	**OM810412**	**ON033778**	**ON009015**
	SDBR-CMU377	**On living leaf of *Coccinia grandis***	**Thailand**	**MZ373244**	**OM810413**	**ON033779**	**ON009016**
	SDBR-CMU381	**On living leaf of *Commelina benghalensis***	**Thailand**	**MZ373253**	**OM810417**	**ON033780**	**ON009017**
	SDBR-CMU384	**On living leaf of *Vigna mungo***	**Thailand**	**MZ373256**	**OM810419**	**ON033786**	–
	SDBR-CMU385	**On living leaf of *Vigna* sp.**	**Thailand**	**MZ373257**	**OM810420**	**ON033787**	**ON009018**
	SDBR-CMU386	**On living leaf of *V. unguiculata***	**Thailand**	**MZ373258**	**OM810421**	**ON033788**	**ON009019**
P. viridisporum	CBS 873.85 [T]	From soil	Turkey	KU846308	KU846278	KU846369	KU846425
	CBS 125835	Rhizosphere soils of bunchgrass	USA	KU846310	KU846280	KU846371	KU846427
Striaticonidium cinctum	CBS 932.69 [T]	From agricultural soil	The Netherlands	KU847239	KU847216	KU847290	KU847329
S. humicola	CBS 388.97	From soil in tropical forest	Papua New Guinea	KU847241	KU847217	KU847291	KU847331
S. synnematum	CBS 479.85 [T]	From leaf of unknown palm	Japan	KU847242	KU847218	KU847292	KU847332
Tangerinosporium thalictricola	CBS 317.61 [T] = IMI 034815	On *Thalictrum flavum*	UK	KU847243	KU847219	–	KU847333
Xenomyrothecium tongaense	CBS 598.80 [T]	On dead thallus of *Halimeda* sp.	Tonga	KU847246	KU847221	KU847295	KU847336

Note: CBS: Culture collection of the Centraalbureau voor Schimmelcultures, Fungal Biodiversity Centre, Utrecht, The Netherlands; CGMCC: China General Microbiological Culture Collection Center; CPC: Collection of P.W. Crous; DAOMC: The Canadian Collection of Fungal Cultures; GUCC: Guizhou University Culture Collection, Guiyang, China; HGUP: Herbarium of Guizhou University, Plant Pathology, China; IMI: International Mycological Institute, CABI-Bioscience, Egham, Bakeham Lane; INIFAT: INIFAT Fungus Collection, Ministerio de Agricultura Habana; KKFC: Kasetsart.Kamphaengsaen Fungal Collection, Thailand; LC: Collection of Lei Cai, Institute of Microbiology, Chinese Academy of Sciences, Beijing, China; MUCL: Mycothèque de l'Université Catholique de Louvian, Belgium; SDBR-CMU: the Culture Collection of the Sustainable Development of Biological Resources Laboratory, Faculty of Science, Chiang Mai University, Chiang Mai, Thailand; TBRC: Thailand Bioresource Research Center, Thailand. Species obtained in this study are in bold. Superscript "[T]" indicates type species and "–" represents the absence of sequence data in GenBank.

5. Conclusions

Leaf spots caused by *Paramyrothecium* spp. were isolated from commercially important plants (*Coffea arabica*, *Tectona grandis*, *Vigna mungo*, and *V. unguiculata*), and non-commercial plants (*Aristolochia* sp., *Centrosema* sp., *Coccinia grandis*, *Commelina benghalensis*, *Lablab purpureus*, *Oroxylum indicum*, *Psophocarpus* sp., *Solanum virginianum*, *Spilanthes* sp., and *Vigna* sp.) in northern Thailand. Based on morphology and concatenated (ITS, *cmdA*, *rpb2*, and *tub2*) phylogeny, *P. vignicola*, *P. breviseta*, *P. eichhorniae*, and *P. foliicola* were identified. The pathogenicity of each isolate was proven using Koch's postulates. The pathogenicity assay revealed that all the isolates can cause the leaf spot disease. Interestingly, cross pathogenicity assay proved the ability of all 16 isolates to cause the disease on a wide range of hosts.

Author Contributions: Conceptualization, R.C. and M.C.S.; methodology, P.W., S.H., N.T., C.S., A.K., P.S. and P.P.; validation, R.C., M.C.S., N.S., J.K. and P.W.J.T.; investigation, P.W., S.H., N.T., C.S., A.K. and P.P.; writing—original draft preparation, P.W., S.H. and C.S.; writing—review and editing, all authors; supervision, R.C.; funding acquisition, R.C. All authors have read and agreed to the published version of the manuscript.

Funding: This research project was supported by Fundamental Fund 2022 (FF65/002), Chiang Mai University.

Institutional Review Board Statement: Not applicable.

Informed Consent Statement: Not applicable.

Data Availability Statement: Publicly available datasets were analyzed in this study. These data can be found here: https://www.ncbi.nlm.nih.gov/ (access on 30 June 2022).

Acknowledgments: This research work was partially supported by Graduate School Chiang Mai University, TA/RA Scholarship, and Chiang Mai University. We thank Hiran A. Ariyawansa for his valuable suggestions and help.

Conflicts of Interest: The authors declare no conflict of interest.

References

1. Ristaino, J.B.; Anderson, P.K.; Bebber, D.P.; Brauman, K.A.; Cunniffe, N.J.; Fedoroff, N.V.; Finegold, C.; Garrett, K.A.; Gilligan, C.A.; Jones, C.M.; et al. The persistent threat of emerging plant disease pandemics to global food security. *Proc. Natl. Acad. Sci. USA* **2021**, *118*, e2022239118. [CrossRef] [PubMed]
2. Jain, A.; Sarsaiya, S.; Wu, Q.; Lu, Y.; Shi, J. A review of plant leaf fungal diseases and its environment speciation. *Bioengineered* **2019**, *10*, 409–424. [CrossRef] [PubMed]
3. Iqbal, Z.; Khan, M.A.; Sharif, M.; Shah, J.H.; ur Rehman, M.H.; Javed, K. An automated detection and classification of citrus plant diseases using image processing techniques: A review. *Comput. Electron. Agric.* **2018**, *153*, 12–32. [CrossRef]
4. Lombard, L.; Houbraken, J.; Decock, C.; Samson, R.A.; Meijer, M.; Réblová, M.; Groenewald, J.Z.; Crous, P.W. Generic hyper-diversity in Stachybotriaceae. *Persoonia* **2016**, *36*, 156–246. [CrossRef] [PubMed]
5. Soliman, M.S. Characterization of *Paramyrothecium roridum* (Basionym *Myrothecium roridum*) Causing Leaf Spot of Strawberry. *J. Plant Prot. Res.* **2020**, *60*, 141–149. [CrossRef]
6. Farr, D.F.; Rossman, A.Y. Fungal Databases, U.S. National Fungus Collections, ARS, USDA. Available online: https://nt.ars-grin.gov/fungaldatabases/ (accessed on 25 March 2022).
7. Matić, S.; Gilardi, G.; Gullino, M.L.; Garibaldi, A. Emergence of Leaf Spot Disease on Leafy Vegetable and Ornamental Crops Caused by *Paramyrothecium* and *Albifimbria* Species. *Phytopathology* **2019**, *109*, 1053–1061. [CrossRef]
8. Haudenshield, J.S.; Miranda, C.; Hartman, G.L. First Report of *Paramyrothecium roridum* (Basionym *Myrothecium roridum*) Causing *Myrothecium* Leaf Spot on Soybean in Africa. *Plant Dis.* **2018**, *102*, 2638. [CrossRef]
9. Chen, Z.D.; Li, P.L.; Chai, A.L.; Guo, W.T.; Shi, Y.X.; Xie, X.W.; Li, B.J. Crown canker caused by *Paramyrothecium roridum* on greenhouse muskmelon (*Cucumis Melo*) in China. *Can. J. Plant Pathol.* **2018**, *40*, 115–120. [CrossRef]
10. Azizi, R.; Ghosta, Y.; Ahmadpour, A. New fungal canker pathogens of apple trees in Iran. *J. Crop Prot.* **2020**, *9*, 669–681.
11. Rennberger, G.; Keinath, A.P. Stachybotriaceae on Cucurbits Demystified: Genetic Diversity and Pathogenicity of Ink Spot Pathogens. *Plant Dis.* **2020**, *104*, 2242–2251. [CrossRef]
12. Aumentado, H.D.; Balendres, M.A. Identification of *Paramyrothecium foliicola* causing crater rot in eggplant and its potential hosts under controlled conditions. *J. Phytopathol.* **2022**, *170*, 148–157. [CrossRef]

13. Huo, J.F.; Yao, Y.R.; Ben, H.Y.; Tian, T.; Yang, L.J.; Wang, Y.; Bai, P.H.; Hao, Y.J.; Wang, W.L. First Report of *Paramyrothecium foliicola* Causing Stem Canker of Cucumber (*Cucumis Sativus*) Seedlings in China. *Plant Dis.* **2021**, *105*, 1859. [CrossRef] [PubMed]
14. Piyaboon, O.; Pawongrat, R.; Unartngam, J.; Chinawong, S.; Unartngam, A. Pathogenicity, host range and activities of a secondary metabolite and enzyme from *Myrothecium roridum* on water hyacinth from Thailand. *Weed Biol. Manag.* **2016**, *16*, 132–144. [CrossRef]
15. Hassan, F.R.; Ghaffar, N.M.; Assaf, L.H.; Abdullah, S.K. Pathogenicity of endogenous isolate of *Paramyrothecium (=Myrothecium) roridum* (Tode) L. Lombard & Crous against the squash beetle *Epilachna chrysomelina* (F.). *J. Plant Prot. Res.* **2021**, *61*, 110–116. [CrossRef]
16. Liu, H.X.; Liu, W.Z.; Chen, Y.C.; Sun, Z.H.; Tan, Y.Z.; Li, H.H.; Zhang, W.M. Cytotoxic trichothecene macrolides from the endophyte fungus *Myrothecium roridum. J. Asian Nat. Prod. Res.* **2016**, *18*, 684–689. [CrossRef]
17. Basnet, B.B.; Liu, L.; Chen, B.; Suleimen, Y.M.; Yu, H.; Guo, S.; Bao, L.; Ren, J.; Liu, H. Four New Cytotoxic Arborinane–Type Triterpenes from the Endolichenic Fungus *Myrothecium inundatum. Planta Med.* **2019**, *85*, 701–707. [CrossRef]
18. Elkhateeb, W.A.; Daba, G.M. *Myrothecium* as Promising Model for Biotechnological Applications, Potentials and Challenges. *Biomed. J. Sci. Tech. Res.* **2019**, *16*, 12126–12131. [CrossRef]
19. Pinruan, U.; Unartngam, J.; Unartngam, A.; Piyaboon, O.; Sommai, S.; Khamsuntorn, P. *Paramyrothecium eichhorniae* sp. nov., Causing Leaf Blight Disease of Water Hyacinth from Thailand. *Mycobiology* **2022**, *50*, 12–19. [CrossRef]
20. Seifert, K.A.; Louis–Seize, G.; Sampson, G. *Myrothecium acadiense*, a new hyphomycete isolated from the weed *Tussilago farfara. Mycotaxon* **2003**, *87*, 317–327.
21. Krisai-Greilhuber, I.; Chen, Y.; Jabeen, S.; Madrid, H.; Marincowitz, S.; Kazaq, A.; Ševčiková, H.; Voglmayr, H.; Yazici, K.; Aptroot, A.; et al. Fungal Systematics and Evolution: FUSE 3. *Sydowia* **2017**, *69*, 229–264. [CrossRef]
22. Crous, P.W.; Osieck, E.R.; Jurjevi, Ž.; Boers, J.; van Iperen, A.L.; Starink–Willemse, M.; Dima, B.; Balashov, S.; Bulgakov, T.S.; Johnston, P.R.; et al. Fungal Planet description sheets: 1284–1382. *Persoonia* **2021**, *47*, 178–374. [CrossRef]
23. Crous, P.W.; Cowan, D.A.; Maggs–Kölling, G.; Yilmaz, N.; Larsson, E.; Angelini, C.; Brandrud, T.E.; Dearnaley, J.D.W.; Dima, B.; Dovana, F.; et al. Fungal Planet description sheets: 1112–1181. *Persoonia* **2020**, *45*, 251–409. [CrossRef] [PubMed]
24. Crous, P.W.; Cowan, D.A.; Maggs-Kölling, G.; Yilmaz, N.; Thangavel, R.; Wingfield, M.J.; Noordeloos, M.E.; Dima, B.; Brandrud, T.E.; Jansen, G.M.; et al. Fungal Planet description sheets: 1182–1283. *Persoonia* **2021**, *46*, 313–528. [CrossRef]
25. Liang, J.; Li, G.; Zhou, S.; Zhao, M.; Cai, L. Myrothecium-like new species from turfgrasses and associated rhizosphere. *MycoKeys* **2019**, *51*, 29–53. [CrossRef] [PubMed]
26. Potts, S.G.; Imperatriz-Fonseca, V.; Ngo, H.T.; Aizen, M.A.; Biesmeijer, J.C.; Breeze, T.D.; Dicks, L.V.; Garibaldi, L.A.; Hill, R.; Settele, J.; et al. Safeguarding pollinators and their values to human well-being. *Nature* **2016**, *540*, 220–229. [CrossRef]
27. Wanasinghe, D.N.; Wijayawardene, N.N.; Xu, J.; Cheewangkoon, R.; Mortimer, P.E. Taxonomic novelties in Magnolia–associated pleosporalean fungi in the Kunming Botanical Gardens (Yunnan, China). *PLoS ONE* **2020**, *15*, e0235855. [CrossRef]
28. White, T.J.; Bruns, T.; Lee, S.; Taylor, J. Amplification and Direct Sequencing of Fungal Ribosomal RNA Genes for Phylogenetics. In *PCR Protocols: A Guide to Methods and Applications*; Innis, M.A., Gelfand, D.H., Sninsky, J.J., White, T.J., Eds.; Academic Press: San Diego, CA, USA, 1990; pp. 315–322.
29. Carbone, I.; Kohn, L.M. A method for designing primer sets for speciation studies in filamentous ascomycetes. *Mycologia* **1999**, *91*, 553–556. [CrossRef]
30. Groenewald, J.Z.; Nakashima, C.; Nishikawa, J.; Shin, H.D.; Park, J.H.; Jama, A.N.; Groenewald, M.; Braun, U.; Crous, P.W. Species concepts in *Cercospora*: Spotting the weeds among the roses. *Stud. Mycol.* **2013**, *75*, 115–170. [CrossRef]
31. O'Donnell, K.; Sarver, B.A.J.; Brandt, M.; Chang, D.C.; Noble-Wang, J.; Park, B.J.; Sutton, D.A.; Benjamin, L.; Lindsley, M.; Padhye, A.; et al. Phylogenetic Diversity and Microsphere Array-Based Genotyping of Human Pathogenic Fusaria, Including Isolates from the Multistate Contact Lens-Associated U.S. Keratitis Outbreaks of 2005 and 2006. *J. Clin. Microbiol.* **2007**, *45*, 2235–2248. [CrossRef]
32. Glass, N.L.; Donaldson, G.C. Development of primer sets designed for use with PCR to amplify conserved genes from filamentous ascomycetes. *Appl. Environ. Microbiol.* **1995**, *61*, 1323–1330. [CrossRef]
33. Katoh, K.; Rozewicki, J.; Yamada, K.D. MAFFT online service: Multiple sequence alignment, interactive sequence choice and visualization. *Brief. Bioinform.* **2019**, *20*, 1160–1166. [CrossRef] [PubMed]
34. Hall, T.A. BioEdit: A user-friendly biological sequence alignment editor and analysis program for Windows 95/98/NT. *Nucleic Acids Symp. Ser.* **1999**, *41*, 95–98.
35. Stamatakis, A.; Hoover, P.; Rougemont, J. A rapid bootstrap algorithm for the RAxML Web Servers. *Syst. Biol.* **2008**, *57*, 758–771. [CrossRef] [PubMed]
36. Stamatakis, A. RAxML version 8: A tool for phylogenetic analysis and post-analysis of large phylogenies. *Bioinformatics* **2014**, *30*, 1312–1313. [CrossRef]
37. Miller, M.A.; Pfeiffer, W.; Schwartz, T. Creating the CIPRES science gateway for inference of large phylogenetic trees. In Proceedings of the Gateway Computing Environments Workshop (GCE), New Orleans, LA, USA, 14 November 2010; pp. 1–8.
38. Huelsenbeck, J.P.; Ronquist, F. MRBAYES: Bayesian inference of phylogenetic trees. *Bioinformatics* **2001**, *17*, 754–755. [CrossRef]
39. Zhaxybayeva, O.; Gogarten, J.P. Bootstrap, Bayesian probability and maximum likelihood mapping: Exploring new tools for comparative genome analyses. *BMC Genom.* **2002**, *3*, 4. [CrossRef]
40. Rannala, B.; Yang, Z. Probability distribution of molecular evolutionary trees: A new method of phylogenetic inference. *J. Mol. Evol.* **1996**, *43*, 304–311. [CrossRef]
41. Rambaut, A. *FigTree. Version 1.4.3*; Institute of Evolutionary Biology, University of Edinburgh: Edinburgh, UK, 2016.

plants

MDPI

Article

Characterization and Control of *Thielaviopsis punctulata* on Date Palm in Saudi Arabia

Khalid A. Alhudaib [1,2], Sherif M. El-Ganainy [1,2,3], Mustafa I. Almaghasla [1,2] and Muhammad N. Sattar [4,*]

[1] Department of Arid Land Agriculture, College of Agriculture and Food Sciences, King Faisal University, P.O. Box 420, Al-Ahsa 31982, Saudi Arabia; kalhudaib@kfu.edu.sa (K.A.A.); salganainy@kfu.edu.sa (S.M.E.-G.); malmghasla@kfu.edu.sa (M.I.A.)
[2] Plant Pests, and Diseases Unit, College of Agriculture and Food Sciences, King Faisal University, P.O. Box 420, Al-Ahsa 31982, Saudi Arabia
[3] Vegetable Diseases Research Department, Plant Pathology Research Institute, ARC, Giza 12619, Egypt
[4] Central Laboratories, King Faisal University, P.O. Box 420, Al-Ahsa 31982, Saudi Arabia
* Correspondence: mnsattar@kfu.edu.sa

Abstract: Date palm (*Phoenix dactylifera* L.) is the most important edible fruit crop in Saudi Arabia. Date palm cultivation and productivity are severely affected by various fungal diseases in date palm-producing countries. In recent years, black scorch disease has emerged as a devastating disease affecting date palm cultivation in the Arabian Peninsula. In the current survey, leaves and root samples were collected from deteriorated date palm trees showing variable symptoms of neck bending, leaf drying, tissue necrosis, wilting, and mortality of the entire tree in the Al-Ahsa region of Saudi Arabia. During microscopic examination, the fungus isolates growing on potato dextrose agar (PDA) media produced thick-walled chlamydospores and endoconidia. The morphological characterization confirmed the presence of *Thielaviopsis punctulata* in the date palm plant samples as the potential agent of black scorch disease. The results were further confirmed by polymerase chain reaction (PCR), sequencing, and phylogenetic dendrograms of partial regions of the ITS, TEF1-α, and β-tubulin genes. The nucleotide sequence comparison showed that the *T. punctulata* isolates were 99.9–100% identical to each other and to the *T. punctulata* isolate identified from Iraq-infecting date palm trees. The pathogenicity of the three selected *T. punctulata* isolates was also confirmed on date palm plants of Khalas cultivar. The morphological, molecular, and pathogenicity results confirmed that *T. punctulata* causes black scorch disease in symptomatic date palm plants in Saudi Arabia. Furthermore, seven commercially available fungicides were also tested for their potential efficacy to control black scorch disease. The in vitro application of the three fungicides Aliette, Score, and Tachigazole reduced the fungal growth zone by 86–100%, respectively, whereas the in vivo studies determined that the fungicides Aliette and Score significantly impeded the mycelial progression of *T. punctulata* with 40% and 73% efficiency, respectively. These fungicides can be used in integrated disease management (IDM) strategies to curb black scorch disease.

Keywords: *Thielaviopsis punctulata*; date palm; black scorch disease; multi-locus phylogeny

Citation: Alhudaib, K.A.; El-Ganainy, S.M.; Almaghasla, M.I.; Sattar, M.N. Characterization and Control of *Thielaviopsis punctulata* on Date Palm in Saudi Arabia. *Plants* **2022**, *11*, 250. https://doi.org/10.3390/plants11030250

Academic Editor: Alessandro Vitale

Received: 20 December 2021
Accepted: 10 January 2022
Published: 18 January 2022

1. Introduction

Date palm (*Phoenix dactylifera* L.) is the most extensively cultivated fruit tree plant in extreme arid and semi-arid regions due to its socio-economic and significant nutritional value [1]. Saudi Arabia is ranked second in date palm production, with an area of 157,000 ha being dedicated to it and 1.1 million tons of dates produced annually [2]. In Saudi Arabia, approximately 450 date palm cultivars have been cultivated in different regions [3]. Soil-borne pathogenic fungi cause serious infections and adversely affect the date quality and production [4,5].

The genus *Thielaviopsis* (family *Ceratocystidaceae*) (former *Ceratocystis pro parte*) includes six species: *T. cerberus*, *T. punctulata*, *T. ethacetica*, *T. musarum*, *T. euricoi*, and *T. paradoxa* [6].

Previously, these species were collectively called Ceratocystis paradoxa complex [7] and have been re-assigned by de Beer et al. [6] into the genus *Thielaviopsis* (type species *T. ethacetica*). Among them, *T. paradoxa* (anamorph of *C. paradoxa*) and *T. punctulata* (anamorph of *C. radicicola*) have been reported to be the causal agents of neck bending, wilting, black scorch, rhizosis, and chlorosis in young leaves not only in the Arabian Peninsula but also in other major date palm-producing countries [8,9]. These fungi infect a wide range of host plants, including coconut, palms, pineapple, and sugarcane [10,11]. Naturally, both species are soil-borne wound pathogens, which can confront any part of the date palm trees at any stage during their life cycle. Black scorch disease can cause economic losses to the date palm industry and may result in losses of newly planted off-shoots of ~50% [4]. This disease has a wide ecological distribution, with *T. paradoxa* being the major disease-causing agent on date palm plantations in Iran [12], Iraq [13], Italy [14], Kuwait [15], Oman [8], Qatar [16], Saudi Arabia [17,18], and the United States [19]. Moreover, *T. punctulata* has also been found to cause black scorch infestation on date palm trees in Oman [20], Qatar [16], and more recently, in the United Arab Emirates [5].

Once they have penetrated any vegetative part of the plant, these fungi cause severe rotting to occur in the buds, heart, inflorescence, leaves, and/or trunk of the plant [8]. Symptoms are characterized by tissue necrosis, the appearance of charcoal-like black and hard tissue lesions, the bending of the terminal bud or heart, and ultimately, sudden decline of the whole tree. Both *Thielaviopsis* species have been found to cause black scorch disease either independently or in combination with secondary fungal pathogens, such as *Alternaria* and *Phoma* spp. [8]. The disease severity may be exacerbated during abiotic stresses or due to poor horticultural practices [21].

The application of chemical control methods has been a major strategy to control black scorch disease despite the broad-spectrum negative impact of fungicides on the surrounding environment as well as their effects on human welfare. It has been found that applying difenoconazole provides effective control against black scorch disease in date palm plants [5]. Nevertheless, adopting integrated disease management (IDM) approaches may enormously reduce the use of chemicals and fungicides in crop plants [22,23]. The use of traditional management practices is also common to control black scorch disease. These may include avoiding wounds, cutting off diseased plant parts or the whole tree, and the integrated use of irrigation and fertilization [24,25]. However, an alternative to fungicides is the use of biological control agents (BCAs), which can provide fair control over the fungal population and can suppress their activity in plant tissues [26]. Thus, exploring native BCAs and/or their natural antagonistic products can also be a potential substitute for conventional fungicides to curtail black scorch disease. However, as a part of the long-term IDM approach, more recent genome editing and biotechnological approaches can provide fair control and can target fungal pathogens in date palm [27,28].

During a survey in the Al-Ahsa region, we observed withering, rhizosis (rapid decline), leaf wilting, neck bending, the yellowing of young leaves, and root necrosis in date palm plantations. The most frequently isolated pathogen from both necrotic roots and symptomatic leaves was *T. punctulata*. However, an extensive survey followed by the complete morphological, biological, and molecular characterization of the causal organism was required. Thus, this study was designed to study black scorch disease in date palm. The outcome of the proposed research project broadens our knowledge of the economic impact that *T. punctulata* has in the agro-ecological regions of Saudi Arabia. The study further explored the potential of various commercially available fungicides to develop an effective IDM strategy against *T. punctulata* in date palm.

2. Results

Date palm trees with progressed fungal infection and that were showing typical black scorch disease symptoms were found in the eastern region of Saudi Arabia. Different parts of the infected date palm trees were affected by the pathogen infection. The plant leaves developed a black charcoal-like hard appearance (Figure 1), while terminal bud infection resulted in severe head bending and in the complete drying of the infected tree (Figure 1A). During the survey of specific fields, entire date palm plants were observed to have black scorch disease symptoms. Other symptoms associated with the black scorch disease that were observed in the date palms were the drying of all of the leaves and, ultimately, absolute destruction of plants during the later stage of disease development (Figure 1). Three root samples (TP1, TP2, and TP3) were collected from three symptomatic date palm trees and were transferred to the laboratory for the isolation and characterization of the potential causal agents.

Figure 1. Black scorch disease symptoms and microscopic confirmation of the causal agent *Thielaviopsis punctulata*. Date palm plants showing neck bending (**A**,**B**) leaf drying under natural field conditions. (**C**) Mycelial growth of *T. punctulata* and conidiogenous cells showing aleuroconidia and phialoconidia. The arrows in green, red, blue, and turquoise represent *T. punctulata* septate hyphae, aleuroconidia, conidiophores, and phialoconidia, respectively. (**D**) Colony morphology of the three *T. punctulata* isolates (TP1, TP2, and TP3) after 2, 3, 4, and 5 days, at 21 °C on potato dextrose agar (PDA), respectively.

2.1. Morphological Characterization of *Thielaviopsis punctulata* as Causing Agent of Black Scorch Disease

The microscopic examination showed optimum mycelial growth and the formation of two types of spores: aleuroconidia or chlamydospores and phialoconidia or endoconidial spores (Figure 1C,D). The thick-walled aleuroconidia were light to dark brown in color, oval-shaped, and were borne singly on the top of short hyphae, whereas the phialoconidia were hyaline to pale brown in color, were cylindrical-shaped, and were formed lengthwise in chains (Figure 1C). The dimensions of the aleuroconidia and phialoconidia were also measured to be $17.0 \pm 1.73 \times 10.0 \pm 1.13$ and $8.0 \pm 1.01 \times 4.0 \pm 0.82$ μm in size, respectively. The microscopic assessment of the morphology of 34 fungal isolates was suggestive of the presence of *T. punctulata* as the causal agent of black scorch in the infected samples. Finally, one purified fungal isolate from each sample (TP1, TP2, and TP3) was selected to perform pathogenicity tests by inoculating the tissue-cultured greenhouse-grown date palm plants (Figure 2). The inoculated plants developed black scorch disease symptoms, which initially presented as the browning of the stem tissues and branches, especially surrounding the inoculation site within 4 weeks of post inoculation (WPI). Later, as the infection progressed, the plants started showing black scorched leaves with leaf malformations, partial or complete tissue necrosis, and wilting at 6 WPI (Figure 2). To establish Koch's postulates, the pathogen was re-isolated from the symptomatic tissues of date palm plants, and from the morphological data, it was confirmed that the pathogen was *T. punctulata*. After 6 WPI, all of the inoculated plants were severely discolored, and severe infections were observed on each part of the plants. However, the control plants did not show any symptoms.

2.2. Molecular Characterization of *Thielaviopsis punctulata*

The successful polymerase chain reaction (PCR) amplification of the three genes *internal transcribed spacer* (ITS) region of the nuclear ribosomal DNA (rDNA), *partial β-tubulin*, and *partial transcription elongation factor* 1-α (TEF1-α) from all of the date palm samples confirmed that the black scorch disease-causing *T. punctulata* fungi was frequently found in all samples.

No DNA sequence data for the *T. punctulata* species are available from Saudi Arabia in the GenBank and/or in the available literature. Therefore, to ensure sequence relatedness and the evolutionary relationships of these *T. punctulata* isolates to other *Thielaviopsis* spp., a detailed sequence comparison was performed, and a phylogenetic dendrogram was constructed, including the available nucleotide (nt) sequences of the aforementioned three genes, which represent the genes that are most closely related to *Thielaviopsis* spp. (Table 1, Figure 3). The nt sequences of the *ITS*, *β-tubulin*, and *TEF1-α* genes from the three isolates were deposited in the GenBank to acquire their respective accession numbers: MZ701784-MZ701786, MZ703651-MZ703653, and MZ703648-MZ703650, respectively (Table 1). A combined phylogenetic dendrogram was constructed for all of the isolates, and it was confirmed that the three isolates in our study were grouped into a well-supported clade (100% bootstrap value) with the *T. punctulata* isolates that had been reported from Iraq, Mauritania, and the United States (Figure 3). The nt sequence comparison showed that our isolates shared 99.9–100% nt sequence identities with each other and with the *T. punctulata* isolate (IMI 316225) identified from the infected date palm trees in Iraq. Whereas, with the other two *T. punctulata* isolates reported from the United States (CBS 114.47) and Mauritania (CBS 167.67), our isolates shared nt sequence identities that were 99.8 and 99.7–99.8% similar, respectively. The nt sequence comparisons and the phylogenetic dendrogram supported that the fungal isolates that were identified from Saudi Arabia were members of the *T. punctulata* species.

Figure 2. Pathogenicity test to fulfil Koch's postulates for three *T. punctulata* isolates on date palm plants of the Khalas cultivar. The plants were inoculated with (**A**) double distilled water (control) and the *T. punctulata* isolates (**B**) TP1, (**C**) TP2, (**D**) and TP3, respectively. Typical symptoms as they appear on the heart (**upper panel**) and rachis (**lower panel**) are shown.

Table 1. List of *Thielaviopsis* spp. for which isolates or sequences were included in this study.

Species Name	* Isolate	GenBank Accession Numbers			Host	Origin
		ITS	β-Tubulin	TEF1-α		
Thielaviopsis cerberus	CBS 130763	JX518355	JX518387	JX518323	*Theobroma cacao*	Cameroon
	CMW 35024	JX518356	JX518388	JX518324	*T. cacao*	Cameroon
	CMW 36641	JX518345	JX518377	JX518313	*Elaeis guineensis*	Cameroon
	CBS 130764	JX518349	JX518381	JX518317	*E. guineensis*	Cameroon
	CBS 130765	JX518348	JX518380	JX518316	*E. guineensis*	Cameroon
T. ethacetica	CBS 374.83	JX518329	JX518361	JX518297	*Phoenix canariensis*	Spain
	CBS 601.70	JX518331	JX518363	JX518299	*Ananas comosus*	Brazil
	CBS 453.66	JX518332	JX518364	JX518300	*Cocos nucifera*	Nigeria
	CMW 36662	JX518353	JX518385	JX518321	*E. guineensis*	Cameroon
	CMW 36771	JX518330	JX518362	JX518298	*Saccharum* sp.	South Africa
	IMI 50560	JX518341	JX518373	JX518309	*A. comosus*	Malaysia
	IMI 344082	JX518339	JX518371	JX518307	*C. nucifera*	Tanzania
	IMI 378943	JX518340	JX518372	JX518308	*E. guineensis*	Papua New Guinea

Table 1. *Cont.*

Species Name	* Isolate	GenBank Accession Numbers			Host	Origin
		ITS	β-Tubulin	TEF1-α		
T. euricoi	CMW 8788	JX518326	JX518358	JX518294	*C. nucifera*	Indonesia
	CMW 8790	JX518327	JX518359	JX518295	*C. nucifera*	Indonesia
	CBS 893.70	JX518335	JX518367	JX518303	*C. nucifera*	Brazil
T. paradoxa.	CBS 130760	JX518346	JX518378	JX518314	*E. guineensis*	Cameroon
	CBS 130762	JX518352	JX518384	JX518320	*E. guineensis*	Cameroon
	CBS 130761	JX518342	JX518374	JX518310	*T. cacao*	Cameroon
	CMW 36754	JX518344	JX518376	JX518312	*E. guineensis*	Cameroon
	CBS 101054	JX518333	JX518365	JX518301	*Rosa* sp.	Netherlands
	CBS 116770	JX518334	JX518366	JX518302	Palm sp.	Ecuador
T. musarum	CMW 1546	JX518325	JX518357	JX518293	*Musa* sp.	New Zealand
T. punctulata	CBS 114.47	KF612023	KF612025	KF612024	*P. dactylifera*	USA
	CBS 167.67	KF953932	KF953931	KF917202	*Lawsonia inermis*	Mauritania
	IMI 316225	JX518338	JX518370	JX518306	*P. dactylifera*	Iraq
	TP1	MZ701784	MZ703651	MZ703648		
	TP2	MZ701785	MZ703652	MZ703649	*P. dactylifera*	Saudi Arabia
	TP3	MZ701786	MZ703653	MZ703650		
Davidsoniella virescens	CMW 11164	AY528984	AY528990	AY528991	*Quercus* sp.	USA

* CBS: Centraalbureau voor Schimmelcultures. CMW: Culture collection of the Forestry, and Agricultural Biotechnology Institute (FABI), University of Pretoria. IMI: International Mycological Institute. TP1-TP3: Three *Thielaviopsis punctulata* isolates (bold text) identified in the current study.

2.3. In Vitro and In Vivo Growth Inhibition Efficiency of Fungicides against Thielaviopsis punctulata

The seven tested fungicides showed variable efficiency to curb the mycelial growth of the three *T. punctulata* isolates. The data showed that there were significant differences among all of the fungicides in their ability to inhibit *T. punctulata* mycelium growth (Table 2). The results further revealed that Aliette, Score, and Tachigazole successfully prevented the growth and sporulation of the three isolates. The results were further confirmed when mycelium growth inhibition was measured. A clear fungal inhibition zone was observed (~100%) when the potato dextrose agar (PDA) media was supplemented with Aliette, Score, and Tachigazole. An exception was observed in the case of the TP1 isolate when Tachigazole was applied with a fungal inhibition zone >86%. There was a significant reduction in the mycelium inhibition zone, which showed an 8 to 24% reduction with the supplementation of the Telder fungicide, whereas no significant difference was observed between Ridomil, Gold, and Uniform because virtually no mycelium growth or sporulation inhibition were observed (Table 2, and Figure 4A).

Table 2. Percentage colony growth inhibition of three isolates of *Thielaviopsis punctulata* using seven fungicides.

Fungicide	TP1	TP2	TP3
Aliette- 80% WG	100 a	100 a	100 a
Infinito 687.5 SC	3.33 d	36.67 b	6.00 bc
Ridomil Gold- 480 SL	0.00 d	41.11 b	5.33 bc
Score 250 EC	100 a	100 a	100 a
Tachigazol- 30% SL	86.67 b	100 a	100 a
Teldor 50 SC	23.78 c	1.78 c	8.22 b
Uniform-446 SE	0.00 d	0.00 d	2.44 cd
Control	0.00 d	0.00 d	0.00 d

Values with similar letters show non-significant differences at $p < 0.05$.

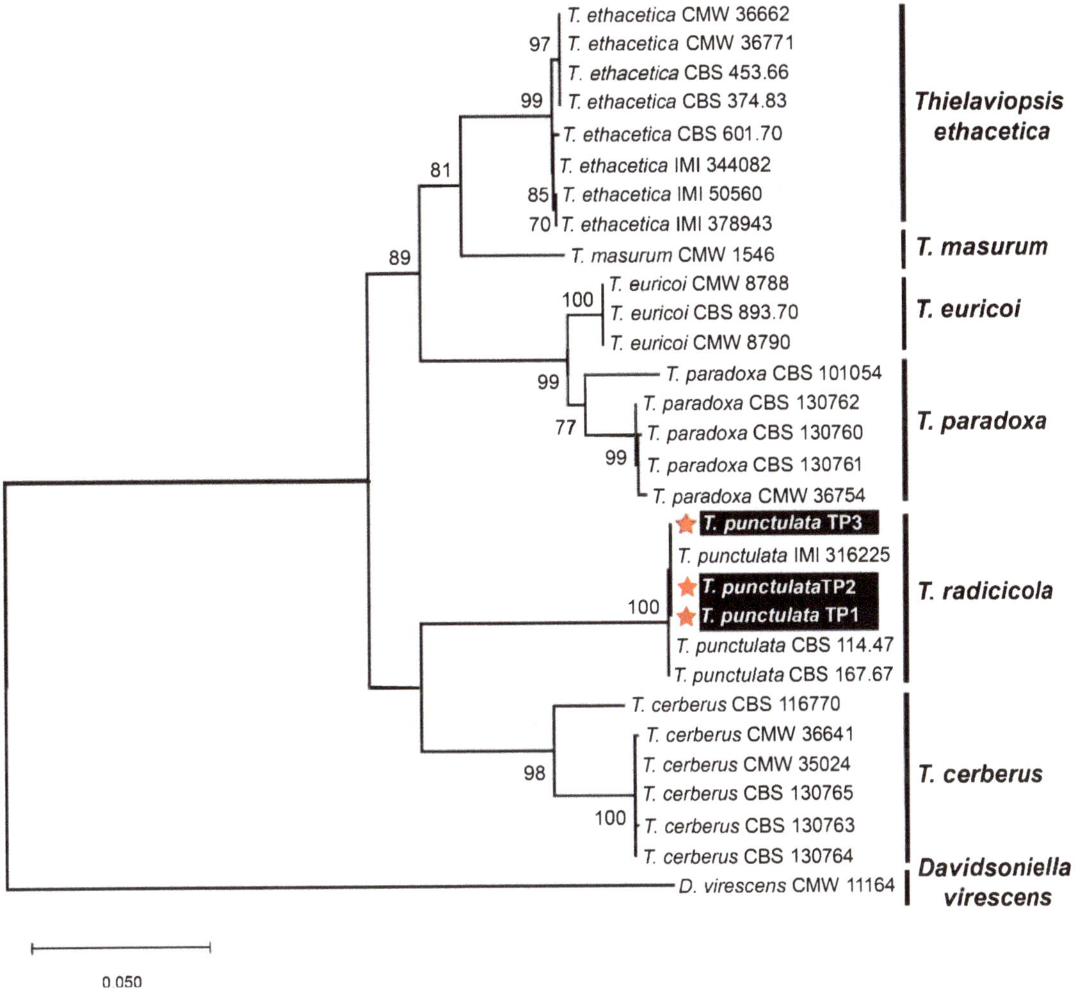

Figure 3. Molecular identification of three *Thielaviopsis punctulata* isolates based upon nucleotide sequences from the internal transcribed spacer (ITS), β-tubulin, and transcription elongation factor 1-α region from the selected isolates of different *Thielaviopsis* species. A maximum-likelihood (ML) algorithm was employed to construct a combined phylogenetic dendrogram by combining the ITS, β-tubulin, and TEF1-α sequences using the best-fit Kimura 2-parameter model (T93+G) in MEGA X software. The isolates that were identified from Saudi Arabia in this study are highlighted in white text on black background and with red asterisks. Only bootstrap values (numeric values at the branch nodes) ≥70 were shown to support the evolutionary relatedness of the isolates. The descriptors for all fungal isolates were named according to de Beer et al. [6].

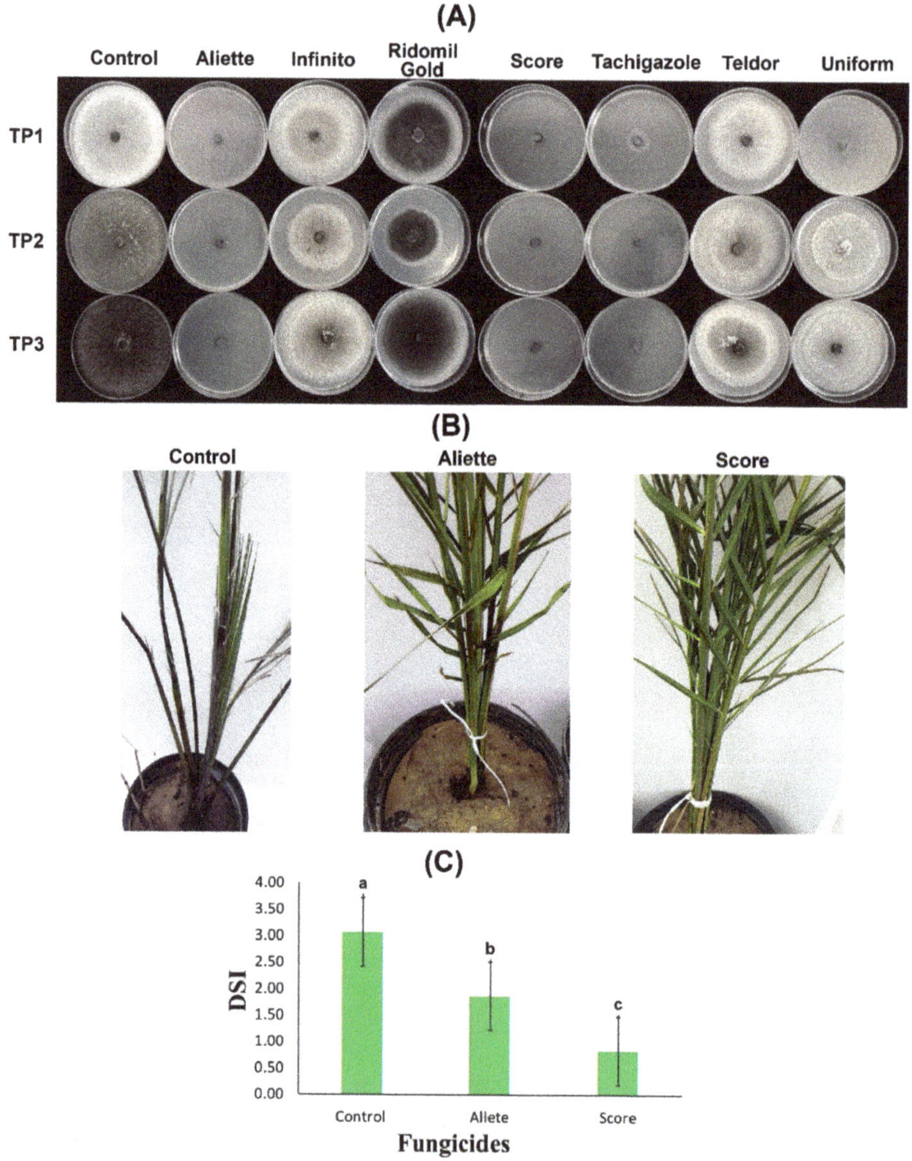

Figure 4. In vitro and in vivo evaluation of fungicides against *Thielaviopsis punctulata*. (**A**) Percentage growth inhibition of mycelium from eight fungicides: Aliette, Infinito, Ridomil Gold, Score, Tachigazole, Teldor, and Uniform. The efficiency of each fungicide was tested on potato dextrose agar (PDA) media against three *T. punctulata* isolates: TP1, TP2, and TP3. (**B**) Date palm plants of the Khalas cultivar were first inoculated with the *T. punctulata* isolate TP2, and at 2 WPI, the plants were treated with the fungicides Aliette and Score. The healthy control was mock-inoculated using double distilled water. (**C**) Disease severity index (DSI) of the infected date palm plants at 8 WPI treated with the Aliette and Score fungicides. The significant values at $p > 0.05$ are shown as different lowercase letters.

To further confirm our in vitro results, two fungicides, Aliette and Score, were sprayed on the date palm Khalas cultivar plants infected with the TP2 isolate of *T. punctulata* under greenhouse conditions, and the efficiency of both fungicides was observed for 8 WPI. Before fungicide spraying, the plants started showing black scorch symptoms at 2 WPI. However, in the plants treated with Score at 4 WPI, the disease progression was significantly reduced, and the fungicide efficiency was 73% compared to the control at 8 WPI (Figure 4B). Although the disease progression was reduced with 40% efficiency with Aliette, it was not a significant reduction compared to the reduction observed for Score. The date palm plants treated with the Score fungicide also showed the emergence of new leaves from the heart. Contrarily, the untreated plants (control) started to dry. Moreover, the disease severity index (DSI) was drastically reduced to 0.83 and 1.87 at 8 WPI compared to 3.07 in the control plants (Figure 4C). Thus, the results of the Score fungicide application were highly promising against all three isolates of *T. punctulata* in our study.

3. Discussion

The date palm has been traditionally cultivated for over 5000 years in arid and semi-arid regions not only because of its edible significance but also due to its social and cultural value [29]. Saudi Arabia contributes ~9 million tons to the global date production with a share of about 17% (https://www.fao.org/faostat/en/#data/QC last accessed 10 December 2021). Among the major biotic stresses, diseases caused by soil-borne fungal pathogens are a major menace to date palm production in Saudi Arabia and include leaf spot, leaf blight, sudden decline, and black scorch disease [18]. In Saudi Arabia, *Fusarium proliferatum*, *F. solani*, *F. brachygibbosum*, *F. oxysporum* and *F. verticillioides* have been recently characterized as causing different fungal-associated diseases [30,31]. Black scorch is another fungal disease that can affect date palm cultivation and is caused by *T. paradoxa* and *T. punctulata* in different parts of the world. Likewise, in Saudi Arabia, both *T. paradoxa* and *T. punctulata* have been found to be associated with black scorch disease in date palm [17,18,32]. However, these studies only focused on the morphological identification of the fungal pathogens and could not determine the molecular anomalies of *Thielaviopsis* spp. in Saudi Arabia. In field surveys, it is difficult to discriminate between black scorch and sudden decline disease due to their similar symptoms. Thus, PCR-based molecular diagnostic techniques can help to differentiate between sudden decline and black scorch disease and can aid in the development of effective control strategies [33]. Black scorch has not been established as an epidemic to date in palm cultivation; however, if established, it may lead to heavy losses in date palm cultivation. Therefore, comprehensive research investigating this important date palm disease is crucial. With this in mind, the current study was designed to pinpoint the exact causal agent of black scorch disease in Saudi Arabia and aimed to find a potential solution to control this devastating pathogen.

The primary morphological data were suggestive of *T. punctulata* being a potential causal agent of the black scorch disease in the collected field samples. It was evident from the microscopic investigation that the pathogen produced abundant endoconidia in the PDA media. The physiological data were based on those acquired from previous studies in which thick-walled chlamydospores appeared singly on short hyphal branches and were light to dark brown in color. Moreover, we observed hyaline to pale brown phialoconidia formed in chains. The length and width of aleuroconidia and phialoconidia were measured as being $17.0 \pm 1.73 \times 10.0 \pm 1.13$ and $8.0 \pm 1.01 \times 4.0 \pm 0.82$ μm in size, respectively. Our results were in accordance with previous reports that aleuroconidia are often larger than phialoconidia in *Thielaviopsis* species [5,17,34]. Although it is difficult to identify fungal species based on spore morphology, it is crucial to study spore morphology because of its critical role in the dispersal, pathogenicity, and survival of the fungal pathogen under study [5]. To elucidate the nature and pathogenicity of the potential pathogen, we inoculated tissue-cultured greenhouse-grown date palm plants with the three isolates of the pathogen. Our data were very similar to the data obtained from previous pathogenicity studies of *T. paradoxa* on *Dracaena marginata* [35], *Butia capitate* [36], *Hyophorbe lagenicaulis* [37], and *Cocos nucifera* [38]. Hence, based upon the morphology and sporulation characteristics, it is difficult to distinguish both *Thielaviopsis* spp., and a DNA-based molecular characterization of the most conserved regions of these fungal pathogens can be an alternative species-specific detection method for black scorch disease in date palm.

Based upon the previous literature, specific genomic regions corresponding to the *ITS*, *β-tubulin*, and *TEF1-α* genes were PCR-amplified and were subsequently sequenced. A combined phylogenetic dendrogram and nt comparison of the three corresponding genes from different *Thielaviopsis* spp. showed that our three isolates were the most closely related to the *T. punctulata* isolate IMI 316225 (nt sequence identity 100%) identified in Iraq. Although the nt sequence identity was shown to be 99.7–99.8% with two other *T. punctulata* isolates CBS 114.47 and CBS 167.67 from the USA and Mauritania, all of the isolates showed a similar evolutionary lineage, as evident from the phylogenetic dendrogram (Figure 3). None of the isolates were grouped with the *T. paradoxa* in the phylogenetic analysis. Thus, based upon our detailed investigation and in contrast to previous morphological studies [17,18,32], we may speculate that the causal agent of black scorch disease in Saudi Arabia is *T. punctulata* and not *T. paradoxa*. Our conclusion is also supported by other reports from the Arabian Peninsula, where it has been shown that *T. punctulata* is the black scorch disease-causing agent in this region [5,8,16,39]. However, an extensive survey across the major date palm-producing areas in the country can better answer this speculation. Molecular monitoring of the soil-borne fungal pathogens has become an important management tool [40]. Thus, application of a high-throughput, precise, and more sensitive molecular diagnostic method to discriminate between *T. punctulata* and *T. paradoxa* species can help to devise species-specific control strategies in date palm. Furthermore, we extended our investigation to find a better control strategy to circumscribe black scorch disease in date palm.

Over-reliance on chemical-based pesticides and fungicides is hazardous to an agroecological ecosystem and may result in the development of pathogen resistance [23]. However, studies on the integrated reduced use of conventional fungicides against *T. punctulata* are very limited [5,25]. Therefore, we continued our work through both *in vitro* and in vivo studies to determine the fungicide that is the most efficient in providing stable control against *T. punctulata* in Saudi Arabia. We selected seven systemic fungicides (Aliette, Infinito, Ridomil Gold, Score, Tachigazole, Teldor, and Uniform) to test their efficacy against black scorch disease. The in vitro investigation using PDA showed that Aliette and Score were the most efficient fungicides at the tested concentration of 300 ppm, providing 100% mycelium inhibition against the three tested *T. punctulata* isolates (Figure 4A), whereas Tachigazole showed 87–100% mycelium inhibition against the three *T. punctulata* isolates. The rest of the fungicides Infinito, Ridomil Gold, Teldor, and Uniform could not produce promising results against the mycelial growth of *T. punctulata*. Henceforth, Aliette (Fosetyl-Al) and Score (Difenoconazole) appear to be the best candidate fungicides for the control of *T. punctulata* in date palm followed by Tachigazole (Hymexazol). However, an empirical demonstration was necessary to test their efficacy against *T. punctulata* under controlled environmental conditions. Our in vivo results on date palm seedlings showed that Score was the most efficient fungicide against *T. punctulata* with ~73% efficiency and 0.83 DSI compared to Aliette, with nearly 40% efficiency and 1.87 DSI against black scorch disease (Figure 4B,C). In a previous study, Croft [41] showed that difenoconazole (Score) was unable to accelerate sugarcane seed germination with *T. paradoxa*. In contrast, our results were similar to the findings of Saeed et al. [5], who found that the difenoconazole-based Score fungicide showed >91% mycelium inhibition against *T. punctulata* in PDA media. In the same study, Saeed et al. [5] reported highly significant inhibition of *T. punctulata* in inoculated date palm seedlings.

4. Materials and Methods

4.1. Collection of Symptomatic Date Palm Plant Samples

The root samples were collected from symptomatic date palm trees in three different locations in the Al-Ahsa province of Saudi Arabia (Figure 1A,B). The collected root samples were washed with tap water to remove any soil particles that had adhered to the roots. The clean roots were fused for further fungal isolation and characterization.

4.2. Isolation of the Root Rot Fungal Pathogens

Tissues were excised from the infected lesions or roots, surface-sterilized using 2% sodium hypochlorite for 2 min followed by rinsing two times with sterilized distilled water, and were dried between pieces of filter paper for 10 min at room temperature. The clean and sterilized pieces of each sample were overlaid on five PDA plates supplemented with 100 ppm ampicillin to avoid any bacterial contamination. Petri plates were incubated at 27 °C for 3–5 days in complete darkness. At least 2–3 emerging fungal colonies were individually subcultured from each plate on PDA for further identification. The morphology of the mycelium and spores was observed under a light microscope (Leica, DM 25000 LED) to characterize different mycological structures (Figure 1C). The spore dimensions were measured using Leica application suite X (LasX). Photographs were taken using a Flexacam C1 camera.

4.3. Pathogenicity Test and Disease Assays

The virulence of the selected isolates was performed to determine their aggressiveness on tissue-cultured plants of the date palm cultivar Khalas grown in soil pots according to the method used by Saeed et al. [5]. After the plant surface had been sterilized with 70% ethanol, the wounded parts of the plants were sprayed with a 5×10^5 mL^{-1} spore suspension of *T. punctulata* (wounds were made with a sterile needle) at the leaf base of the plants to facilitate infection. Moreover, the soil pots of the same plants were also infested with *T. punctulata* by drenching 50 mL of potato broth medium (PBM) with 1×10^6 mL^{-1} spore suspension. The control plants were only sprayed with sterilized distilled water, and soil was drenched by 50 mL of PBM only. Plastic bags were used to cover the inoculated plants to provide sufficient humidity for infection, and the samples kept at 27 °C under controlled environmental conditions. The plants were examined closely for any black scorch disease symptoms for 6 weeks. The disease severity index was calculated using the scale from 0 to 5, as previously described [5,32].

4.4. Molecular Characterization of the Purified Isolates of *Thielaviopsis punctulata*

Total genomic DNA was extracted from the dried mycelium that had been cultured on PDA media following the Dellaporta extraction method [42], with some minor modifications. The extracted DNA was used either directly for a PCR or was frozen for further experiments. The PCR reactions were performed to amplify the ITS region, partial β-tubulin, and partial TEF1-α using their respective primers (Table 3). PCR amplification was carried out as previously described [7] using the ESCO Swift Maxi Thermal Cycler. In a 25 μL total reaction volume, ~40 ng of fungal DNA were mixed with 2.5 μL of 10 × Taq Polymerase buffer, 2 mM MgCl2, 1.5 μL of 10 μM primers, 2.5 μL of 10 mM dNTPs, 0.3 μL of 5U Taq DNA Polymerase, and a final reaction volume with nuclease-free water. The PCR was performed as initial denaturation at 95 °C for 2 min, followed by 35 cycles of 95 °C for 30 s, 52–58 °C for 30–60 s, and 72 °C for 30 s, and the final elongation cycle performed at 72 °C for 10 min. The resultant PCR amplicons were further confirmed by agarose gel electrophoresis. The confirmed PCR products were purified using the CloneJet PCR cloning kit (ThermoFisher Scientific, Waltham, MA, USA) and were completely sequenced at Macrogen Inc., (Seoul, Korea).

Table 3. Nucleotide sequence of the primers used in the study.

Locus Name	Primer	Sequence (5′-3′)	PCR Program *	Reference
ITS region	ITS4 ITS5	TCCTCCGCTTATTGATATGC GGAAGTAAAAGTCGTAACAAGG	35 cycles at 94 °C for 30 s, 52 °C for 30 s, 72 °C for 30 s	[43]
TEF1-α	EF1F EF1R	TGCGGTGGTATCGACAAGCGT AGCATGTTGTCGCCGTT GAAG	35 cycles at 94 °C for 30 s, 58 °C for 60 s, 72 °C for 90 s	[44]
β-tubulin	Bt1a Bt1b	TTCCCCCGTCTCCACTTCTTCATG GACGAGATCGTTCATGTTGAACTC	34 cycles at 94 °C for 60 s, 58 °C for 60 s, 72 °C for 60 s	[45]

* Initial denaturation at 95 °C for 2 min and the final elongation cycle performed at 72 °C for 10 min, respectively.

The obtained sequences for ITS, β-tubulin, and TEF1-α were initially compared to their respective sequences using BLASTn in the NCBI GenBank database for their primary identification. The highly similar sequences for each region were retrieved from the NCBI. The sequences of each gene in the study were aligned separately with the respective sequences of the most closely related *Thielaviopsis* spp. isolates using the ClustalW algorithm in MEGA X [46]. Thus, three individual datasets including sequences from each and the relevant sequences retrieved from the NCBI were constructed. Finally, a combined dataset of all three genes was generated to construct a phylogenetic dendrogram and to infer the evolutionary relationships. Furthermore, the pairwise nt sequence identities of all of the fungal isolates were calculated using the Muscle algorithm that is available in the species demarcation tool (SDTv1.2) software (University of Cape Town, Cape Town, South Africa).

4.5. In Vitro Evaluation of Fungicides against Thielaviopsis punctulata

The antifungal evaluation of seven fungicides: Aliette (80% WG), Infinito (687.5 SC), Ridomil Gold (480 SL), Score (250 EC), Tachigazole (30% SL), Teldor (50 SC), and Uniform (446 SE), was performed as previously described [47] (Table 4). The fungicide solution of each fungicide contained a 300 ppm final concentration of each respective fungicide. These fungicide solutions were then aseptically introduced into the sterilized PDA media supplemented with ampicillin at 55 °C to avoid any bacterial growth. The media was carefully poured into the Petri plates. The fungal plug ~5 mm of each *T. punctulata* isolate was transferred to the control and the treatment plates followed by incubation at 25 °C for six days according to the method used by Jonathan et al. [47]. After six days, the radial growth of each fungal isolate was measured in the dishes to determine the growth inhibition efficiency of each fungicide. The percentage mycelium growth inhibition was measured as follows:

$$MI\% = 100 \times (Mc - Mt)/Mc$$

where Mc is the diameter of the mycelium growth on the medium without fungicide, while Mt is mycelium growth on the medium with each fungicide.

Table 4. List of fungicides and their characteristics used to curtail black scorch disease in date palm.

Fungicides	Active Ingredient	Chemical Group	Dose/L
Aliette- 80% WG	Fosetyl-Al	Organophosphate	2.5 g
INFINITO 687.5 SC	Fluopicolide + Propamocarb HCL	Acylpicolide + Carbamate	1.5 mL
RIDOMIL GOLD- 480 SL	Metalaxyl-M	Phenylamide	2 mL
Score 250 EC	Difenoconazole	Triazole	0.5
TACHIGAZOL- 30% SL	Hymexazol	Oxazoles	1.5 mL
Teldor 50 SC	Fenhexamid	Anilide	0.5 mL
UNIFORM-446 SE	Metalaxyl-M + Azoxystrobin	Phenylamide + Strobilurin	0.5 mL

4.6. In Vivo Evaluation of Fungicides to Control Black Scorch Disease

Two-year-old tissue-cultured plants from the date palm cultivar Khalas grown under greenhouse conditions were used to assess the efficacy of two fungicides, i.e., Aliette and Score, against black scorch disease. The plants were inoculated at the leaf base region with *T. punctulata* isolate TP2, as described above. The inoculated plants were kept in greenhouse conditions and were covered with transparent plastic bags for 5 days to facilitate fungal infection. After 2 WPI, the plants were sprayed with the commercially recommended doses of the respective fungicides. The control plants were only sprayed with double distilled water. The experiment was repeated three times with five replicates for each treatment.

4.7. Data Analysis

The data for the in vitro experiments were analyzed using a two-way analysis of variance (ANOVA) approach with a treatment effect using the general linear model (GLM) procedure. The least significant difference (LSD) test was applied to separate the means at the 5% significance level.

The in vivo evaluation of two fungicides on the infected date palm plants was performed in a completely randomized design (CRD) under greenhouse conditions. For the disease assay and in vivo assessment of the Aliette and Score fungicides, three replicates with five plants per replicate were used. The significance was determined using the LSD test with the statistical difference set at $p < 0.05$. The collected data were statistically analyzed using the MSTAT-C program (v 2.10). All statistical analyses were carried out using SAS/STAT® 9.3 software (SAS, Cary, NC, USA)

5. Conclusions

Our study confirmed *T. punctulata* as the black scorch disease-causing agent in Saudi Arabia. Although black scorch disease has already been reported from Saudi Arabia, previous investigations were based upon morphological characterizations of the pathogen. Our research represents a detailed study on the morphology, pathogenicity, biology, evolutionary relatedness, and potential control mechanisms of *T. punctulata* in Saudi Arabia. However, further research should explore the complete infection cycle of *T. punctulata* and the development of an effective IDM approach.

Author Contributions: Conceptualization, S.M.E.-G. and K.A.A.; methodology, S.M.E.-G. and M.I.A.; software, M.N.S.; validation, S.M.E.-G. and M.N.S.; formal analysis, S.M.E.-G.; investigation, S.M.E.-G., M.I.A. and M.N.S.; data curation, M.N.S.; writing—original draft preparation, S.M.E.-G. and M.N.S.; writing—review and editing, K.A.A. and M.N.S.; supervision, K.A.A. and M.N.S.; project administration, K.A.A., M.N.S. and S.M.E.-G.; funding acquisition, K.A.A., M.N.S., and S.M.E.-G. All authors have read and agreed to the published version of the manuscript.

Funding: This research work was funded by the Deputyship for Research and Innovation, Ministry of Education in Saudi Arabia, through project number IFT 20005.

Institutional Review Board Statement: Not Applicable.

Informed Consent Statement: Not Applicable.

Data Availability Statement: Not Applicable.

Acknowledgments: The authors extend their appreciation to the Deputyship for Research and Innovation, Ministry of Education in Saudi Arabia for funding this research work through project number IFT 20005. Furthermore, the authors are grateful to the anonymous reviewers and editors for their comments that helped improve the manuscript.

Conflicts of Interest: The authors declare no conflict of interest.

References

1. Iqbal, Z.; Sattar, M.N.; Al-Khayri, J.M. Whole-Genome Mapping of Date Palm (*Phoenix dactylifera* L.). In *The Date Palm Genome*; Springer: Cham, Switzerland, 2021; Volume 1, pp. 181–199.
2. Krueger, R.R. Date Palm (*Phoenix dactylifera* L.) Biology and Utilization. In *The Date Palm Genome*; Springer: Cham, Switzerland, 2021; Volume 1, pp. 3–28.
3. Saboori, S.; Noormohammadi, Z.; Sheidai, M.; Marashi, S. Insight into date palm diversity: Genetic and morphological investigations. *Plant Mol. Biol. Rep.* **2021**, *39*, 137–145. [CrossRef]
4. Abdelmonem, A.; Rasmy, M. Major diseases of date palm and their control. *Commun. Inst. For. Bohem* **2007**, *23*, 9–23.
5. Saeed, E.E.; Sham, A.; El-Tarabily, K.; Abu Elsamen, F.; Iratni, R.; AbuQamar, S.F. Chemical control of black scorch disease on date palm caused by the fungal pathogen *Thielaviopsis punctulata* in United Arab Emirates. *Plant Dis.* **2016**, *100*, 2370–2376. [CrossRef]
6. De Beer, Z.W.; Duong, T.A.; Barnes, I.; Wingfield, B.D.; Wingfield, M.J. Redefining Ceratocystis and allied genera. *Stud. Mycol.* **2014**, *79*, 187–219. [CrossRef] [PubMed]
7. Mbenoun, M.; De Beer, Z.W.; Wingfield, M.J.; Wingfield, B.D.; Roux, J. Reconsidering species boundaries in the *Ceratocystis paradoxa* complex, including a new species from oil palm and cacao in Cameroon. *Mycologia* **2014**, *106*, 757–784. [CrossRef] [PubMed]
8. Al-Raisi, Y.; B'Chir, M.; Al-Mandhari, A.; Deadman, M.; Gowen, S. First report of *Ceratocystis radicicola* associated with date palm disease in Oman. *New Dis. Rep.* **2011**, *23*, 23. [CrossRef]
9. Al-Sadi, A. Phylogenetic and population genetic analysis of *Ceratocystis radicicola* infecting date palms. *J. Plant Pathol.* **2013**, *95*, 49–57.
10. Demiray, S.T.; Akcali, E.; Uysal, A.; Kurt, S. First report of *Thielaviopsis paradoxa* causing main stalk rot on banana in Turkey. *Plant Dis.* **2020**, *104*, 2733–2734. [CrossRef]
11. Gaitan-Chaparro, S.; Navia-Rodriguez, E.; Romero, H.M. Assessment of inoculation methods of *Thielaviopsis paradoxa* (de seynes) Hohn into oil palm seedlings under greenhouse conditions. *J. Fungi* **2021**, *7*, 910. [CrossRef]
12. Mirzaee, M.; Tajali, H.; Javadmosavi, S. *Thielaviopsis paradoxa* causing neck bending disease of date palm in Iran. *J. Plant Pathol.* **2014**, *96*, 4–122.
13. Abbas, I.; Al-Izi, M.; Aboud, H.; Saleh, H. Neck bending: A new disease affecting date palm in Iraq. In *Proceedings of the Sixth Arab Congress of Plant Protection*; Arab Plant Protection Society: Beirut, Lebanon, 1997.
14. Polizzi, G.; Castello, I.; Vitale, A.; Catara, V.; Sofia, V.S.; Tamburino, V. First report of *Thielaviopsis* trunk rot of date palm in Italy. *Plant Dis.* **2006**, *90*, 972. [CrossRef] [PubMed]
15. Mubarak, H.; Riaz, M.; As-Saeed, I.; Hameed, J. Physiological studies and chemical control of black scorch disease of date palm caused by *Thielaviopsis* (*Ceratocystis*) *paradoxa* in Kuwait. *Pak. J. Phytopathol.* **1994**, *6*, 7–12.
16. Al-Naemi, F.A.; Nishad, R.; Ahmed, T.A.; Radwan, O. First report of *Thielaviopsis punctulata* causing black scorch disease on date palm in Qatar. *Plant Dis.* **2014**, *98*, 1437. [CrossRef] [PubMed]
17. Al-Sharidy, A.; Molan, Y. Survey of fungi associated with black scorch and leaf spots of date palm in Riyadh Area. *Saudi J. Biol. Sci.* **2008**, *15*, 113–118.
18. Ammar, M.; El-Naggar, M. Date palm (*Phoenix dactylifera* L.) fungal diseases in Najran, Saudi Arabia. *Int. J. Plant Pathol.* **2011**, *2*, 126–135. [CrossRef]
19. Garofalo, J.F.; McMillan, R.T. Thielaviopsis diseases of palms. In *Proceedings of the Florida State Horticultural Society*; Florida State Horticultural Society: Sarasota, FL, USA, 2004; pp. 324–325.
20. Al-Sadi, A.M.; Al-Jabri, A.H.; Al-Mazroui, S.S.; Al-Mahmooli, I.H. Characterization and pathogenicity of fungi and oomycetes associated with root diseases of date palms in Oman. *Crop Protect.* **2012**, *37*, 1–6. [CrossRef]
21. Suleman, P.; Al-Musallam, A.; Menezes, C. The effect of biofungicide Mycostop on *Ceratocystis radicicola*, the causal agent of black scorch on date palm. *BioControl* **2002**, *47*, 207–216. [CrossRef]
22. Percival, G.C.; Graham, S. The potential of resistance inducers and synthetic fungicide combinations for management of foliar diseases of nursery stock. *Crop Protect.* **2021**, *145*, 105636. [CrossRef]
23. Prasertsan, P.; Nutongkaew, T.; Leamdum, C.; Suyotha, W.; Boukaew, S. Direct biotransformation of oil palm frond juice to ethanol and acetic acid by simultaneous fermentation of co-cultures and the efficacy of its culture filtrate as an antifungal agent against black seed rot disease. *Biomass Convers. Biorefin.* **2021**, 1–10. [CrossRef]
24. El Bouhssini, M. *Date Palm Pests and Diseases: Integrated Management Guide*; International Center for Agricultural Research in the Dry Areas (ICARDA): Beirut, Lebanon, 2018.
25. Dharmaputra, O.S.; Hasbullah, R.; Fransiscus, J. Use of calcium chloride and chitosan to control *Thielaviopsis paradoxa* in salak pondoh fruit during storage. *J. Fitopatol. Indones.* **2021**, *17*, 131–140. [CrossRef]
26. Degani, O.; Dor, S. Trichoderma biological control to protect sensitive maize hybrids against late wilt disease in the field. *J. Fungi* **2021**, *7*, 315. [CrossRef]
27. Sattar, M.N.; Iqbal, Z.; Al-Khayri, J.M.; Jain, S.M. Induced genetic variations in fruit trees using new breeding tools: Food security and climate resilience. *Plants* **2021**, *10*, 1347. [CrossRef]
28. Sattar, M.N.; Iqbal, Z.; Tahir, M.N.; Shahid, M.S.; Khurshid, M.; Al-Khateeb, A.A.; Al-Khateeb, S.A. CRISPR/Cas9: A practical approach in date palm genome editing. *Front. Plant Sci.* **2017**, *8*, 1469. [CrossRef] [PubMed]

29. Sattar, M.N.; Iqbal, Z.; Al-Khayri, J.M. CRISPR-Cas Based Precision Breeding in Date Palm: Future Applications. In *The Date Palm Genome*; Springer: Cham, Switzerland, 2021; Volume 2, pp. 169–199.
30. Saleh, A.A.; Sharafaddin, A.H.; El Komy, M.H.; Ibrahim, Y.E.; Hamad, Y.K. Molecular and physiological characterization of Fusarium strains associated with different diseases in date palm. *PLoS ONE* **2021**, *16*, e0254170. [CrossRef] [PubMed]
31. Saleh, A.A.; Sharafaddin, A.H.; El Komy, M.H.; Ibrahim, Y.E.; Hamad, Y.K.; Molan, Y.Y. Fusarium species associated with date palm in Saudi Arabia. *Eur. J. Plant Pathol.* **2017**, *148*, 367–377. [CrossRef]
32. Molan, Y.; Al-Obeed, R.; Harhash, M.; El-Husseini, S. Decline of date-palm offshoots with *Chalara paradoxa* in Riyadh region. *J. King Saud. Univ.* **2004**, *16*, 79–86.
33. Alwahshi, K.J.; Saeed, E.E.; Sham, A.; Alblooshi, A.A.; Alblooshi, M.M.; El-Tarabily, K.A.; AbuQamar, S.F. Molecular identification and disease management of date palm sudden decline syndrome in the United Arab Emirates. *Int. J. Mol. Sci.* **2019**, *20*, 923. [CrossRef] [PubMed]
34. Abdullah, S.K.; Lorca, L.; Jansson, H. Diseases of date palms (*Phoenix dactylifera* L.). *Basrah J. Date Palm Res* **2010**, *9*, 1–44.
35. Santos, Á.F.D.; Inácio, C.A.; Guedes, M.V.; Tomaz, R. First report of *Thielaviopsis paradoxa* causing stem rot in *Dracaena marginata* in Brazil. *Summa Phytopathol.* **2012**, *38*, 345–346. [CrossRef]
36. Gepp, V.; Hernández, L.; Alaniz, S.; Zaccari, F. First report of *Thielaviopsis paradoxa* causing palm fruit rot of Butia capitata in Uruguay. *New Dis. Rep* **2013**, *27*, 12. [CrossRef]
37. Soytong, K.; Pongak, W.; Kasiolarn, H. Biological control of Thielaviopsis bud rot of *Hyophorbe lagenicaulis* in the field. *J. Agric. Technol.* **2005**, *1*, 235–245.
38. Pinho, D.B.; Dutra, D.C.; Pereira, O.L. Notes on *Ceratocystis paradoxa* causing internal post-harvest rot disease on immature coconut in Brazil. *Trop. Plant Pathol.* **2013**, *38*, 152–157. [CrossRef]
39. Saeed, E.E.; Sham, A.; Salmin, Z.; Abdelmowla, Y.; Iratni, R.; El-Tarabily, K.; AbuQamar, S. *Streptomyces globosus* UAE1, a potential effective biocontrol agent for black scorch disease in date palm plantations. *Front. Microbiol.* **2017**, *8*, 1455. [CrossRef]
40. Tzelepis, G.; Bejai, S.; Sattar, M.N.; Schwelm, A.; Ilbäck, J.; Fogelqvist, J.; Dixelius, C. Detection of Verticillium species in Swedish soils using real-time PCR. *Arch. Microbiol.* **2017**, *199*, 1383–1389. [CrossRef]
41. Croft, A.R. Improving germination of sugarcane and the control of pineapple disease. *Proc. Aust. Soc. Sugarcane Technol.* **1998**, *20*, 300–306.
42. Dellaporta, S.; Wood, J.; Hicks, J. A rapid method for DNA extraction from plant tissue. *Plant Mol. Biol. Rep.* **1983**, *1*, 19–21. [CrossRef]
43. White, T.J.; Bruns, T.; Lee, S.; Taylor, J. Amplification and direct sequencing of fungal ribosomal RNA genes for phylogenetics. *PCR Protoc. Guide Methods Appl.* **1990**, *18*, 315–322.
44. Jacobs, K.; Bergdahl, D.R.; Wingfield, M.J.; Halik, S.; Seifert, K.A.; Bright, D.E.; Wingfield, B.D. *Leptographium wingfieldii* introduced into North America and found associated with exotic *Tomicus piniperda* and native bark beetles. *Mycol. Res.* **2004**, *108*, 411–418. [CrossRef]
45. Glass, N.L.; Donaldson, G.C. Development of primer sets designed for use with the PCR to amplify conserved genes from filamentous ascomycetes. *Appl. Environ. Microbiol.* **1995**, *61*, 1323–1330. [CrossRef] [PubMed]
46. Kumar, S.; Stecher, G.; Li, M.; Knyaz, C.; Tamura, K. MEGA X: Molecular evolutionary genetics analysis across computing platforms. *Mol. Biol. Evol.* **2018**, *35*, 1547. [CrossRef]
47. Jonathan, S.; Udoh, E.; Ojomo, E.; Olawuyi, O.; Babalola, B. Efficacy of Jatropha curcas Linn. as fungicides in the control of Ceratocystis paradoxa (Chalara anamorph) IMI 501775 associated with bole rot of Cocos nucifera Linn. seedlings. *Rep. Opin* **2012**, *4*, 48–60.

plants

Article

YOLO-JD: A Deep Learning Network for Jute Diseases and Pests Detection from Images

Dawei Li [1,2,3,†], Foysal Ahmed [1,†], Nailong Wu [1,3,*] and Arlin I. Sethi [4]

1 College of Information Sciences and Technology, Donghua University, Shanghai 201620, China; daweili@dhu.edu.cn (D.L.); foysal.9@outlook.com (F.A.)
2 State Key Laboratory for Modification of Chemical Fibers and Polymer Materials, Donghua University, Shanghai 201620, China
3 Engineering Research Center of Digitized Textile and Fashion Technology, Ministry of Education, Donghua University, Shanghai 201620, China
4 Department of Chemistry, Faculty of Science, National University of Bangladesh, Gazipur, Dhaka 1704, Bangladesh; arlinishrat10@gmail.com
* Correspondence: nathan_wu@dhu.edu.cn
† These authors contributed equally to this work.

Abstract: Recently, disease prevention in jute plants has become an urgent topic as a result of the growing demand for finer quality fiber. This research presents a deep learning network called YOLO-JD for detecting jute diseases from images. In the main architecture of YOLO-JD, we integrated three new modules such as Sand Clock Feature Extraction Module (SCFEM), Deep Sand Clock Feature Extraction Module (DSCFEM), and Spatial Pyramid Pooling Module (SPPM) to extract image features effectively. We also built a new large-scale image dataset for jute diseases and pests with ten classes. Compared with other state-of-the-art experiments, YOLO-JD has achieved the best detection accuracy, with an average mAP of 96.63%.

Keywords: Jute; disease detection; deep learning; YOLO-JD; image processing

Citation: Li, D.; Ahmed, F.; Wu, N.; Sethi, A.I. YOLO-JD: A Deep Learning Network for Jute Diseases and Pests Detection from Images. *Plants* **2022**, *11*, 937. https://doi.org/10.3390/plants11070937

Academic Editor: Alessandro Vitale

Received: 26 February 2022
Accepted: 27 March 2022
Published: 30 March 2022

1. Introduction

Jute (*Corchorus olitorius* L. or *C. capsularis* L.) is one of the most important fiber crops and an inexpensive fiber source of high quality. It is also referred to as the "golden fiber" crop. Jute is a strong fiber that is soft, lustrous, and relatively lengthy. In the Indo-Bangladesh subcontinent, commercial jute farming is mostly limited to the latitudes of 80°18′ E–92° E and 21°24′ N–26°30′ N [1]. Jute is a herbaceous annual that may grow to a height of 10 to 12 feet (3 to 3.6 m) and has a cylindrical stalk about the thickness of a finger. It is said to have originated in the Indian subcontinent. The main differences between the two jute species cultivated for fiber lie in the form of their seed pods, growth habits, and fiber qualities. Most types of jute prefer well-drained sandy loam in warm and humid areas with at least 3 to 4 inches (7.5 to 10 cm) of monthly rainfall during the growing season. The light green leaves of the plant are usually 4 to 6 inches (10 to 15 cm) long, 2 inches (5 cm) broad, serrated on the margins, and taper to a point.

Jute is a biodegradable natural polymer that decomposes quickly in the environment. On the other hand, although synthetic polymers including polystyrene, polyethylene, polypropylene, and polyvinyl chloride offer better mechanical qualities, sustainability, and durability than natural polymers for producing plastics, they are not bio-degradable and can seriously pollute the environment [2]. Plastic pollution is one of the biggest environmental issues nowadays. In 2019, plastic manufacturing and incineration produced more than 850 million metric tons of greenhouse gases, the equivalent of the emissions from 189 coal power plants with a 500 megawatt capacity [3]. The most serious issue with plastic is that it does not decompose in the environment and has accumulated for decades in streams, agricultural soils, rivers, and the ocean [3]. To protect the environment, it is necessary to substitute synthetic polymers with bio-degradable and ecologically friendly polymers. Therefore, natural jute polymer has become increasingly popular in both domestic and international markets. The process of jute fiber production comprises several steps (shown in Figure 1). In addition, due to the importance of jute production, the disease prevention in the jute species has become an urgent task of precision agriculture.

Figure 1. Demonstration of the full process of jute manufacturing. The production process comprises at least six steps, starting from harvesting to the binding of jute fiber.

Plant diseases and pests are a global threat to crop yields, and they may be even more destructive for smallholder farmers whose livelihoods depend heavily on healthy harvests. Unfortunately, jute is still usually cultivated by smallholder farmers. Disease symptoms appear on leaves, fruits, buds, and young branches on jute plants. Jute diseases come in a variety of types, each of which can result in big economic loss. Recently, disease prevention has become increasingly significant as a result of the demand for finer quality fiber [4]. Precision plant protection offers a non-destructive means of managing plant diseases based on the concept of spatio-temporal variability [5,6], and those works have inspired us to transplant new technology from the computer vision and artificial intelligence fields to detect and manage plant diseases.

In this scenario, early and precise detection of plant diseases and pests is critical for avoiding losses in agricultural production. Traditionally, the detection of plant diseases and pest is manually performed by experts such as botanists and agricultural engineers. The disease investigation usually begins with a visual assessment and then a laboratory test. Detection of plant diseases and pests by visual inspection is extremely beneficial for new farmers. Traditional approaches are typically time consuming and need complex procedures, as well as some specialized knowledge. Therefore, during the past several years, researchers have used image processing and machine learning techniques to detect or classify plant diseases. For example, Maniyath et al. [7] proposed a classification architecture using a machine learning approach to detect plant diseases and pests. Gavhale et al. [8] proposed a framework using K-means clustering to recognize the defects and areas of disease on plant leaves. Hossain et al. [9] used a Support Vector Machine (SVM) to recognize the diseases on tea leaves.

Deep learning (DL) has previously been proven to be successful for real-life object identification, recognition, and classification [10]. The agricultural industry has resorted to DL-based models for the solution. State-of-the-art outcomes have been achieved using deep learning approaches on tasks such as plant identification, fruit harvesting, and crop/weed classification. Recent research has also concentrated on the detection of plant disease [11]. Convolutional neural networks such as YOLOv3 [12], YOLOv4 [13], Faster R-CNN [14], Mask R-CNN [15], and SSD [16] were successfully applied in crop disease detection. For example, Hammad et al. [17] realized image-based plant disease identification by meta-architectures based on deep learning. Chowdhury et al. [18] proposed a deep learning model based on EfficientNet and they used 18,161 tomato leaf images to classify tomato diseases. A previous article by Mohanty et al. [19] using AlexNet and GoogleNet models was able to identify 14 crop species and 26 diseases from images. They used a dataset comprising 54,306 images of both diseased and healthy plant leaves. Görlich et al. [20] proposed a UAV-Based classification of Cercospora leaf spot disease on RGB images. Chen et al. [21] designed a model that could automatically detect rubber tree diseases on images using an improved YOLOv5 model. Arsenovic et al. [22] proposed a new deep learning approach to detect 13 different types of plant diseases. Vishnoi et al. [23] developed two different DL approaches to detect diseases in the PlantVillage dataset. Wagle et al. [24] proposed a CNN model with transfer learning from AlexNet to detect nine species of plants from the PlantVillage dataset.

There are still several existing limitations in disease identification for jute plants; e.g., (i) the lack of a dataset for jute diseases; (ii) the majority of the current approaches on plant disease detection are based on traditional machine learning methods, generating unsatisfactory performances; (iii) the research on jute disease detection via image processing is rare; and (iv) it is difficult to implement multi-class disease detection because different diseases have very diversified appearances.

To transcend the mentioned above limitations, we formulate the objectives of this research as follows. The first objective of this research is to establish a brand new image dataset for jute diseases and pests with accurate manual labels, which should not only include several thousands of images captured at different environments and weather conditions, but should also incorporate multiple disease/pest classes. To the best of our knowledge, the dataset will be the first published jute disease and pest image dataset in the field. The second objective is to explore new network architectures and modules under deep learning that is fit for crop diseases detection, especially for jute diseases. The new network is also expected to outcompete some popular networks designed for object detection. The third objective is to validate the application feasibility of the deep learning models in YOLO-family on detection (or recognition) of jute diseases and pests, and provide guidance for scientists working on both agricultural engineering and artificial intelligence.

The content of this paper is structured as follows. The Jute diseases and pests dataset and the architecture of our detection model YOLO-JD are specified in Section 2. Experimental results with the ablation study are provided in Section 3. Discussion of the results takes place in Section 4. Conclusions are drawn in the last section.

2. Material and Methods

2.1. Dataset

The images of jute diseases and pests were collected at Jamalpur and Narail districts in Bangladesh in July 2021. To diversify the dataset, the images were captured over the course of a single day under both sunny and cloudy weather. The images were captured by a Canon Powershot G16 camera and the camera of a Samsung Galaxy S10 with different viewing angles and different distances (0.3–0.5 m). In total, 4418 images in multiple jute disease and pest classes were obtained. The light intensity and background circumstance of the images vary greatly in the dataset. Though the image sizes are not uniform in our dataset, we prepare a normalization step at the beginning of the network to unify all images to a fixed resolution of 640×640. Eight common diseases including stem rot, anthracnose, black band, soft rot, tip blight, dieback, jute mosaic, and jute chlorosis, as well as two pests—*Jute Hairy Caterpillar*, and *Comophila sabulifers*—are incorporated into our dataset. Some of the sample images are displayed in Figure 2, and the symptomatic patterns and causes [25] of all Jute diseases and pests are listed in Table 1, respectively.

Figure 2. Some sample images from our jute diseases and pests dataset. (**a**) the stem rot disease. (**b**) the anthracnose disease. (**c**) the black band disease. (**d**) the soft rot disease. (**e**) the tip blight disease. (**f**) the dieback disease. (**g**) the jute mosaic disease. (**h**) the jute chlorosis disease. (**i**) the *Cosmophila sabulifera* caterpillar on a jute leaf. (**j**) some *Hairy Caterpillars* on a jute leaf.

Table 1. The indices and causes of Jute diseases and pests.

Index	Name of Disease/Pest	Cause	Causal Organism
D-1	Stem rot	Fungal	*Macrophomina phaseolina* (Tassi) Goid.
D-2	Anthracnose	Fungal	*Colletotrichum corchorum* Ikata and Tanaka; *C. gloeosporioides* (Penz.) Penz and Sacc.
D-3	Black band	Fungal	*Botryodiplodia theobromae* (Pat.) Griff and Maubl.
D-4	Soft rot	Fungal	*Sclerotium rolfsii* Sacc. (*Athelia rolfsii*)
D-5	Tip blight	Fungal	*Curvularia subulata* (Nees ex Fr.) Boedijn
D-6	Die back	Fungal	*Diplodia corchori* Syd. and P. Syd.
D-7	Jute mosaic	Viral	A Begomovirus of the Geminiviridae family, vector: *Bemisia tabaci* Genn. (Whitefly).
D-8	Jute Chlorosis	Viral	A member of Tobravirusgenus
P-1	*Cosmophila sabulifera*	Pest	
P-2	*Hairy Caterpillar*	Pest	

Stem rot usually causes long and blackened rotted areas on the main stem of jutes. This disease is economically the most serious disease for jute. Stem rot reduces the yield of fiber both quantitatively and qualitatively, and even produces infected seeds to the next generation. The symptoms of anthracnose disease are sunken spots of various colors on different parts of plants, usually observed on stems. Irregular spots of anthracnose disease often cause deep necrosis spots on stems, and may further result in cracks on the fiber, and even the withering of the infected plant. The disease can also infect jute seeds; the infected ones are lighter in color, with shrunken shapes and poor germination. Black band was a minor disease in the past, but now it has become more prevalent due to climate change. Black band disease causes dark-colored areas on the infected stem, together with the defoliation of plants. Initially, it may often be confused with Stem rot because the infected areas are both spot-like. Soft rot disease is still a common fungal disease on jute plants. The disease may appear on all growing areas of a jute plant but the intensity of the disease is usually low. Attack of soft rot happens when the Jute crop reaches 80–90 days old. The fungus grows from soil and later slowly infects fallen leaves of jute, and from there it goes up to the stem base and then travels to other parts of the plant. In the past, tip blight was a minor disease but now it has developed into different new varieties. The disease causes the blighting of newly emerged sprouts at the tip of young plants. The infected sprouts turn from green to black and then slowly become rotten in high humidity. Dieback disease is relatively rare. The dieback disease usually happens at the top of the plant, and leaves begin to droop and wither, they later become dried up. Infected branches slowly turn brown and later black, and remain attached as dead and dry parts. Jute mosaic is a common disease nowadays. The disease creates small yellow dots (like flakes) on the leaf lamina in the early stage, then gradually the dots enlarge themselves to become yellow mosaics on leaves. Chlorosis of jute causes yellow chlorotic spots with sharp margins on the leaves. The *Cosmophila sabulifera* and the *Hairy Caterpillar* are two common pests on jute leaves. The first caterpillar is much larger than the latter one, and individuals of *Hairy Caterpillar* usually are inclined to aggregate into a flock.

The entire preparation procedure of our jute disease and pest dataset is as follows. First, we apply image pre-processing methods such as brightness correction andimage filtering on sample images to enhance the quality of the dataset. In the dataset, 556 images were selected to form the testing dataset, and the rest of the 3862 images were used to form the training set. Then, an annotation software called 'LabelImg' [26] was used to draw the ground truth bounding boxes of the disease or pests in all images.

2.2. Overall Architecture

Based on YOLOv4, the YOLOv5 improved in terms of both detection performance and computational complexity, making it to be perhaps the most popular solution for object detection tasks nowadays. Despite its popularity, the standard YOLOv5 still has a problem in generalization and domain adaptation (e.g., a performance decline can be observed on our jute dataset when applying YOLOv5). Inspired from YOLOv5, this research proposes a unique model: YOLO-JD, for detection and recognition of jute disease and pests by evaluating the architecture. Figure 3 shows the overall architecture of the proposed YOLO-JD, which can be divided into three main components—(i) the head (backbone) component, a backbone network that uses the Sand Clock Feature Extraction Module (SCFEM), Spatial Pyramid Pooling Module (SPPM), and the Deep Sand Clock Feature Extraction Module (DSCFEM) to extract features at different levels; (ii) the neck component, that collects cross-stage features extracted from three different layers of the head component, and then generates three different high-level feature maps; and (iii) the detection component, that incorporates anchor results under different scales to create an aggregated detection box. The full architecture of YOLO-JD also contains several kinds of compact operations and calculation steps such as CBL (Conv2D + Batch Normalization + Leaky ReLU activation), NMS (Non-max Suppression), Up-sampling (Us), and Concatenation.

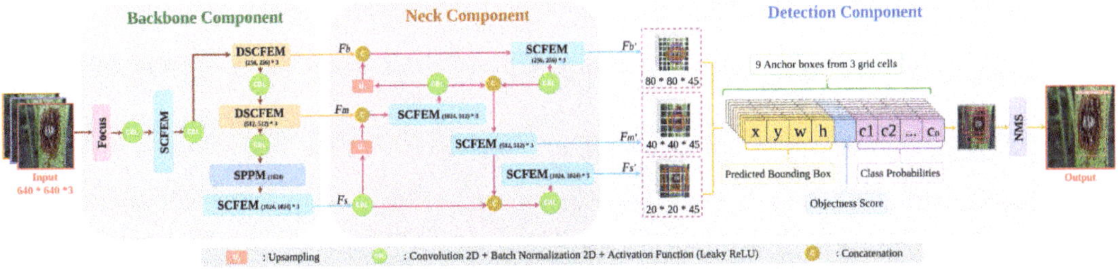

Figure 3. The overall architecture of YOLO-JD. The architecture contains three components—the head (backbone) component, the neck component, and the detection component. Structures of the three new modules: SPPM, SCFEM, and DSCFEM are detailed in Figure 4.

In the head component, the input feature dimension changes from $640 \times 640 \times 3$ to $320 \times 320 \times 32$ after the focus module with a shuffling scheme shown in Figure 3. The features then pass through several different operations and modules such as SCFEM, CBL, DSCFEM, and SPPM, and generate a multi-level output for the neck component. The multi-level output includes three feature maps, two of which are the outputs of DSCFEMs, and the other one is the output of SCFEM. In each CBL operation, we sequentially carry out Conv2D, batch normalization, and the Leaky ReLU activation. The DSCFEMs are used in the first component of YOLO-JD network to collect important low-level image features. The SCFEMs are applied mainly in the middle component for the extraction of mid-level features.

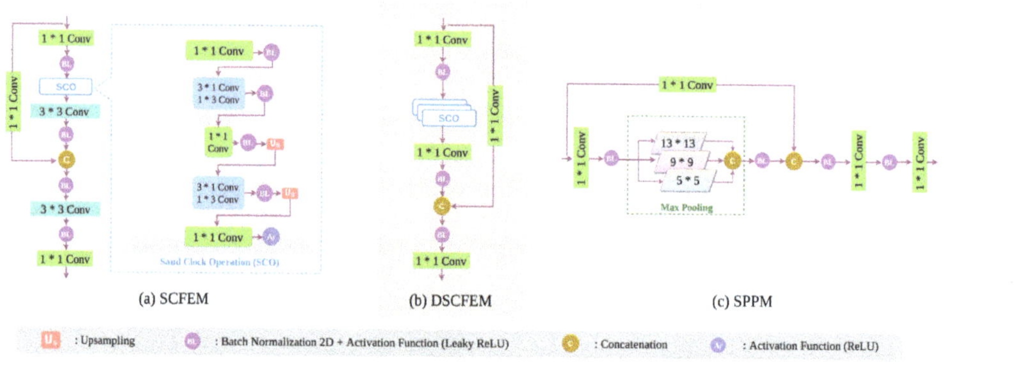

Figure 4. The detailed demonstration of several key modules in YOLO-JD. (**a**) Shows the architecture of the Sand Clock Feature Extraction Module (SCFEM). (**b**) Shows the details of the Deep Sand Clock Feature Extraction Module (DSCFEM). (**c**) Shows the Spatial Pyramid Pooling Module (SPPM).

Finally, the detection component is a standard scheme inherited from the YOLO family and it creates multi-scale grids for detecting objects with different sizes (e.g., grid size 8×8 for detecting small objects, grid size 16×16 for detecting medium objects, and grid size of 32×32 for detecting big objects). After that, we use 1×1 convolution to combine all feature maps to create 9 different anchor results and carry out K-means clustering to combine 9 anchor boxes on the feature map output from the previous layer. In the final stage, the Non-max Suppression (NMS) operation was applied to select only one bounding box out of many overlapping ones as the final detection.

2.2.1. Sand Clock Feature Extraction Module (SCFEM)

The Sand Clock Feature Extraction Module (SCFEM) is designed to extract high-quality mid-level features from the input image. The detailed architecture of SCFEM is given in Figure 4a. The backbone component of YOLO-JD contains two SCFEM modules, and we insert four SCFEM modules in the neck component of YOLO-JD. The SCFEM module contains an important feature extraction block called the Sand Clock Operation (SCO). In SCO there are five steps. The first step is a 1×1 conv followed by a BL operation (batch normalization + Leaky ReLU activation), then the second step uses two spatially separable convolutions (3×1 conv + 1×3 conv) followed by BL to abstract features. The third step has a 1×1 conv followed by a BL operation, which creates a "thin" feature map. The fourth step is similar to the second one. And the last step of SCO is still a 1×1 conv followed by an activation function such as ReLU. When passing through the calculation of the five steps above, the feature maps first gradually become small and then become bigger, taking the shape of a sand clock. Thus, the block is named SCO. The step of two spatially separable convolutions (3×1 conv + 1×3 conv) is used to replace the traditional 3×3 conv because the former has fewer parameters to compute. In SCFEM, we also use multiple 1×1 convs and 3×3 convs, as well as a skip connection, to enhance its feature extraction ability. In the neck component of YOLO-JD, three SCFEMs are applied to generate three high-level feature layers with different scales to serve the following detection purposes, respectively.

2.2.2. Deep Sand Clock Feature Extraction Module (DSCFEM)

The biggest difference between Deep Sand Clock Feature Extraction Module (DSCFEM) and SCFEM is that there are three consecutive SCO blocks in the DSCFEM (Figure 4b). This structure makes the structure of DSCFEM much deeper than SCFEM, and also explains why the name begins with "deep". The second difference between DSCFEM and SCFEM is that we only use 1×1 convs in the part outside SCOs. As the DSCFEM has a deeper design than SCFEM, the wide usage of 1×1 convs rather than 3×3 convs can reduce the network parameters, and shortens the training and referencing time. DSCFEM is applied two times in the backbone component of YOLO-JD. The module is responsible for the efficient abstraction of low-level image features.

2.2.3. Spatial Pyramid Pooling Module (SPPM)

The Spatial Pyramid Pooling Module (SPPM) works only once in the backbone component and its feature output becomes the input of the SCFEM with the highest resolution. The SPPM (shown in Figure 4c) first creates a feature pyramid and then uses convolutions to integrate the features under different scales. In SPPM, the network learns the object features with different receptive fields, and feature maps with multiple receptive fields are naturally fused to create an effective feature embedding that has both local and global focuses. In implementation, we use three different convolution kernels to generate multi-layer pyramid maps, which are then separately max-pooled and concatenated to generate the output. Due to its multi-scale feature extraction, SPPM may enhance the recognition of objects at varied sizes.

2.3. Loss Functions

A comprehensive loss function is designed for training YOLO-JD, and this loss function contains several different sub-losses.

The first sub-loss is the Intersection over Union (*IoU*) loss, which descends from a basic criterion used frequently in target detection and tracking. *IoU* is defined as the ratio of the intersection of the prediction box B^p and its Ground Truth (GT) box B^{gt} to the union of the prediction and its GT, and *IoU* loss is given as:

$$L_{IoU}(B^p, B^{gt}) = 1 - \frac{|B^p \cap B^{gt}|}{|B^p \cup B^{gt}|}. \tag{1}$$

In most cases, *IoU* can reasonably evaluate the detection performance. However, when there is no intersection between a prediction and its GT, the *IoU* loss reaches the maximum value 1.0. It then becomes impossible to distinguish the relative distance between the area of GT and the predicted area since a very bad prediction (very far away) and a not so bad prediction (near but still no intersection) are punished with the same L_{IoU}. To improve the training, we resort to *CIoU* loss (*Complete IoU*) [27], a generalized *IoU* sub-loss that takes three geometric factors into account—i.e., the overlapping factor (the standard *IoU* loss), the distance factor D, and the aspect ratio factor V. The L_{CIoU} can then be defined as follows,

$$L_{CIoU} = L_{IoU}(B^p, B^{gt}) + D(B^p, B^{gt}) + V(B^p, B^{gt}), \tag{2}$$

where D and V denote the distance factor and the aspect ratio factor, respectively. The distance factor $D(B^p, B^{gt})$ is also a binary function that accepts the information of the GT area and the prediction. The equation of the distance factor can be written as

$$D(B^p, B^{gt}) = \frac{(b^p - b^{gt})^2}{c^2}, \tag{3}$$

in which b^p and b^{gt} are of the central point coordinates of boxes B^p and B^{gt}, respectively. In addition, c is the diagonal length of the bounding box that circumscribes B^p and B^{gt}. The distance factor $D(B^p, B^{gt})$ can be regarded as a normalized distance between two boxes.

The aspect ratio factor can be calculated via the following equation,

$$V = \frac{4}{\pi^2}\left(\arctan\frac{\omega^{gt}}{h^{gt}} - \arctan\frac{\omega^p}{h^p}\right)^2, \tag{4}$$

where h and ω represent the width and height of a bounding box, respectively. The lower the L_{CIoU} is, the better the predicted box approximates the ground truth.

3. Experiments

3.1. Evaluation Metrics

For our jute disease dataset, each detected bounding box can be categorized into three cases. The true positive (*TP*) indicates that the detected box has an *IoU* value (defined as $|B^p \cap B^{gt}| / |B^p \cup B^{gt}|$) higher than 50% against its ground truth box. The false positive (*FP*) indicates that the detected box has an *IoU* value lower than 50%. The false negative (*FN*) indicates a ground truth box that is not covered by any detection. Based on *TP*, *FP*, and *FN*, we define *Precision* (Prec), *Recall* (Rec), and F1-measure (*F*1). Precision reflects the correctness of a model in all detected boxes. It is defined as the ratio of the number of TPs to the number of all detected bounding boxes:

$$Precision = \frac{TPs}{TPs + FPs}. \tag{5}$$

Recall reflects the ability of a model to cover all ground truth bounding boxes. It is defined as the ratio of the number of *TPs* to the number of all bounding boxes in ground truth:

$$Pecall = \frac{TPs}{TPs + FNs}. \tag{6}$$

*F*1-measure is a combination of *Precision* and *Recall*, and it is defined as follows,

$$F1 = 2 \times \frac{Precision \times Recall}{(Precision + Pecall)}. \tag{7}$$

The mean average precision (*mAP*) is defined as:

$$mAP = \frac{1}{C} \sum_{i=1}^{C} Precision(i). \tag{8}$$

In Equation (8), *C* is the number of total disease classes, and $Precision(i)$ (shown by (5)) stands for the precision of each disease class.

3.2. Training Details

We implemented YOLO-JD by PyTorch and trained it on a single GPU (Nvidia RTX 2080). The YOLO-JD model runs on a computer with an AMD 3700x CPU under the Ubuntu 18.04 operating system. We used a learning rate of 0.02 for the first 100 epochs and then 0.01 for the last 50 epochs with a mini-batch size of 8. We trained our network using Adam optimizer (SGD momentum rate at 0.937, weight decay rate at 0.005, epoch's warm-up rate at 3.0, and warm-up initial momentum rate at 0.8).

3.3. Quantitative Results

To prove the effectiveness of YOLO-JD, we compare it with only detection models from the YOLO family. This is because nowadays the YOLO family holds the best image object detection performance in various applications. The contrasted models include YOLOv3 [12], YOLOv3(tiny) [12], YOLOv4 [13], YOLOv4(tiny) [14], YOLOv5-s [28], YOLOv5-m [28], YOLOv5-l [28], YOLOv5-x [28]. Except for YOLO-JD, we obtained pre-trained models on the COCO dataset [29] for all other methods compared, and we then conduct transferred learning on the jute dataset to speed up the convergence, respectively. Different from all others, our YOLO-JD was trained directly on the jute dataset. Table 2 reports the quantitative results on our jute disease test images. Our YOLO-JD achieved the best performance on all four metrics: *Prec*, *Rec*, *F*1, and *mAP*.

Table 2. The quantitative comparison of several methods including YOLO-JD on the Jute disease test dataset. The best measures are in boldface.

Measures	Methods	D-1	D-2	D-3	D-4	D-5	D-6	D-7	D-8	P-1	P-2	Mean
	YOLOv3	73.43	78.71	78.51	73.81	83.22	97.81	98.32	94.40	80.13	74.23	83.26
	YOLOv4	83.33	81.94	80.03	85.71	84.64	97.77	97.31	91.07	82.89	76.10	86.08
	YOLOv3 Tiny	69.23	71.13	68.01	75.60	88.81	94.72	98.03	96.29	79.01	71.94	81.27
	YOLOv4 Tiny	67.74	82.85	79.16	81.25	82.98	97.43	44.54	92.59	82.89	77.77	78.92
Prec (%)	YOLOv5s	83.82	82.91	71.22	75.92	89.41	**98.91**	96.21	97.31	84.40	82.12	86.22
	YOLOv5m	69.13	65.34	69.54	74.52	85.55	97.84	96.44	**98.80**	80.33	70.33	80.78
	YOLOv5l	69.51	71.75	73.10	81.52	88.62	98.34	96.15	96.22	85.75	78.41	83.93
	YOLOv5x	68.24	66.62	72.71	78.21	89.70	98.82	96.71	98.15	83.31	78.15	83.06
	YOLO-JD (ours)	**98.34**	**96.21**	**95.82**	**97.10**	**95.31**	98.10	**98.65**	97.90	**91.60**	**92.90**	**96.19**
	YOLOv3	81.73	83.98	71.74	85.46	97.53	92.72	96.11	92.74	83.76	86.73	87.25
	YOLOv4	85.64	90.17	68.93	**94.87**	87.64	91.76	94.15	90.92	75.36	76.98	85.64
	YOLOv3 Tiny	78.64	87.53	73.51	91.24	93.74	93.83	95.81	89.83	86.36	89.36	87.98
	YOLOv4 Tiny	74.63	89.74	77.93	83.76	78.54	94.62	**97.33**	81.02	88.36	73.72	83.96
Rec (%)	YOLOv5s	78.83	84.61	59.70	82.91	97.33	95.90	95.71	91.22	92.51	82.91	86.16
	YOLOv5m	84.81	89.43	74.11	89.72	**98.75**	91.33	92.24	95.35	96.23	89.30	90.12
	YOLOv5l	78.84	85.92	75.93	91.33	97.95	91.85	93.91	89.20	93.40	83.65	88.19
	YOLOv5x	78.12	88.55	73.91	92.94	97.71	91.89	91.73	89.41	93.61	84.21	88.20
	YOLO-JD (ours)	**92.41**	**98.62**	**86.92**	93.22	98.10	**96.75**	96.71	**96.21**	**97.31**	**95.20**	**95.14**
	YOLOv3	77.11	81.25	74.96	79.20	89.79	95.19	**96.98**	93.56	81.88	79.97	84.99
	YOLOv4	84.46	85.85	74.05	82.68	86.11	94.66	95.54	90.99	78.94	76.53	84.98
	YOLOv3 Tiny	73.63	78.48	70.64	90.05	91.20	93.91	97.47	92.94	82.52	79.70	85.05
	YOLOv4 Tiny	71.01	86.15	78.54	82.48	80.69	96.00	60.53	86.41	85.53	75.69	80.30
F1 (%)	YOLOv5s	81.22	83.74	64.94	79.24	93.18	97.85	95.59	94.15	88.26	82.49	86.06
	YOLOv5m	76.14	75.47	71.72	81.10	91.75	94.43	94.05	97.60	87.53	78.66	85.84
	YOLOv5l	73.85	77.74	74.47	85.98	93.01	95.40	94.05	92.56	89.00	80.91	85.69
	YOLOv5x	72.81	76.00	73.29	84.54	93.14	95.41	94.57	93.54	88.15	81.03	85.24
	YOLO-JD (ours)	**95.25**	**97.38**	**92.55**	**94.04**	**96.11**	**97.94**	96.83	**98.01**	**95.29**	**94.03**	**95.74**
	YOLOv3	89.23	88.42	85.43	96.13	96.41	97.72	98.53	95.72	90.10	84.75	92.24
	YOLOv4	86.10	91.44	83.33	97.55	96.55	96.91	97.51	98.33	94.74	89.25	93.17
	YOLOv3 Tiny	89.42	88.51	85.52	96.52	96.72	98.52	98.72	95.91	85.92	85.91	92.16
	YOLOv4 Tiny	78.53	84.71	82.51	91.71	90.33	97.88	76.22	94.90	92.77	80.33	86.98
mAP (%)	YOLOv5s	79.71	87.34	63.57	96.31	97.85	96.40	98.51	93.25	93.35	81.50	88.78
	YOLOv5m	79.83	87.55	68.60	95.72	98.40	96.21	98.73	97.41	92.55	83.14	89.82
	YOLOv5l	79.94	86.73	75.61	96.54	98.31	96.33	98.75	93.92	93.41	84.35	90.39
	YOLOv5x	80.22	88.61	72.23	**96.91**	**98.80**	94.35	**98.90**	95.95	94.40	84.31	90.47
	YOLO-JD (ours)	**97.21**	**96.10**	**89.40**	94.23	98.61	**98.50**	98.10	**98.70**	**96.90**	**97.30**	**96.63**

3.4. Qualitative Results

The qualitative comparison across YOLO-JD and other models is given in Figure 5. We choose one example from each disease category of the testing dataset. The first column of Figure 5 shows the input test images; the second column gives the ground truth images. The results of our YOLO-JD are given by the third column. The other eight state-of-the-art methods are shown in the fourth to the eleventh columns, respectively. Each row of Figure 5 stands for a type of disease, and the rows are arranged in the same order as in Table 1. For example, the first row of Figure 5 shows detection results of the stem rot disease of all models, in which YOLO-JD is the most similar to the ground truth bounding box. The 2nd row of Figure 5 shows detection results of another disease—anthracnose; YOLOv3, YOLO-v3(tiny), YOLOv5-m, and YOLOv5-l all have an extra false detection box, and our YOLO-JD detects the disease area with high accuracy. The 3rd row shows the results of detecting the disease of black band, our YOLO-JD avoids false positives and is almost identical to ground truth. On the fourth, fifth, sixth, seventh, and eighth rows, our method is still the closest to the ground truth across all models compared. The ninth and tenth rows show two test images of jute pests, respectively. YOLO-JD successfully detected the accurate pest areas, and our results are the closest to the ground truth across all models compared.

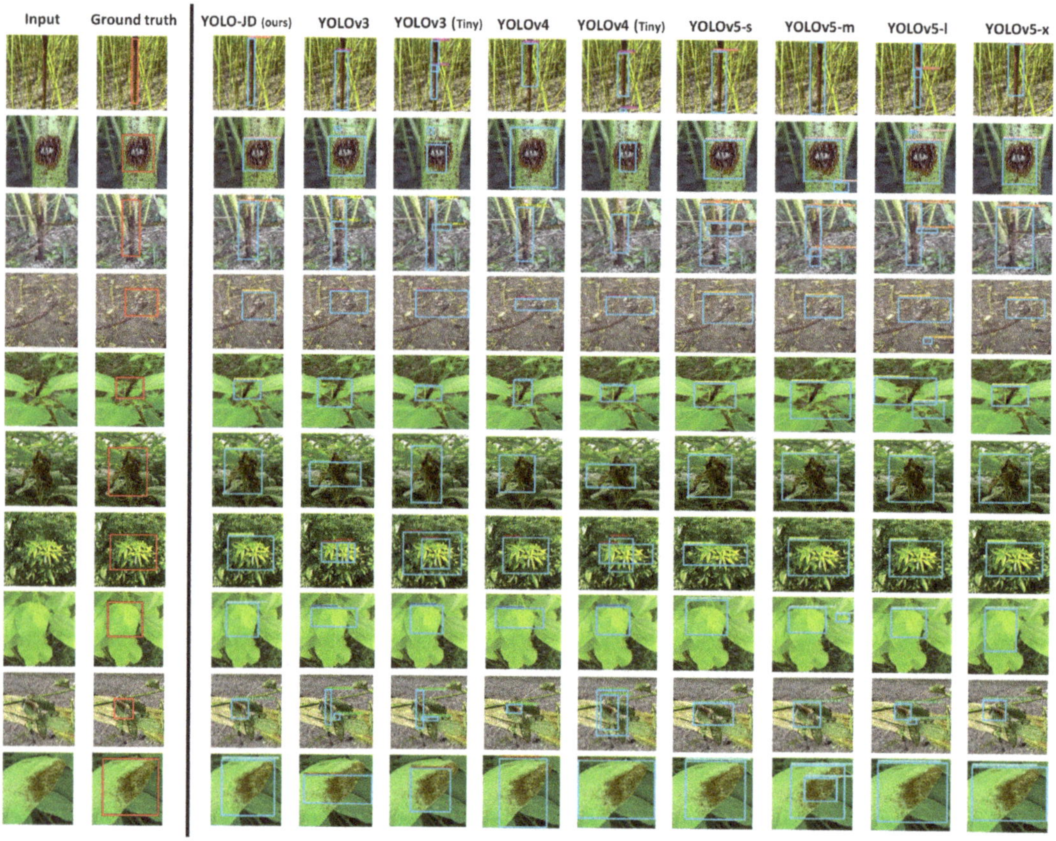

Figure 5. Qualitative comparison between our YOLO-JD and eight other models on Jute disease detection.

YOLO-JD has ability to locate multiple instances of a disease on the same image, and is also able to detect multiple classes of diseases and pests on the same image. Figure 6 shows that our YOLO-JD successfully detects multiple disease cases from three images in the testing dataset. Figure 6a contains 1 case of D-2 and 3 cases of P-1. Figure 6b contains 1 case of D-2 and 1 case of P-1. Figure 6c contains 1 case of D-6 and 1 case of P-1.

3.5. Ablation Analysis

To prove the independent contribution of each module to the total performance of the new modules added in YOLO-JD, we perform a simple but effective ablation analysis on the Jute disease dataset. The results of all ablation cases are shown in Table 3. In the "A1" version of our model, we replaced the SPPM module with the original Spatial Pyramid Pooling structure in the standard YOLOv5 while remaining all other parts unchanged. In "A2" version of our model, we replace the DSCFEM with the "C3" module in the original YOLOv5. In "A3" version of our model, we replace the SCFEM with the "BottleneckCSP" module in the original YOLOv5. The "C" version in Table 3 means the complete YOLO-JD model. We compared the complete YOLO-JD model with the "A1", "A2", and "A3" models using *Precision*, *Recall*, *F*1 on the same training scheme and the same dataset. The fully-deployed YOLO-JD has the best performance in the ablation comparison.

Figure 6. YOLO-JD detection on images that have multiple instances of the same disease and that have multiple classes of diseases and pests on the same image. (**a**,**b**) are two Jute stem images both contain Anthracnose disease and the *Cosmophila sabulifera* pest at the same time. (**c**) is an image contains the die back disease and the *Cosmophila sabulifera* pest at the same time.

Table 3. YOLO-JD the results of the peeling test of the network on the Object Detection task. The best measures are in boldface. The "√" sign means deployment in network.

	Ver	SCFEM	DSCFEM	SPPM	D-1	D-2	D-3	D-4	D-5	D-6	D-7	D-8	P-1	P-2	Mean
Prec (%)	A1	√	√		**98.4**	88.1	90.0	91.9	93.8	99.4	98.0	99.4	91.1	91.1	94.1
	A2	√		√	91.6	85.3	87.0	92.4	92.1	99.1	96.2	99.1	89.7	89.7	92.2
	A3		√	√	91.9	78.6	86.8	87.2	87.8	97.3	83.3	97.4	84.2	84.2	87.8
	C	√	√	√	98.3	**96.2**	**98.0**	**97.1**	**95.3**	**99.5**	**98.3**	**99.7**	**91.6**	**91.6**	**96.5**
Rec (%)	A1	√	√		**93.7**	97.4	83.3	99.1	98.3	92.7	99.2	89.5	94.8	94.8	94.2
	A2	√		√	89.4	97.1	79.6	98.7	99.3	91.7	99.1	89.0	93.4	93.4	93.1
	A3		√	√	86.2	89.5	85.1	93.8	99.1	89.5	98.3	90.9	95.7	95.7	92.3
	C	√	√	√	92.4	**98.6**	**86.9**	**99.3**	**99.5**	**98.1**	**99.5**	**96.2**	**99.3**	**99.3**	**96.9**
F1 (%)	A1	√	√		**96.0**	92.5	86.5	94.9	95.9	95.9	98.5	94.1	92.9	92.9	94.1
	A2	√		√	90.5	90.8	83.1	95.4	95.5	95.2	97.6	93.7	91.5	91.5	92.4
	A3		√	√	88.9	83.6	85.9	90.3	93.1	93.2	90.1	94.0	89.6	89.5	89.8
	C	√	√	√	95.2	**97.4**	**92.1**	**98.1**	**97.3**	**98.7**	**98.8**	**97.9**	**95.3**	**95.2**	**96.6**
mAP (%)	A1	√	√		97.1	97.4	87.6	99.3	99.5	94.4	99.1	91.7	97.5	97.5	96.1
	A2	√		√	93.6	97.8	87.0	99.1	99.3	**98.7**	98.6	**98.7**	97.5	97.1	96.7
	A3		√	√	89.0	90.9	89.2	98.8	99.1	92.3	98.2	93.5	96.4	96.4	94.4
	C	√	√	√	**97.2**	**98.1**	**89.4**	**99.7**	**99.6**	98.5	**99.3**	**98.7**	**99.4**	**99.4**	**97.9**

4. Discussion

In this discussion, we will highlight studies that used deep learning models to detect or recognize different crop diseases with high accuracies. Lee et al. [30] used a Convolution Neural Network (CNN) to process the plant images, and after removing the background and leaving only the potato leaves in the image to judge the symptoms of diseases for potato plants, achieved an accuracy of around 99%. On the other hand, Islam et al. [31] used transfer learning with VGG16 network to detect potato diseases from images with an accuracy of 99.43%. Likewise, Olivares et al. [32–34] used machine learning algorithms such as Random Forest to accurately identify soil properties associated with disease symptoms of tropical diseases of bananas. In our work, YOLO-JD achieved an average mAP of 96.63%

for multiple diseases and pests for jute plants. Together with our YOLO-JD, the above works disclose the popularity and the broad application prospects of machine learning and deep learning on disease prevention for precision agriculture. Therefore, it is possible to effectively use the CNN-like or YOLO-like architectures to detect crop diseases from images, and provide highly accurate results. Though the size of the architectures and the number of parameters may vary from task to task, the creation of suitable models for various types of human-machine interfaces are possible and even straightforward.

5. Conclusions

In this paper, we present a new model YOLO-JD for detecting jute diseases and pests. The main contributions of this paper are threefold: (i) we built a new image dataset with accurate manual labels for jute diseases, and the dataset contains ten classes (eight in diseases, and two in pests); (ii) in this study, we integrated three new modules into the YOLO-JD architecture and achieved an average mAP at 96.63% and *F*1-score at 95.83% for all disease classes; and (iii) YOLO-JD outcompeted several other state-of-the-art methods from the YOLO family both qualitatively and quantitatively.

In the future, we will continue to optimize YOLO-JD for better performance. We are also going to update the jute disease dataset, and try to accommodate YOLO-JD to light-weight applications (such as apps for mobile devices).

Author Contributions: Conceptualization, D.L. and F.A.; methodology, D.L. and F.A.; software, F.A.; validation, N.W.; investigation, A.I.S.; data acquisition, F.A. and A.I.S.; writing—original draft preparation, F.A.; writing—review and editing, D.L., N.W. and F.A.; visualization, F.A.; supervision, D.L.; funding acquisition, D.L. and N.W. All authors have read and agreed to the published version of the manuscript.

Funding: This work was supported in part by the Shanghai Rising-Star Program (No. 21QA1400100), Shanghai Natural Science Foundation (No. 20ZR1400800), and in part by the National Natural Science Foundation of China (No. 52101346).

Institutional Review Board Statement: Not applicable.

Informed Consent Statement: Not applicable.

Data Availability Statement: Data and code are available upon request.

Conflicts of Interest: The authors declare no conflict of interest.

References

1. Mahapatra, B.S.; Mitra, S.; Ramasubramanian, T.; Sinha, M.K. Research on jute (*Corchorus olitorius* and *C. capsularis*) and kenaf (*Hibiscus cannabinus* and *H. sabdariffa*): Present status and future perspective. *Indian J. Agric. Sci.* **2009**, *79*, 951–967.
2. Miah, M.J.; Khan, M.A.; Khan, R.A. Fabrication and Characterization of Jute Fiber Reinforced Low Density Polyethylene Based Composites: Effects of Chemical Treatment. *J. Sci. Res.* **2011**, *3*, 249–259. [CrossRef]
3. CIEL; EIP; FracTracker Alliance; GAIA; 5Gyres; Breakfreefromplastic. *Plastic & Climate: The Hidden Costs of a Plastic Planet*; CIEL: Washington, DC, USA, 2019; pp. 1–108. Available online: https://www.ciel.org/wp-content/uploads/2019/05/Plastic-and-Climate-FINAL-2019.pdf (accessed on 26 March 2022).
4. Barkoula, N.M.; Alcock, B.; Cabrera, N.O.; Peijs, T. Flame-Retardancy Properties of Intumescent Ammonium Poly(Phosphate) and Mineral Filler Magnesium Hydroxide in Combination with Graphene. *Polym. Polym. Compos.* **2008**, *16*, 101–113.
5. Balasundram, S.K.; Golhani, K.; Shamshiri, R.R.; Vadamalai, G. Precision agriculture technologies for management of plant diseases. In *Plant Disease Management Strategies for Sustainable Agriculture through Traditional and Modern Approaches*; Springer: Cham, Switzerland, 2020; pp. 259–278.
6. Traversari, S.; Cacini, S.; Galieni, A.; Nesi, B.; Nicastro, N.; Pane, C. Precision agriculture digital technologies for sustainable fungal disease management of ornamental plants. *Sustainability* **2021**, *13*, 3707. [CrossRef]
7. Maniyath, S.R.; Vinod, P.V.; Niveditha, M.; Pooja, R.; Prasad Bhat, N.; Shashank, N.; Hebbar, R. Plant disease detection using machine learning. In Proceedings of the 2018 International Conference on Design Innovations for 3Cs Compute Communicate Control, ICDI3C 2018, Bangalore, India, 25–26 April 2018; pp. 41–45. [CrossRef]
8. Gavhale, K.R.; Gawande, U.; Hajari, K.O. Unhealthy region of citrus leaf detection using image processing techniques. In Proceedings of the International Conference for Convergence for Technology—2014, Pune, India, 6–8 April 2014; pp. 2–7. [CrossRef]

9. Hossain, M.S.; Mou, R.M.; Hasan, M.M.; Chakraborty, S.; Abdur Razzak, M. Recognition and detection of tea leaf's diseases using support vector machine. In Proceedings of the 2018 IEEE 14th International Colloquium on Signal Processing & Its Applications (CSPA), Penang, Malaysia, 9–10 March 2018; pp. 150–154. [CrossRef]
10. Jiao, L.; Zhang, F.; Liu, F.; Yang, S.; Li, L.; Feng, Z.; Qu, R. A survey of deep learning-based object detection. *IEEE Access* **2019**, *7*, 128837–128868. [CrossRef]
11. Saleem, M.H.; Potgieter, J.; Arif, K.M. Plant disease classification: A comparative evaluation of convolutional neural networks and deep learning optimizers. *Plants* **2020**, *9*, 1319. [CrossRef] [PubMed]
12. Redmon, J.; Farhadi, A. YOLOv3: An Incremental Improvement. 2018. Available online: http://arxiv.org/abs/1804.02767 (accessed on 8 April 2018).
13. Bochkovskiy, A.; Wang, C.-Y.; Liao, H.-Y.M. YOLOv4: Optimal Speed and Accuracy of Object Detection. 2020. Available online: http://arxiv.org/abs/2004.10934 (accessed on 23 April 2020).
14. Ren, S.; He, K.; Girshick, R.; Sun, J. Faster R-CNN: Towards Real-Time Object Detection with Region Proposal Networks. *IEEE Trans. Pattern Anal. Mach. Intell.* **2017**, *39*, 1137–1149. [CrossRef] [PubMed]
15. He, K.; Gkioxari, G.; Dollár, P.; Girshick, R. Mask R-CNN. *IEEE Trans. Pattern Anal. Mach. Intell.* **2020**, *42*, 386–397. [CrossRef] [PubMed]
16. Liu, W.; Anguelov, D.; Erhan, D.; Szegedy, C.; Reed, S.; Fu, C.Y.; Berg, A.C. *SSD: Single Shot Multibox Detector*; Computer Vision-ECCV 2016. Lecture Notes in Computer Science (including Subser. Lect. Notes Artif. Intell. Lect. Notes Bioinformatics); Springer: Cham, Switzerland, 2016; Volume 9905, pp. 21–37. Available online: https://doi.org/10.1007/978-3-319-46448-0_2 (accessed on 29 December 2016). [CrossRef]
17. Hammad Saleem, M.; Khanchi, S.; Potgieter, J.; Mahmood Arif, K. Image-based plant disease identification by deep learning meta-architectures. *Plants* **2020**, *9*, 1451. [CrossRef] [PubMed]
18. Chowdhury, M.E.H.; Rahman, T.; Khandakar, A.; Ayari, M.A.; Khan, A.U.; Khan, M.S.; Al-Emadi, N.; Reaz, M.B.I.; Islam, M.T.; Ali, S.H.M. Automatic and Reliable Leaf Disease Detection Using Deep Learning Techniques. *AgriEngineering* **2021**, 3, 294–312. [CrossRef]
19. Mohanty, S.P.; Hughes, D.P.; Salathé, M. Using deep learning for image-based plant disease detection. *Front. Plant Sci.* **2016**, *7*, 1419. [CrossRef] [PubMed]
20. Görlich, F.; Marks, E.; Mahlein, A.K.; König, K.; Lottes, P.; Stachniss, C. Uav-based classification of cercospora leaf spot using rgb images. *Drones* **2021**, *5*, 34. [CrossRef]
21. Chen, Z.; Wu, R.; Lin, Y.; Li, C.; Chen, S.; Yuan, Z.; Chen, S.; Zou, X. Plant Disease Recognition Model Based on Improved YOLOv5. *Agronomy* **2022**, *12*, 365. [CrossRef]
22. Arsenovic, M.; Karanovic, M.; Sladojevic, S.; Anderla, A.; Stefanovic, D. Solving current limitations of deep learning based approaches for plant disease detection. *Symmetry* **2019**, *11*, 939. [CrossRef]
23. Vishnoi, V.K.; Kumar, K.; Kumar, B. *Plant Disease Detection Using Computational Intelligence and Image Processing*; Springer: Berlin/Heidelberg, Germany, 2021; Volume 128.
24. Wagle, S.A.; Harikrishnan, R.; Ali, S.H.M.; Faseehuddin, M. Classification of plant leaves using new compact convolutional neural network models. *Plants* **2022**, *11*, 24. [CrossRef] [PubMed]
25. De, R.K. *Jute Diseases: Diagnosis and Management*; ICAR-Center Research Institute for Jute and Allied Fibres (Indian Council of Agricultural Research): Kolkata, India, 2019; ISBN 9789353822149. Available online: http://www.crijaf.org.in/ (accessed on 1 April 2019).
26. Tzutalin. labelImg. 2015. Available online: https://github.com/tzutalin/labelImg (accessed on 27 July 2015).
27. Zheng, Z.; Wang, P.; Ren, D.; Liu, W.; Ye, R.; Hu, Q.; Zuo, W. Enhancing Geometric Factors in Model Learning and Inference for Object Detection and Instance Segmentation. *IEEE Trans. Cybern.* **2021**, *20*, 1–13. [CrossRef] [PubMed]
28. Ultralytics. YOLOv5. 2020. Available online: https://github.com/ultralytics/yolov5 (accessed on 25 June 2020).
29. Lin, T.Y.; Maire, M.; Belongie, S.; Hays, J.; Perona, P.; Ramanan, D.; Dollár, P.; Zitnick, C.L. *Microsoft COCO: Common Objects in Context*; Computer Vision-ECCV 2014. Lecture Notes in Computer Science (including Subser. Lect. Notes Artif. Intell. Lect. Notes Bioinformatics); Springer: Cham, Swizerland, 2014; Volume 8693, pp. 740–755. [CrossRef]
30. Lee, T.Y.; Yu, J.Y.; Chang, Y.C.; Yang, J.M. Health Detection for Potato Leaf with Convolutional Neural Network. In Proceedings of the 2020 Indo—Taiwan 2nd International Conference on Computing, Analytics and Networks (Indo-Taiwan ICAN), Rajpura, India, 7–15 February 2020; pp. 289–293. [CrossRef]
31. Islam, F.; Hoq, M.N.; Rahman, C.M. Application of transfer learning to detect potato disease from leaf image. In Proceedings of the IEEE International Conference on Robotics, Automation, Artificial-Intelligence and Internet-of-Things (RAAICON), Dhaka, Bangladesh, 29 November–1 December 2019; pp. 127–130. [CrossRef]
32. Olivares, B.O.; Rey, J.C.; Lobo, D.; Navas-Cortés, J.A.; Gómez, J.A.; Landa, B.B. Fusarium wilt of bananas: A review of agro-environmental factors in the venezuelan production system affecting its development. *Agronomy* **2021**, *11*, 986. [CrossRef]
33. Olivares, B.O.; Paredes, F.; Rey, J.C.; Lobo, D.; Galvis-Causil, S. The relationship between the normalized difference vegetation index, rainfall, and potential evapotranspiration in a banana plantation of Venezuela. *Soc. Psychol. Soc.* **2021**, *12*, 58–64. [CrossRef]
34. Olivares, B. *Determination of the Potential Influence of Soil in the Differentiation of Productivity and in the Classification of Susceptible Areas to Banana wilt in Venezuela*; UCOPress: Córdoba, Spain, 2022; pp. 89–111.

plants

MDPI

Article

Fusarium nirenbergiae (*Fusarium oxysporum* Species Complex) Causing the Wilting of Passion Fruit in Italy

Dalia Aiello, Alberto Fiorenza, Giuseppa Rosaria Leonardi, Alessandro Vitale * and Giancarlo Polizzi

Dipartimento di Agricoltura, Alimentazione e Ambiente, sez. Patologia Vegetale, University of Catania, Via S. Sofia 100, 95123 Catania, Italy; dalia.aiello@unict.it (D.A.); alberto.fiorenza.93@gmail.com (A.F.); leonardigiusi@outlook.it (G.R.L.); gpolizzi@unict.it (G.P.)
* Correspondence: alevital@unict.it

Abstract: Passion fruit (*Passiflora edulis* Sims.) is an ever-increasing interest crop in Italy because it is mainly cultivated for its edible fruit and, secondly, as an ornamental evergreen climber. During the summer of 2020, two-year-old plants of purple passion fruit in one of the most important expanding production areas of Sicily (southern Italy) showed symptoms of yellowing, wilting, and vascular discoloration. *Fusarium*-like fungal colonies were consistently yielded from symptomatic crown and stem tissues. Five representative isolates were characterized by a morphological and molecular analysis based on a multilocus phylogeny using RNA polymerase's second largest subunit (RPB2) and translation elongation factor 1-alpha (EF-1α) genes, as *Fusarium nirenbergiae* (*Fusarium oxysporum* species complex). Pathogenicity tests conducted on healthy 1-year-old passion fruit cuttings revealed symptoms similar to those observed in the field. To our knowledge, this is the first report of Fusarium wilt on passion fruit caused by *Fusarium nirenbergiae*. This report focuses on the phytopathological implications of this fungal pathogen, which may represent a future significant threat for the expanding passion fruit production in Italy and Europe.

Keywords: wilt; passion fruit; *Fusarium oxysporum* species complex

Citation: Aiello, D.; Fiorenza, A.; Leonardi, G.R.; Vitale, A.; Polizzi, G. *Fusarium nirenbergiae* (*Fusarium oxysporum* Species Complex) Causing the Wilting of Passion Fruit in Italy. *Plants* **2021**, *10*, 2011. https://doi.org/10.3390/plants10102011

Academic Editor: Pablo Castillo

Received: 1 September 2021
Accepted: 24 September 2021
Published: 26 September 2021

1. Introduction

In recent years, tropical fruit production increased worldwide due to the increasing demand of global markets and more efficient transportation and storage techniques [1,2]. Most of the tropical fruit is destined for fresh consumption or industrial transformation. Among these, passion fruit (*Passiflora edulis* Sims.) is one of the most exported and consumed fruit commodities. It originated in tropical and subtropical America [3], and it is now extensively cultivated worldwide, including Australia, New Zealand, India, Africa, and South America [4,5]. Passion fruit is mainly cultivated for its edible fruit but secondarily also for its attractive flowers on ornamental evergreen vines.

In Italy, the cultivation of *P. edulis* (also known as purple passion fruit) as a fruit crop in some regions characterized by a Mediterranean climate (e.g., Sicily and Calabria) is gaining growing interest by local farmers, and it is carried out under greenhouse and, to a lower extent, open field conditions. Indeed, although the crop is well adapted to a wide rainfall range (1000–2500 mm for crop season), minimum temperatures below 5 °C should be avoided because they seriously compromise the plant growth and nutrient uptake [6–8]. In this regard, it should be noted that a process of reconversion of protected tomato and vegetable crops into tropical fruit plantations is currently taking place in southern Italy and Sicily.

Unfortunately, this species is affected by many diseases during its different growth stages, and this reduces the yield and the farmers' income [9]. One of the most widely reported fungal pathogens affecting passion fruit is *Fusarium oxysporum* f. sp. *passiflorae*, which causes the Fusarium wilt. It was first reported in Australia [10] but is nowadays spread worldwide [11–13]. Among Fusarium diseases, *Neocosmospora solani* (=*Fusarium solani*) is responsible for the basal stem rot [14–16]. According to Viana & Costa [17], the species *F. oxysporum* f. sp. *passiflorae* and *N. solani* are the most damaging ones to passion fruit crops. Minor diseases have been reported on passion fruit, such as the damping-off of seedlings and collar and root rot in adult plants caused by *Rhizoctonia solani* [18] and collar rot caused by *Phytophthora* spp. [19]. During a recent survey performed in Sicily, young passion fruit plants showing symptoms of general yellowing and wilting were observed in some of the most representative production areas. Given the increasing interest of local growers in expanding passion fruit cultivation, the aim of this study was to characterize the fungal species associated with those symptoms and test their pathogenicity, in order to better understand the syndrome's aetiology.

2. Results

The symptoms observed in the greenhouse consisted of leaf yellowing and wilting (Figure 1a,b), external crown and root rot and wood discoloration moving upward to the canopy (Figure 1c). The disease incidence reached 10% of the cultivated plants. Colonies with white or light purple aerial mycelia and violet pigmentation on the underside of the cultures developed after 14 days on PDA, being firstly identified as *Fusarium*-like. Sporodochial macroconidia with 2 to 5 septa, grown on OA, measured (23.09–) 28.76 $\pm$ 3.06 (–35.48) μm × (1.99–) 3.84 $\pm$ 0.58 (–4.75) μm (Figure 1f,g). Oval, unicellular microconidia developed on short monophialides, grown on OA, measured (3.1–) 5.17 $\pm$ 1.35 (–9.17) μm × (1.3–) 1.98 $\pm$ 0.37 (–2.9) μm (Figure 1i).

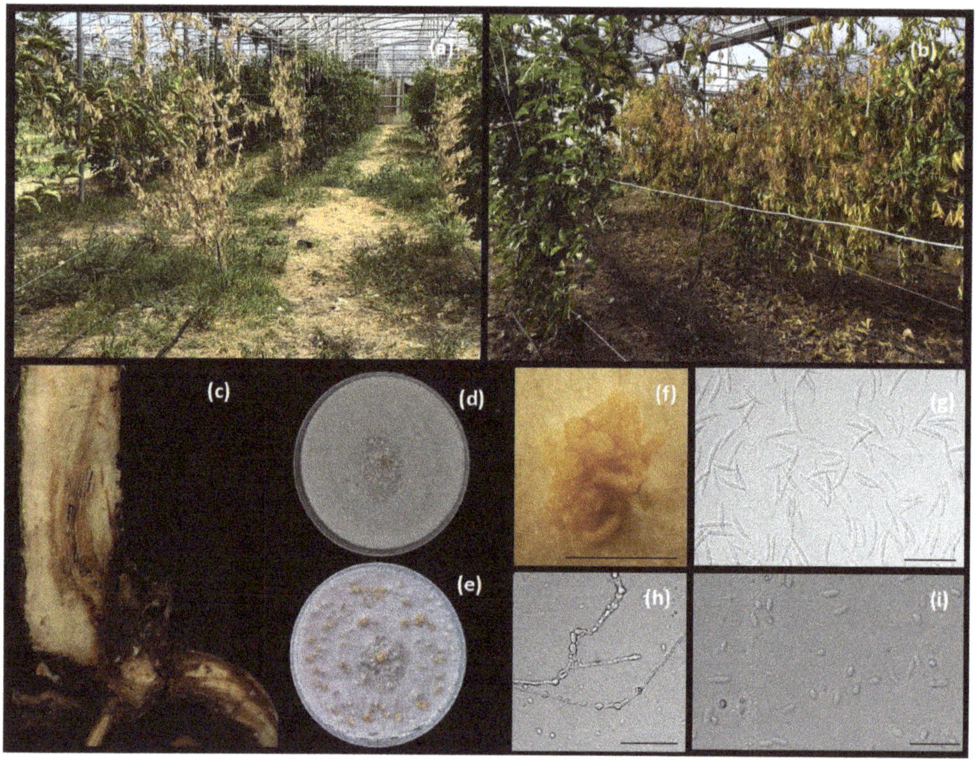

Figure 1. Disease symptoms and *Fusarium nirenbergiae* features: (**a**,**b**), yellowing and wilting of passion fruit plants in greenhouse; (**c**), vascular discoloration on a collar portion; (**d**,**e**), *F. nirenbergiae* (Di3A-Pef1 isolate) grown on 7 day-old (up) and 14 day-old (down) OA; (**f**), sporodochia on OA; (**g**), sporodochial conidia (macroconidia); (**h**), chlamydospores on SNA; (**i**), aerial conidia (microconidia). Scale bars, (**f**): 2 mm; (**g**–**i**): 50 μm.

PCR edit amplicons resulted in 528 bp for the partial ITS region, 287 bp for EF-1α and 953 bp for RPB2. The sequences were registered in GenBank as follows: MZ398141, MZ398142, MZ398143, MZ398144, MZ398145 for ITS, MZ408109, MZ408110, MZ408111, MZ408112, MZ408113 for RPB2 and MZ408114, MZ408115, MZ408116, MZ408117, MZ408118 for EF1-α. A GenBank BLASTn analysis and a pairwise sequence alignment on the MLST database indicated that all the isolates from passion fruit belonged to the *Fusarium oxysporum* species complex (FOSC). In particular, the MLST search resulted in high identity values (96–100%) (Acc. number MH582354) for the EF1-α gene and 98% (Acc. number MH582140) for the RBP2 gene with a *F. oxysporum* species complex (FOSC). The MP heuristic search resulted in 83 parsimony-informative characters, while 109 were variable and parsimony-uninformative and 1412 were constant. A maximum of 320 equally most parsimonious trees were retained (Tree length = 249, CI = 0.851, RI = 0.898 and RC = 0.765).

The bootstrap support values from the parsimony analysis are shown close to the branch node. The group of representative isolates Di3A-Pef1-5 clustered with the reference strain of *F. nirenbergiae*, as shown in Figure 2, and were clearly separated by the other sequences provided in the study by Lombard et al. [20]. The isolates were then identified as *Fusarium nirenbergiae* L. Lombard & Crous.

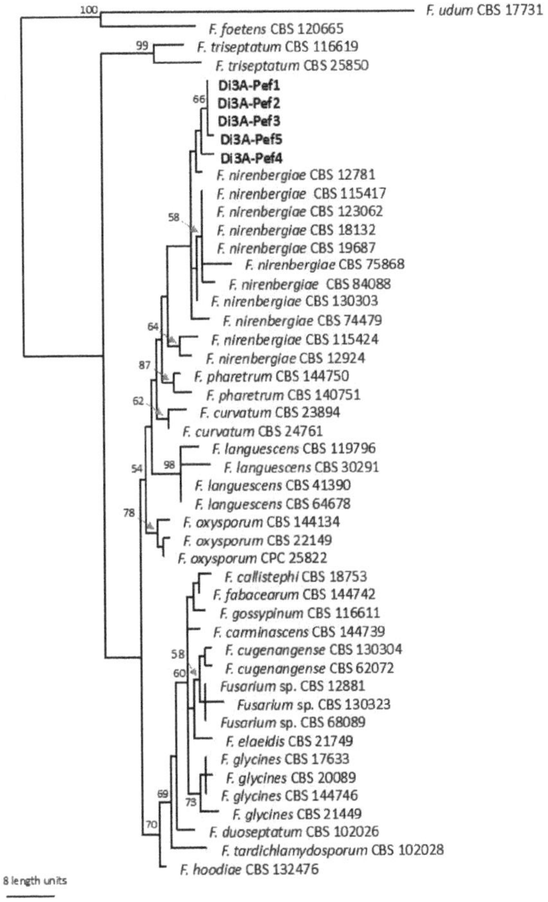

Figure 2. Single most parsimonious phylogenetic tree resulting from the MP analysis of combined *EF1-α* and *rbp2* sequence data. The isolates in bold were sequenced in this study. The numbers represent MP bootstrap values.

The inoculated isolate after five months caused symptoms similar to those observed under greenhouse conditions in all inoculated plants. The symptoms consisted of leaf yellowing and wilting. After 7 months all plants died. A longitudinal section of the inoculated plants reveals the internal discolorations moving upward to the canopy. The control remains symptomless. From the symptomatic tissues, *F. nirembergiae* was always re-isolated, and it was characterized as previously described.

3. Discussion and Conclusions

To the best of our knowledge, this paper represents the first report of *F. nirembergiae*, belonging to the FOSC complex, as a causal agent of Fusarium wilt of passion fruit. In this regard, both the morphological characterization and the analysis of the ITS, EF1-α and RBP2 sequences allowed us to correctly allocate a representative number of detected strains within the *F. nirenbergiae* group, being distinctly separated by the other taxa, as recently shown by Lombard et al. [20] and Crous et al. [21]. Based on the present findings, *F. nirenbergiae* was strongly grouped in a separated subclade of FOSC, phylogenetically close to *F. curvatum*. Although little information regarding *F. nirembergiae*'s pathogenicity and host range is currently available, except for the study by Zhao et al. [22] on *Acer negundo*, this species (belonging to FOSC) is able to colonize and infect host vascular tissues; for this reason, it is reported worldwide as responsible for Fusarium wilt [14]. The first symptoms consist of leaf yellowing and wilt, followed by the plant's collapse. This disease is observed in adult and young plants under favorable conditions for the infection development, such as high temperature and humidity and a high potential inoculum in the soil [5,23]. Once this fungal pathogen is established in the field, its control is very difficult, since fungicide application does not result in a significant reduction of the disease amount, and the pathogen can persist in the soil for many years in the absence of the host [14]. Hence, the incidence data are very worrying as regards the nature of the fungal pathogen and dissemination ability of *F. nirenbergiae* under greenhouse conditions. If, on the one hand, protected systems could facilitate the cultivation of purple passion fruit, on the other hand they could aggravate the consequences of this phytopathological issue. Indeed, this could represent a future threat for the expansion of this tropical crop, which is replacing protected tomato cultivation in different areas of southern Italy. Therefore, disease management should be focused mainly on preventative and pathogen exclusion measures, avoiding plantation in areas with a severe history of Fusarium wilt infections or selecting healthy propagation material in combination with adequate agronomic practices. Additionally, other sustainable strategies should include the use of resistant cultivars, as recommended by several authors [24,25]. Comprehensively, the increasing trend of tropical plantations in Italy leads us to focus more on fungal diseases that could represent limiting factors for future production. According to presented data combined with recent findings [20,21], it cannot be excluded that some past reports of *F. oxysporum* f. sp. *passiflorae* could confirm that *F. nirembergiae* is a causal agent of Fusarium wilt. However, further surveys should be performed on *P. edulis* orchards in Italy and worldwide to confirm the new aetiology of the Fusarium wilt of passion fruit and its real diffusion.

4. Materials and Methods

4.1. Field Survey, Isolations and Morphological Characterization

In July of 2020, 50 two-year-old 'purple' passion fruit plants cultivated in a greenhouse in the Syracuse province (Sicily, Italy) appeared stunted, defoliated and severely wilted. Diseased vascular tissues (0.5 cm^2) were surface-disinfected for 1 min in a 1.2% sodium hypochlorite (NaOCl) solution, rinsed in sterile water, placed on a potato dextrose agar (PDA, Lickson, Vicari, Italy) amended with 0.1 g/L of streptomycin sulphate (Sigma-Aldrich, St. Louis, MO, USA), to prevent bacterial growth, and then incubated at 25 $\pm$ 1 °C until fungal colonies were observed. Single-spore isolates were obtained from pure cultures grown on APDA. To induce sporulation, five representative single-spore isolates (named Di3A-Pef1, Di3A-Pef2, Di3A-Pef3, Di3A-Pef4 and Di3A-Pef5) were selected and transferred on a synthetic nutrient-poor agar (SNA) [26], Oatmeal Agar (OA, Difco, Detroit, MI, USA) and PDA for morphological characterization. A total of 50 macro- and micro-conidia were measured (length and width size) using a fluorescence microscope (Olympus-BX61) coupled to an Olympus DP70 digital camera; images and measurements were captured using the software analySIS Image Processing. Conidia sizes are reported as the minimum and maximum in parentheses, and the average is reported with the standard deviation.

4.2. Molecular Characterization and Phylogeny

Genomic DNA of the selected isolates (Di3A-Pef 1-2-3-4-5) was extracted using the Gentra Puregene Yeast/Bact kit (Qiagen, Hilden, Germany) according to the manufacturer's protocol. The internal transcribed spacer of the ribosomal DNA (rDNA-ITS), partial translation elongation factor alpha gene (EF-1α) and RNA polymerase II gene (RPB2) were targeted for PCR amplification and sequencing. The primers used for these regions were: ITS5 and ITS4 for ITS [27], EF1-728F and EF1-986R for EF-1α [28] and 5f2 and 7cr for RPB2 [29]. The PCR products were purified and sequenced in both directions by Macrogen Inc. (Seoul, Korea). The sequences were edited using MEGAX: Molecular Evolutionary Genetics Analysis across computing platforms [30], manual adjustments of alignments were made when necessary and submitted to GenBank. Moreover, the sequences were blasted in the NCBIs GenBank nucleotide database and on the Fusarium MLST database of the Westerdijk Fungal Biodiversity Institute (http://www.westerdijkinstitute.nl/fusarium/, accessed on 21 May 2021). For comparison, 44 additional sequences were selected according to the recent literature [20] (Table 1). The phylogenetic analysis was based on the Maximum Parsimony (MP). The MP analysis was done using PAUP v. 4.0a165 [31]. Phylogenetic relationships were estimated by heuristic searches with 100 random addition sequences. A tree bisection-reconnection was used, with the branch swapping option set to 'best trees' only, with all characters weighted equally and alignment gaps treated as the fifth state. The tree length (TL), consistency index (CI), retention index (RI) and rescaled consistency index (RC) were calculated for parsimony and the bootstrap analyses were based on 1000 replicates [32]. *Fusarium foetens* (CBS 120665) and *F. udum* (CBS 12881) served as outgroups.

Table 1. Characteristics of *Fusarium* isolates included in the phylogenetic analysis.

Species	Culture Accession	Host/Substrates	Special form	Origin	GeneBank Accession	
					rpb2	EF1-α
Fusarium callistephi	CBS 187.53	*Callistephus chinensis*	*callistephi*	The Netherlands	MH484875	MH484966
F. carminascens	CBS 144739	*Zea mays*		South Africa	MH484934	MH485025
F. cugenengense	CBS 620.72	*Crocus* sp.	*gladioli*	Germany	MH484879	MH484970
F. cugenengense	CBS 130304	*Gossypium barbadense*	*vasinfectum*	China	MH484921	MH485012
F.curvatum	CBS 247.61	*Matthiola incana*	*matthiolae*	Germany	MH484876	MH484967
F.curvatum	CBS 238.94	*Beaucarnia* sp.	*meniscoideum*	The Netherlands	MH484893	MH484984
F. duoseptatum	CBS 102026	*Musa sapientum*	*cubense*	Malaysia	MH484896	MH484987
F. elaeidis	CBS 217.49	*Elaeis* sp.	*elaeidis*	Zaire	MH484870	MH484961
F. fabacearum	CBS 144742	*Zea mays*		South Africa	MH484938	MH485029
F. foetens	CBS 120665	*Nicotiana tabacum*		Iran	MH484918	MH485009
F. glycines	CBS 144746	*Glycine max*		South Africa	MH484942	MH485033
F. glycines	CBS 20089	*Ocimum basilicum*	*basilici*	Italy	MH484888	MH484979
F. glycines	CBS 17633	*Linum usitatissium*	*lini*	Unknown	MH484868	MH484959
F. glycines	CBS 21449	Unknown		Argentina	MH484869	MH484960
F. gossypinum	CBS 116611	*Gossypium hirsutum*	*vasinfectum*	Ivory Coast	MH484907	MH484998
F. hoodiae	CBS 132474	*Hoodia gordonii*	*hoodiae*	South Africa	MH484929	MH485020
F. languescens	CBS 41390	*Solanum lycopersicum*	*lycopersici*	Israel	MH484890	MH484981
F. languescens	CBS 119796	*Zea mays*		South Africa	MH484917	MH485008
F. languescens	CBS 30,291	*Solanum lycopersicum*	*lycopersici*	The Netherlands	MH484892	MH484983
F. languescens	CBS 646.78	*Solanum lycopersicum*	*lycopersici*	Morocco	MH484881	MH484972
F. nirembergiae	CBS 744.79	*Passiflora edulis*	*passiflorae*	Brazil	MH484882	MH484973
F. nirembergiae	CBS 115424	*Agothosma betulina*		South Africa	MH484906	MH484997
F. nirembergiae	CBS 12924	*Secale cereale*		Unknown	MH484864	MH484955
F. nirembergiae	CBS 12781	*Chrysanthemum* sp.	*chrysanthemi*	USA	MH484883	MH484974
F. nirembergiae	CBS 130303	*Solanum lycopersicum*	*radicis-lycopersici*	USA	MH484923	MH485014
F. nirembergiae	CBS 115417	*Agothosma betulina*		South Africa	MH484903	MH484994
F. nirembergiae	CBS 19687	*Bouvardia longiflora*	*bouvardiae*	Italy	MH484886	MH484977
F. nirembergiae	CBS 123062	Tulip roots		USA	MH484919	MH485010
F. nirembergiae	CBS 18132	*Solanum tuberosum*		USA	MH484867	MH484958
F. nirembergiae	CBS 75868	*Solanum lycopersicum*	*lycopersici*	The Netherlands	MH484877	MH484968

Table 1. *Cont.*

Species	Culture Accession	Host/Substrates	Special form	Origin	GeneBank Accession	
					rpb2	EF1-α
F. nirembergiae	CBS 840.88	*Dianthus caryophyllus*	*dianthi*	The Netherlands	MH484887	MH484978
F. oxysporum	CBS 221.49	*Camellia sinensis*	*medicaginis*	South East Asia	MH484872	MH484963
F. oxysporum	CPC 25822	*Protea* sp.		South Africa	MH484943	MH485034
F. oxysporum	CBS 144134	*Solanum tuberosum*		Germany	MH484953	MH485044
F. pharetrum	CBS 144751	*Aliodendron dichotomum*		South Africa	MH484952	MH485043
F. pharetrum	CBS 144750	*Aliodendron dichotomum*		South Africa	MH484951	MH485042
F. trachichlamydosporum	CBS 102028	*Musa sapientum*	*cubense*	Malaysia	MH484897	MH484988
F. triseptatum	CBS 258.50	*Ipomea batatas*	*batatas*	USA	MH484873	MH484964
F. triseptatum	CBS 116619	*Gossypium hirsutum*	*vasinfectum*	Ivory Coast	MH484910	MH485001
F. udum	CBS 177.31	*Digitaria ariantha*		South Africa	MH484866	MH484957
Fusarium sp.	CBS 12881	*Chrysanthemum* sp.	*chrysanthemi*	USA	MH484884	MH484975
Fusarium sp.	CBS 130323	Human nail		Australia	MH484927	MH485018
Fusarium sp.	CBS 68089	*Cucumis sativus*	*cucurbitacearum*	The Netherlands	MH484889	MH484980

4.3. Pathogenicity Tests

In order to fulfil Koch's postulates, pathogenicity tests were conducted on one-year-old potted cuttings using the mycelial plug technique. In detail, 18 healthy cuttings were inoculated, removing a piece of bark of the crown root with a scalpel blade and applying a mycelial plug (0.3 cm^2), taken from a 14-day-old Di3A-Pef 1 isolate, upside down on the wound and subsequently covered with soil to prevent desiccation. The controls consisted of sterile PDA plugs applied as described above to the same number of healthy young plants. Re-isolation attempts were performed from representative inoculated plants.

Author Contributions: Conceptualization, D.A., A.V. and G.P.; methodology, D.A., A.F. and G.R.L.; software, A.F.; validation, D.A., A.V. and G.P.; formal analysis, A.V.; investigation, A.F. and G.R.L.; resources, A.V. and G.P.; data curation, D.A., A.F. and A.V.; writing—original draft preparation, D.A., A.F. and A.V.; writing—review and editing, A.V. and G.P.; visualization, A.V.; supervision, G.P.; project administration, G.P.; funding acquisition, A.V. and G.P. All authors have read and agreed to the published version of the manuscript.

Funding: This research was funded by the following grants: Programma Ricerca di Ateneo MEDIT-ECO UNICT 2020–2022 Linea 2-University of Catania (Italy); Starting Grant 2020, University of Catania (Italy); Fondi di Ateneo 2020–2022, University of Catania (Italy), Linea Open Access. Research Project 2016–2018, University of Catania 5A722192134.

Data Availability Statement: The data presented in this study are available on request from the corresponding author.

Acknowledgments: All authors are grateful to all technicians and growers who supported this research.

Conflicts of Interest: The authors declare no conflict of interest.

References

1. Ding, P. Tropical fruits. In *Encyclopedia of Applied Plant Sciences*, 2nd ed.; Thomas, B., Murphy, D.J., Murray, B.G., Eds.; Academic Press: Oxford, UK, 2017; pp. 431–434.
2. Underhill, S. Fruits of Tropical Climates: Commercial and Dietary Importance. In *Encyclopedia of Food Sciences and Nutrition*, 2nd ed.; Caballero, B., Ed.; Academic Press: Oxford, UK, 2003; pp. 2780–2785.
3. Cervi, A.C. *O Gênero Passiflora (Passifloraceae) No Brasil, Es- Pécies Descritas Após o Ano de 1950*; Real Jardin Botánico: Madrid, Spain, 2006.
4. Manicom, B.; Ruggiero, C.; Ploetz, R.C.; de Goes, A. Diseases of passion fruit. In *Diseases of Tropical Fruit Crops*; Ploetz, R.C., Ed.; CAB International: Wallingford, UK, 2003; pp. 413–441.
5. Vanderplank, J. *Passion Flowers*, 2nd ed.; MIT Press: London, UK, 1996; p. 224.
6. Das, M.R.; Hossain, T.; Mia, M.B.; Ahmed, J.; Kariman, A.S.; Hossain, M.M. Fruit setting behavior of passion fruit. *Am. J. Plant Sci.* **2013**, *4*, 1066–1073. [CrossRef]
7. NDA-ARC. *Growing Granadillas*; Institute of Tropical and Subtropical Crops: Pretoria, South Africa, 1999; pp. 1–9.
8. Rao, B.N.; Jha, A.K.; Deo, C.; Kumar, S.; Roy, S.S.; Ngachan, S.V. Effect of irrigation and mulching on growth, yield and quality of passion fruit (*Passiflora edulis* Sims.). *J. Crop Weed* **2013**, *9*, 94–98.

9. Fischer, I.H.; Lourenco, S.A.; Martins, M.C.; Kimati, H.; Amorim, L. Seleção de plantas resistentes e de fungicidas para o controle da podridão do colo do maracujazeiro causada por *Nectria hematococca*. *Fitopatol. Bras.* **2005**, *30*, 250–258. [CrossRef]
10. McKnight, T. A wilt disease of the passion vines (*Passiflora edulis*) caused by a species of *Fusarium*. *Queensl. J. Agric. Sci.* **1951**, *8*, 1–4.
11. Garcia, E.; Paiva, D.; Costa, J.; Portugal, A.; Ares, A. First report of Fusarium wilt caused by *Fusarium oxysporum* f. sp. passiflorae on Passion Fruit in Portugal. *Plant Dis.* **2019**, *103*, 2680.
12. Liberato, J.R.; Costa, H. Doenças fúngicas, bacterianas e fitonematóides. In *Maracujá: Tecnologia de produção, Pós-Colheita, Agroindústria, Mercado*; Cinco Continentes: Porto Alegre, Brazil, 2001; pp. 243–276.
13. Rooney-Latham, S.; Blomquist, C.L.; Scheck, H.J. First report of Fusarium wilt caused by *Fusarium oxysporum* f. sp. passiflorae on Passion fruit in north America. *Plant Dis.* **2011**, *95*, 1478.
14. Fischer, I.H.; Rezende, J.A.M. Diseases of passion flower (*Passiflora* spp.). *Pest Technol.* **2008**, *2*, 1–19.
15. Li, D.F.; Yang, J.Q.; Zhang, X.Y.; Sun, L.F. Identification of the pathogen causing collar rot of passion fruit in Fujian. *Acta Phytopathol. Sin.* **1993**, *23*, 372.
16. Ploetz, R.C. Sudden wilt of passionfruit in southern Florida caused by *Nectria haematococca*. *Plant Dis.* **1991**, *75*, 1071–1073. [CrossRef]
17. Viana, F.M.P.; Costa, A.F. Doenças do maracujazeiro. In *Doenças de Fruteiras Tropicais de Interesse Agroindustrial*; Freire, F.C.O., Cardoso, J.E., Viana, F.M.P., Eds.; Embrapa informação Tecnológica: Brasília, Brazil, 2003; pp. 276–285.
18. Bezerra, J.L.; De Oliveira, M.L. Damping-off of passion fruit caused by *Rhizoctonia* sp. [*Passiflora edulis*]. *Fitopatol. Brasil.* **1984**, *9*, 273–276.
19. Young, B.R. Root rot of passionfruit vine (*Passiflora edulis* Sims.) in the Auckland area, New Zealand. *J Agric. Res.* **1970**, *13*, 119–125.
20. Lombard, L.; Sandoval-Denis, M.; Lamprecht, S.C.; Crous, P.W. Epitypification of *Fusarium oxysporum* clearing the taxonomic chaos. *Persoonia* **2019**, *43*, 1–47. [CrossRef]
21. Crous, P.; Lombard, L.; Sandoval-Denis, M.; Seifert, K.; Schroers, H.-J.; Chaverri, P.; Gené, J.; Guarro, J.; Hirooka, Y.; Bensch, K.; et al. *Fusarium*: More than a node or a foot-shaped basal cell. *Stud. Mycol.* **2021**, *98*, 100116. [CrossRef] [PubMed]
22. Zhao, X.; Li, H.; Zhou, L.; Chen, F.; Chen, F. Wilt of *Acer negundo* L. caused by *Fusarium nirenbergiae* in China. *J. For. Res.* **2020**, *31*, 2013–2022. [CrossRef]
23. Bennett, R.S.; Davis, R.M. Method for rapid production of *Fusarium oxysporum* f. sp. vasinfectum chlamydospores. *J. Cotton Sci.* **2013**, *17*, 52–59.
24. De Carvalho, J.A.; de Jesus, J.G.; Araujo, K.L.; Serafim, M.E.; Gilio, T.A.S.; Neves, L.G. Passion fruit (*Passiflora* spp.) species as sources of resistance to soil phytopathogens *Fusarium solani* and *Fusarium oxysporum* f. sp. passiflorae complex. *Rev. Bras. Frutic.* **2021**, *43*. [CrossRef]
25. Silva, A.S.; Oliveira, E.J.; Haddad, F.; Jesus, O.N.; Oliveira, S.A.S.; Costa, M.A.P. Variação genética em isolados de *Fusarium oxysporum* f. sp. passiflorae com marcadores AFLP. *Sci. Agric.* **2013**, *70*, 108–115.
26. Nirenberg, H.I. A simplified method for identifying *Fusarium* spp. occurring on wheat. *Can. J. Bot.* **1981**, *59*, 1599–1609. [CrossRef]
27. White, T.J.; Bruns, T.; Lee, S.; Taylor, J.L. Amplification and direct sequencing of fungal ribosomal RNA genes for phylogenetics. In *PCR Protocols: A Guide to Methods and Applications*; Innes, M.A., Gelfand, D.H., Sninsky, J.J., White, T.J., Eds.; Academic Press: New York, NY, USA, 1990; pp. 315–322.
28. Carbone, I.; Kohn, L.M. A method for designing primer sets for speciation studies in filamentous ascomycetes. *Mycologia* **1999**, *91*, 553–556. [CrossRef]
29. Liu, Y.J.; Whelen, S.; Hall, B.D. Phylogenetic relationships among ascomycetes: Evidence from an RNA polymerse II subunit. *Mol. Biol. Evol.* **1999**, *16*, 1799–1808. [CrossRef]
30. Kumar, S.; Stecher, G.; Li, M.; Knyaz, C.; Tamura, K. MEGA X: Molecular evolutionary genetics analysis across computing platforms. *Mol. Biol. Evol.* **2018**, *35*, 1547–1549. [CrossRef] [PubMed]
31. Swofford, D.L. *PAUP*: Phylogenetic Analysis Using Parsimony (*and Other Methods), v. 4.0b10*; Sinauer Associates: Sunderland, MA, USA, 2003.
32. Hillis, D.M.; Bull, J.J. An empirical test of bootstrapping as a method for assessing confidence in phylogenetic analysis. *Syst. Biol.* **1993**, *42*, 182–192. [CrossRef]

plants

Article

COS-OGA Applications in Organic Vineyard Manage Major Airborne Diseases and Maintain Postharvest Quality of Wine Grapes

Francesca Calderone [1], Alessandro Vitale [1,*], Salvina Panebianco [2], Monia Federica Lombardo [1] and Gabriella Cirvilleri [1]

[1] Dipartimento di Agricoltura, Alimentazione e Ambiente, University of Catania, 95123 Catania, Italy; francesca.calderone00@gmail.com (F.C.); monia.lombardo@phd.unict.it (M.F.L.); gcirvil@unict.it (G.C.)

[2] Dipartimento di Fisica e Astronomia, University of Catania, Via S. Sofia 64, 95123 Catania, Italy; salvina.panebianco@ct.infn.it

* Correspondence: alevital@unict.it

Abstract: In most wine-growing countries of the world the interest for organic viticulture and eco-friendly grape production processes increased significantly in the last decade. Organic viticulture is currently dependent on the availability of Cu and S compounds, but their massive use over time has led to negative effects on environment health. Consequently, the purpose of this study was to evaluate the effectiveness of alternative and sustainable treatments against powdery mildew, gray mold and sour rot under the field conditions on Nero d'Avola and Inzolia Sicilian cultivars. In detail, the efficacy of COS-OGA, composed by a complex of oligochitosans and oligopectates, and its effects in combination with arbuscular mycorrhizal fungi (AMF) were evaluated to reduce airborne disease infections of grape. COS-OGA combined with AMF induced a significant reduction in powdery mildew severity both on Nero d'Avola and Inzolia with a mean percentage decrease of about 15% and 33%, respectively. Moreover, COS-OGA alone and combined with AMF gave a good protection against gray mold and sour rot with results similar to the Cu–S complex (performance in disease reduction ranging from 65 to 100%) on tested cultivars. Similarly, the COS-OGA and AMF integration provided good performances in enhancing average yield and did not negatively impact quality and microbial communities of wine grape. Overall, COS-OGA alone and in combination could be proposed as a valid and safer option for the sustainable management of the main grapevine pathogens in organic agroecosystems.

Keywords: organic vineyards; sustainable management; powdery mildew; gray mold; sour rot; postharvest quality

Citation: Calderone, F.; Vitale, A.; Panebianco, S.; Lombardo, M.F.; Cirvilleri, G. COS-OGA Applications in Organic Vineyard Manage Major Airborne Diseases and Maintain Postharvest Quality of Wine Grapes. *Plants* **2022**, *11*, 1763. https://doi.org/10.3390/plants11131763

Academic Editor: Inmaculada Pascual

Received: 24 May 2022
Accepted: 29 June 2022
Published: 1 July 2022

1. Introduction

Environmental and food safety issues are driving the wine sector towards innovative systems characterized by eco-friendly and sustainable approaches [1,2]. To this regard, the expansion of the organic viticulture and wine market is increasing more and more and is globally widespread [3]. European countries hold a predominant position in such a scenario, since Spain, France and Italy account for 75% of the world surface destined to organic wine grape production [4]. Italy represents one of the largest organic grape and wine producers, since more than 15% of Italian wine grape cultivation surface is addressed to organic production and covers about 107,143 ha. In detail, Sicily is the first organic viticulture region in Italy with 29,669 ha, corresponding to over 25% of the organic Italian wine-growing surface [4]. Sicilian wine is also well-known for quality and typicality mainly referable to a wide range of indigenous germplasm (i.e., Nero d'Avola, Nerello Mascalese, Nocera, Grillo, Inzolia, Cataratto). European certifications were assigned over time to Sicilian wine production areas, such as 1 "Guaranteed and Controlled Designation of Origin (GCDO)" and 23 "Controlled Designation of Origin (CDOs)". These labels define the different Sicilian terroirs, in which chemical and fertilization inputs should be reduced to preserve biodiversity and maintain balanced agro-ecosystems [1]. The eco-friendly approach, involving a strong reduction in chemical inputs, is also supported by the agro-food industry and the global government measures.

Powdery mildew, gray mold and downy mildew, caused by *Erysiphe necator*, *Botrytis cinerea* and *Plasmopara viticola*, respectively, represent major grapevine diseases, that strongly affect yield and quality worldwide. The sour rot, caused by a complex of bacteria and yeast, can be also considered a serious threat for the grape production and wine-making process [5]. The control of these diseases relies almost exclusively on fungicide applications [6,7], including Cu and S compounds which are extensively used in organic vineyards [8,9]. These compounds are considered mandatory steps for disease protection in organic winegrowing due to lack of available alternatives and are often applied simultaneously to control *P. viticola* and *E. necator* infections [10]. Consequently, the massive use of Cu, and to a lesser extent for S compounds, over the last century has led to negative effects for environment health [11–13], which are conflicting with the principles of organic agriculture. The European Commission has regulated the use of Cu compounds per year [14] being the maximum allowed 28 kg/ha over 7 years (averagely 4 kg ha^{-1} per year) [15]. Consequently, Cu molecule is considered an active substance candidate for impending withdrawal for agricultural purposes [16] as it has already pursued in some European countries, such as Denmark and the Netherlands [11]. Despite the negative impact, Cu is still necessary in organic viticulture, due to its wide activity spectrum, high efficacy against downy mildew and low cost [8].

According to eco-friendly approaches and increasing organic wine demands, the global scientific community focused on developing alternatives to Cu in order to reduce and/or replace it in main crops, such as grapevine. There are many substances under ongoing testing and validation and some are already on the global market, such as inorganic substances (i.e., zeolite, potassium and sodium hydrogen carbonate); biological control agents (BCAs); biostimulants, including also arbuscular mycorrhizal fungi (AMF) and plant growth-promoting rhizobacteria (PGPR), and resistance inducer (RI) products [11,17–19].

For example, many BCAs are already commercially available for winegrowers. Several bioformulates based on *Trichoderma* spp., *Streptomyces* spp., *Aureobasidium pullulans*, *Bacillus subtilis* and *B. amyloliquefaciens* are used to manage gray mold infections, whereas *Ampelomyces quisqualis*-based formulates are applied for the control of the powdery mildew of grape [1,17]. With regard to biostimulants, the most promising and widely used are seaweed extracts, protein hydrolysates, humic and fulvic acids, silicon, AMF and PGPR [18]. Several studies showed that mycorrhizal colonization provides pathogen protection through inducing plant systemic resistance [19,20]. AMF are also considered among the viable alternatives for a sustainable vineyard management being able to establish root symbiotic relationship with grapevine such as it happens for some species as *Glomus* and *Rhizophagus* [19,21,22].

Several RIs have been recently studied focusing on their mode of action and performances versus grapevine pathogens [23]. Some of these provided adequate protection in controlled conditions [24,25] and others in vineyards [8,26,27]. Among these, laminarin and chitosan aroused much interest. Romanazzi et al. [27] reported the effectiveness of laminarin and Saccharomyces extracts mixtures and chitosan alone in reducing downy mildew in vineyards although to a lesser extent if compared to copper compounds. Furthermore, chitosan confirmed similar performances against *P. viticola* both under high and low disease pressures [28].

Among the RIs present on the global market, chitosan oligosaccharides (COS)–oligogalacturonides (OGA) complex has been found to induce resistance against biotrophic pathogens in different crop plants [29]. As a consequence of the favorable opinion expressed by the European Food Safety Authority [30], the European Commission [31] approved it as "the first low-risk active substance". COS-OGA is an active substance resulting from the combination of a complex of COS, which are compounds found in fungal cell walls and crustacean exoskeletons, with pectin-derived OGA originating from plant cell walls. Therefore, chitosan and pectin form a stabilized complex, known as an egg box, which triggers a set of signaling resulting in defense reactions against potential invaders. COS and OGA fragments are detected by plant as non-self and self-molecules, respectively. The coupled danger signals increase the speed and intensity of plant defense response [32]. COS-OGA gave adequate protection against powdery mildew of tomato and cucumber [26,33]. Moreover, COS-OGA showed good activity to prevent potato late blight and root-knot nematode attacks (*Meloidogyne graminicola*) on rice crop [29,34]. Otherwise, little information is available in the literature about COS-OGA performance against grapevine diseases regarding the powdery mildew control in integrated production systems [7,26].

Thus, the aim of this paper was the evaluation of the performance of COS-OGA alone and in combination with AMF—commonly used in organic vineyards of this Sicilian district—on Nero d'Avola and Inzolia cultivars (i) in managing powdery mildew, gray mold and sour rot representing key pathogens for Mediterranean grape production; (ii) in reducing yield losses and maintaining postharvest chemical characteristics of grape; and (iii) in affecting the culturable carposphere microflora.

2. Results

2.1. Field Experiments

The effectiveness of sustainable alternative compounds against powdery mildew, gray mold and sour rot was evaluated through the in-field assessment of symptoms on bunches confirmed by laboratory results. The treatment efficacies were always referred to relative untreated controls calculating Abbott's formula. Sicilian grape cultivars Nero d'Avola and Inzolia were evaluated and disease levels compared at the end of crop cycle to observe different responses to treatments and/or disease susceptibility. Since ranking of treatments efficacies/effects was similar for both trials conducted in the two vineyards, diseases, yield, and chemical and microbiological data were averaged for at least two trials and reported for each cultivar.

2.1.1. Climate Data

Climatic conditions favorable to powdery mildew were detected in the 2020 season. In Figure 1, the mean weekly values of air temperature, relative humidity and rain are reported by averaging raw data from 1 April up to 30 September. Moreover, in this Figure, phenological stages of vine budbreak (from 25 to 31 March), bloom and veraison are indicated.

2.1.2. Efficacy in Controlling Powdery Mildew Infections in Vineyards

Effects of two single factors, treatment and cultivar, were always significant for all the tested parameters, whereas site effects were not significant (Table 1). Therefore, the data clearly showed a different susceptibility to the powdery mildew of grape between the two tested cultivars, being Nero d'Avola more susceptible than Inzolia cultivar (Table 1, Figure 2). All first and second order interactions were not significant versus disease parameters except for treatment × cultivar on DS and I_{MK} variables (Table 1).

Based on the ANOVA results, the trials were analyzed separately for Nero d'Avola and Inzolia cultivars. Post-hoc analyses to establish the ranking of effectiveness at the end of crop cycle are reported in the Table 2.

These data showed that no significant DI differences were observed among different treatments on Nero d'Avola, whereas these differences were detected for Inzolia cultivar, and in this case the Cu–S complex and Cu–S complex plus mycorrhiza combination were the only effective treatments (significant data). Otherwise, significant differences were always observed among DS and I_{MK} values both on Nero d'Avola and Inzolia. In detail, the Cu–S complex and Cu–S complex plus mycorrhiza combination were the most effective treatments being able to significantly reduce powdery mildew DS and I_{MK} values by about 56–62% on Nero d'Avola and about 43–48% on Inzolia. Although to a lesser extent,

DS and I_{MK} were significantly decreased by COS-OGA plus mycorrhiza treatment both on Nero d'Avola and Inzolia with percentage reductions of about 15% and 33%, respectively. COS-OGA applied alone significantly reduced about by 28% DS and I_{MK} values on Inzolia cultivar.

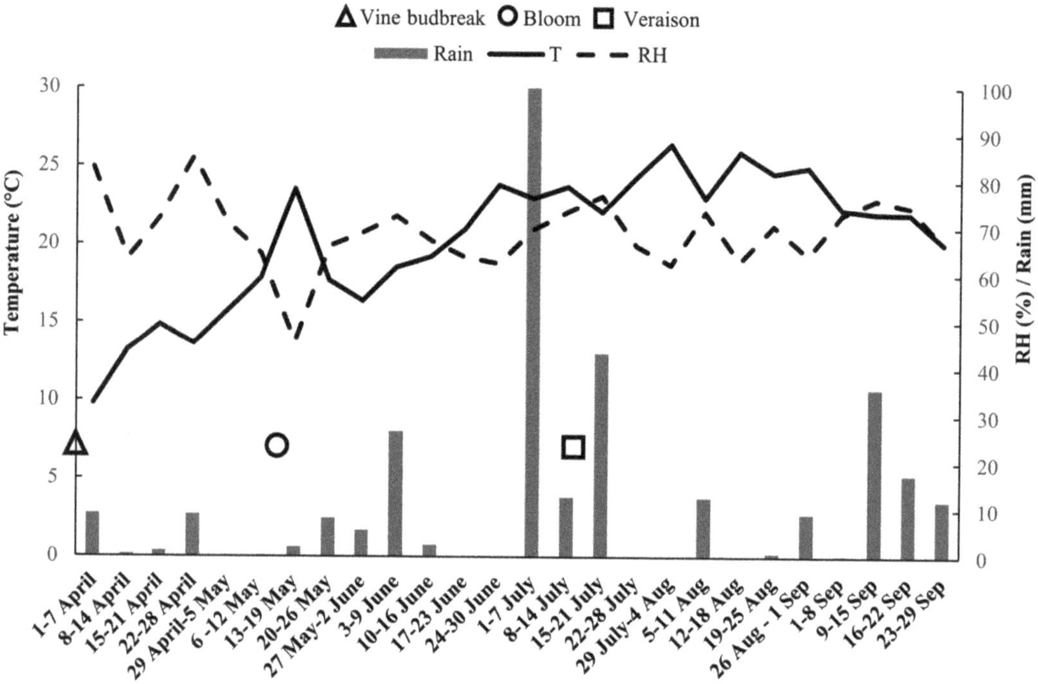

Figure 1. Climate data and main phenological stages of grapevines detected in the 2020 season, from April to September, by the weather stations of Novara di Sicilia and Patti (ME). T = temperature; RH = relative humidity.

Table 1. Effects of single factors and their interactions in ANOVA on the powdery mildew infection on wine grape caused by *Erysiphe necator* over time.

Source of Variation	df	Disease Incidence (DI)		Disease Severity (DS)		McKinney's Index (I_{MK})	
		F	*p*-Value	F	*p*-Value	F	*p*-Value
Treatment	4	4.909	*0.001366*	72.439	*<0.0001*	59.012	*<0.0001*
Cultivar	1	11.879	*0.000908*	206.501	*<0.0001*	173.176	*<0.0001*
Site	1	2.182	0.143575 ns	1.084	0.300866 ns	1.384	0.242915 ns
Treatment × Cultivar	4	0.364	0.833815 ns	17.995	*<0.0001*	15.512	*<0.0001*
Treatment × Site	4	0.061	0.993079 ns	0.024	0.998849 ns	0.093	0.984535 ns
Treatment × Cultivar × Site	4	0.242	0.913439 ns	0.029	0.998354 ns	0.042	0.996549 ns

p-value of fixed effects associated to F test; *ns*: not significant data.

(a) (b)

Figure 2. Symptoms observed on grapes caused by *Erysiphe necator*. Whitish powdery efflorescence (**a**) and berry cracks (**b**) on bunches of Nero d'Avola.

Table 2. Post-hoc analyses of treatment effects on disease incidence (DI), severity (DS) and McKinney's index (I_{MK}) of powdery mildew on Nero d'Avola and Inzolia wine grapes caused by *Erysiphe necator* at the final production stages (on September 8th).

Treatment	Nero d'Avola [a,b]			Inzolia [a,b]		
	DI (%)	DS (0-to-4)	I_{MK} (%)	DI (%)	DS (0-to-4)	I_{MK} (%)
Untreated control	100 ± 0.0 [ns]	3.4 ± 0.11 a	86.2 ± 2.90 a	97.5 ± 2.5 a	2.1 ± 0.15 a	52.5 ± 3.88 a
Cu–S complex	92.5 ± 5.0	1.3 ± 0.11 c	33.1 ± 2.72 c	82.5 ± 5.0 b	1.1 ± 0.11 c	26.9 ± 2.72 c
Cu–S complex + mycorrhiza	92.5 ± 5.0	1.5 ± 0.18 c	37.5 ± 4.41 c	82.5 ± 5.0 b	1.2 ± 0.10 bc	28.8 ± 2.50 bc
COS-OGA	100 ± 0.0	3.2 ± 0.23 ab	81.2 ± 5.85 ab	92.5 ± 5.0 ab	1.5 ± 0.21 b	38.8 ± 5.28 b
COS-OGA + mycorrhiza	100 ± 0.0	2.9 ± 0.18 b	73.1 ± 4.59 b	92.5 ± 5.0 ab	1.4 ± 0.12 bc	35.0 ± 3.03 bc

[a] Data expressed as means of the two trials and followed by standard error of the means (± SEM). Each value derives from 5 replicates, each formed by at least 4 bunches. [b] Arcsine transformation was used on percentage data prior to analysis, whereas untransformed data (%) are presented. DI, DS and I_{MK} values followed by the same letter within each column are not significantly different according to Fisher's least significance differences test ($\alpha = 0.05$). DI = Disease incidence; DS = disease severity; I_{MK} = McKinney's index; ns = not significant data.

2.1.3. Efficacy in Controlling Gray Mold Infections in Vineyards

The effects of single factors, treatment and cultivar were always significant on both DI, DS and I_{MK} parameters, whereas the effects of the site were always not significant. All interactions between factors were not significant on DI, DS and I_{MK} except for treatment × cultivar on DS parameter (Table 3). Therefore, Nero d'Avola revealed higher susceptibility degree to gray mold infection (Figure 3) than Inzolia cultivar (Table 3).

Based on ANOVA results, the trials were analyzed separately for Nero d'Avola and Inzolia cultivar. Post-hoc analyses showed the same ranking of treatment effectiveness for Nero d'Avola and Inzolia cultivars (Table 4).

Table 3. Effects of single factors and their interactions in ANOVA on gray mold caused by *Botrytis cinerea* on wine grape.

Source of Variation	df	Disease Incidence (DI)		Disease Severity (DS)		McKinney's Index (I_{MK})	
		F	*p*-Value	F	*p*-Value	F	*p*-Value
Treatment	4	17.8744	*<0.0001*	35.4128	*<0.0001*	27.6946	*<0.0001*
Cultivar	1	778291	*<0.0001*	95.1193	*<0.0001*	96.7338	*<0.0001*
Site	1	0.1608	0.689486 ns	1.1743	0.281772 ns	0.6645	0.417390 ns
Treatment × Cultivar	4	0.8693	0.486126 ns	12.4587	*<0.0001*	1.6885	0.160808 ns
Treatment × Site	4	0.0352	0.997585 ns	0.0734	0.990027 ns	0.0633	0.992476 ns
Treatment × Cultivar × Site	4	0.0754	0.989510 ns	0.0183	0.999327 ns	0.0609	0.993018 ns

p-value of fixed effects associated to F test; *ns*: not significant data.

Figure 3. Symptoms observed on bunches of Nero d'Avola caused by *Botrytis cinerea*.

Table 4. Post-hoc analyses of treatment effects on disease incidence (DI), severity (DS) and McKinney's index (I_{MK}) of gray mold caused by *Botrytis cinerea* on Nero d'Avola and Inzolia wine grapes.

Treatment	Nero d'Avola [a,b]			Inzolia [a,b]		
	DI (%)	DS (0-to-4)	I_{MK} (%)	DI (%)	DS (0-to-4)	I_{MK} (%)
Untreated control	70 ± 9.35 a	1.1 ± 0.16 a	27.5 ± 4.12 a	22.5 ± 2.5 a	0.2 ± 0.02 a	5.6 ± 0.62 a
Cu–S complex	17.5 ± 7.51 b	0.2 ± 0.07 b	4.4 ± 1.87 b	0.0 ± 0.0 b	0.0 ± 0.0 b	0.0 ± 0.0 b
Cu–S complex + mycorrhiza	22.5 ± 10.0 b	0.2 ± 0.10 b	5.6 ± 2.50 b	0.0 ± 0.0 b	0.0 ± 0.0 b	0.0 ± 0.0 b
COS-OGA	35 ± 10.0 b	0.3 ± 0.08 b	7.5 ± 2.12 b	0.0 ± 0.0 b	0.0 ± 0.0 b	0.0 ± 0.0 b
COS-OGA + mycorrhiza	25 ± 13.69 b	0.2 ± 0.11 b	5.6 ± 2.86 b	0.0 ± 0.0 b	0.0 ± 0.0 b	0.0 ± 0.0 b

[a] Data expressed as means of the two trials and followed by standard error of the means (± SEM). Each value derives from 5 replicates, each formed by at least 4 bunches. [b] Arcsine transformation was used on percentage data prior to analysis, whereas untransformed data (%) are presented. DI, DS and I_{MK} values followed by the same letter within each column are not significantly different according to Fisher's least significance differences test (α = 0.05). DI = Disease incidence; DS = disease severity; I_{MK} = McKinney's index; ns = not significant data.

On Nero d'Avola, significant differences were detected among treatments for DI parameter. In detail, Cu–S complex, Cu–S complex plus mycorrhiza, COS-OGA and COS-OGA plus mycorrhiza treatments significantly reduced the DI variable with values comprised between 50% and 75%. All treatments significantly reduced DS and I_{MK} values from 73% up to 85% if compared to the untreated control according to Abbott's formula. Moreover, on Inzolia grape all treatments inhibited gray mold development.

2.1.4. Efficacy in Controlling Sour Rot Infections in Vineyards

Similar to previous experiments, the effects of the single factors, treatment and cultivar, were significant on DI, DS and I_{MK} parameters, whereas the effects of the site were always not significant. All interactions between factors were not significant on all disease parameters except for treatment × cultivar on DS (Table 5). Nero d'Avola revealed higher susceptibility degree to sour rot infection (Figure 4) than Inzolia cultivar (Table 5).

Table 5. Effects of single factors and their interactions in ANOVA on the sour rot infection caused by phytopathogenic bacteria and yeasts on wine grape.

Source of Variation	df	Disease Incidence		Disease Severity		McKinney's Index (I_{MK})	
		F	*p*-Value	F	*p*-Value	F	*p*-Value
Treatment	4	44.8487	*<0.0001*	34.2016	*<0.0001*	34.1885	*<0.0001*
Cultivar	1	205.5921	*<0.0001*	154.9407	*<0.0001*	192.4708	*<0.0001*
Site	1	1.5921	0.210692 ns	0.7905	0.376611 ns	0.6381 ns	0.426747 ns
Treatment × Cultivar	4	2.2039	0.075924 ns	9.3083	<0.0001	1.9823 ns	0.105118 ns
Treatment × Site	4	0.0461	0.995922 ns	0.0198	0.999221 ns	0.0276 ns	0.998498 ns
Treatment × Cultivar × Site	4	0.0855	0.986682 ns	0.0277	0.998490 ns	0.0599 ns	0.993223 ns

p-value of fixed effects associated to F test; *ns*: not significant data.

Post-hoc analyses revealed a similar ranking of efficacy on two wine grape cultivars except for DS and IMK parameters on Nero d'Avola (Table 6). On this cultivar, all treatments were significantly effective to reduce the number of infected bunches compared to the untreated control. In detail, the most effective treatment was once again the Cu–S complex plus mycorrhiza combination, since it reduced DI, DS and IMK values by approximately 72–82% if compared to untreated controls (Abbott's formula). Although with slightly lower performances, COS-OGA and COS-OGA plus mycorrhiza were able to significantly reduce sour rot decay. On Inzolia, all treatments were significantly effective in reducing or inhibiting sour rot infections if compared to the control.

Figure 4. Symptoms observed on bunches of Nero d'Avola caused by sour rot.

Table 6. Post-hoc analyses of treatment effects on disease incidence (DI), severity (DS) and McKinney's index (I_{MK}) of sour rot caused by phytopathogenic bacteria and yeasts on Nero d'Avola and Inzolia wine grapes.

	Nero d'Avola [a,b]			Inzolia [a,b]		
Treatment	**DI (%)**	**DS (0-to-4)**	**I_{MK} (%)**	**DI (%)**	**DS (0-to-4)**	**I_{MK} (%)**
Untreated control	97.5 ± 2.5 a	1.7 ± 0.10 a	43.1 ± 2.50 a	37.5 ± 5.6 a	0.4 ± 0.07 a	10.6 ± 1.88 a
Cu–S complex	40.0 ± 10.7 b	0.5 ± 0.12 bc	11.2 ± 2.89 bc	0.0 ± 0.0 b	0.0 ± 0.0 b	0.0 ± 0.0 b
Cu–S complex + mycorrhiza	27.5 ± 11.4 b	0.3 ± 0.13 c	8.1 ± 3.37 c	0.0 ± 0.0 b	0.0 ± 0.0 b	0.0 ± 0.0 b
COS-OGA	47.5 ± 6.1 b	0.9 ± 0.25 b	22.5 ± 6.20 b	7.5 ± 7.5 b	0.07 ± 0.07 b	1.9 ± 1.87 b
COS-OGA + mycorrhiza	47.5 ± 6.1 b	0.6 ± 0.19 bc	15.6 ± 4.84 bc	2.5 ± 2.5 b	0.02 ± 0.02 b	0.6 ± 0.62 b

[a] Data expressed as means of the two trials and followed by standard error of the means (± SEM). Each value derives from 5 replicates, each formed by at least 4 bunches. [b] Arcsine transformation was used on percentage data prior to analysis, whereas untransformed data (%) are presented. DI, DS and I_{MK} values followed by the same letter within each column are not significantly different according to Fisher's least significance differences test (α = 0.05). DI = Disease incidence; DS = disease severity; I_{MK} = McKinney's index; ns = not significant data.

2.2. Disease Incidence and Severity Progressions of Powdery Mildew

Powdery mildew disease (DI and DS) progressions over time for each cultivar are shown in Figure 5. On Nero d'Avola, the DI value was averagely very high and it reached maximum level in July (DI = 100%) in untreated control plots, whereas DS value reached the maximum in August (DS = 3.5). Likewise, powdery mildew infections detected on Inzolia grape showed averagely high DI values. In detail, high decay amounts were already observed starting from June and increased at the following evaluations. Unlike Nero d'Avola, on Inzolia the higher level of DI was reached in August (97.5%), whereas DS maximum levels were reached in September (DS = 2.15). Comprehensively, lower values of DS were recorded overt time on Inzolia vineyards if compared to Nero d'Avola, thus confirming the higher disease susceptibility of the red wine grape cultivar (Figure 5).

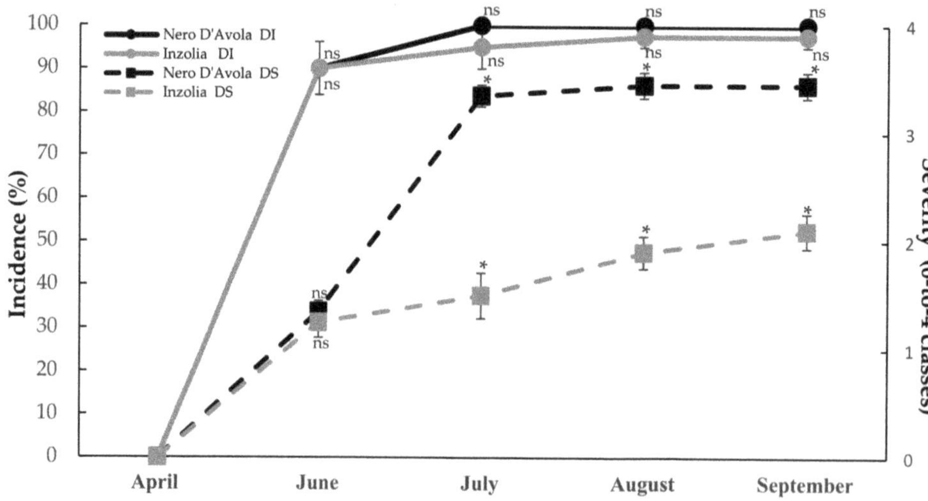

Figure 5. Progression over time of powdery mildew infections caused by *Erysiphe necator* in control plots of vineyards. Continuous lines indicate dual comparisons of DI progression values (percent) over time whereas dotted lines show dual comparisons of DS progression values (0-to-4 scale) over time between Nero d'Avola (black lines) and Inzolia (grey lines). ns = not significant; * = significant differences.

2.3. Yield and Chemical Analysis of Grape Must

The impact of sustainable compounds on grape yield and quality parameters of wine grape, including the total soluble solids (°Brix), total acidity (g L^{-1} of tartaric acid) and pH, was assessed (Table 7).

Table 7. Grape production and oenological parameters of Nero d'Avola and Inzolia.

	Nero d'Avola [a,b]			
Treatment	**Yield (kg plant^{-1})**	**Sugar Content (°Brix)**	**Total Acidity (g L^{-1})**	**pH**
Untreated control	0.6 ± 0.04 d	20.8 ± 0.19 [ns]	6.2 ± 0.08 [ns]	3.1 ± 0.01 [ns]
Cu–S complex	1.4 ± 0.03 a	21.0 ± 0.16	6.2 ± 0.03	3.2 ± 0.01
Cu–S complex + mycorrhiza	1.4 ± 0.03 a	20.6 ± 0.17	6.1 ± 0.07	3.3 ± 0.15
COS-OGA	0.9 ± 0.04 c	20.9 ± 0.22	6.0 ± 0.07	3.2 ± 0.01
COS-OGA + mycorrhiza	1.0 ± 0.05 b	20.8 ± 0.25	6.2 ± 0.05	3.3 ± 0.14
	Inzolia [a,b]			
Untreated control	0.71 ± 0.02 d	19.88 ± 0.09 [ns]	5.12 ± 0.11 [ns]	3.36 ± 0.01 [ns]
Cu–S complex	1.80 ± 0.08 a	19.52 ± 0.09	5.36 ± 0.15	3.46 ± 0.10
Cu–S complex + mycorrhiza	1.80 ± 0.06 a	19.82 ± 0.10	5.18 ± 0.08	3.42 ± 0.05
COS-OGA	1.36 ± 0.03 c	20.04 ± 0.08	5.00 ± 0.03	3.45 ± 0.05
COS-OGA + mycorrhiza	1.51 ± 0.04 b	19.93 ± 0.09	5.30 ± 0.04	3.43 ± 0.05

[a] Data expressed as means of the two trials and followed by standard error of the means (±SEM). Each value derives from 5 replicates, each formed by at least 4 bunches. [b] Values followed by the same letter within each column are not significantly different according to Fisher's least significance differences test ($\alpha = 0.05$). ns = not significant data.

Treatments provided significant effects only on yields of Nero d'Avola and Inzolia wine grapes (Table 7). Nero d'Avola production always increased significantly in all treated plots if compared to the untreated control. Cu–S complex and Cu–S complex plus mycorrhiza combination provided the best performances with the highest production increases (about by 133%) if referred to relative control (Table 7). Similarly, the Cu–S complex and Cu–S complex plus mycorrhiza combination increased wine grape yield by 153% on Inzolia vineyard (Table 7). Although with lower performances, also the COS-OGA and COS-OGA plus mycorrhiza combination significantly increased the average yields of Nero d'Avola (from 50 to 67%) and Inzolia (from 92 to 113%) wine grapes if compared to those of untreated controls. The chemical analysis of the wine grapes showed that the sugar content, total acidity and pH were not significantly influenced by all treatments with respect to those recorded on Nero d'Avola and Inzolia untreated controls (Table 7).

2.4. Microbiological Analysis of Grapes

Four carposphere microbial communities were separately evaluated for each cultivar (fungal and yeast populations, and aerobic bacterial and fluorescent bacteria populations).

The cultivable fungal and yeast populations on Nero d'Avola and Inzolia berries are reported in Figure 6a and 6c, respectively. The fungal load detected on Nero d'Avola carposphere ranged from 3.44 to 3.61 Log_{10} CFU g^{-1}, while the yeast load was comprised between 4.18 and 5.26 Log_{10} CFU g^{-1} throughout all treatments. The fungi load on Inzolia carposphere ranged from 3.29 to 3.44 Log_{10} CFU g^{-1}, while the yeast load from 4.57 to 5.17 Log_{10} CFU g^{-1}. Fungal and yeast loads were not significantly influenced by tested treatments both on Nero d'Avola and Inzolia wine grapes if compared to the relative untreated controls. The cultivable aerobic and fluorescent bacterial populations are reported in Figure 6b,d, respectively, for Nero d'Avola and Inzolia. The aerobic bacteria load on Nero d'Avola carposphere ranged from 4.75 to 5.35 Log_{10} CFU g^{-1}, while the fluorescent bacteria from 2.41 to 3.73 Log_{10} CFU g^{-1} throughout all treatments. The aerobic bacteria load on Inzolia was comprised between 3.44 and 5.33 Log_{10} CFU g^{-1}, while the fluorescent bacteria load ranged from 3.37 to 3.44 Log_{10} CFU g^{-1}. The fluorescent bacteria load was not significantly influenced by the tested treatments both on Nero d'Avola and Inzolia grape.

Otherwise, only for aerobic bacteria load significant differences were detected on different treated Inzolia wine grape (Figure 6d). In particular, in COS-OGA treatments (alone and in combination with mycorrhiza) aerobic bacteria load was similar to untreated control (not significant data); diversely Cu–S complex and Cu–S complex plus mycorrhiza combination significantly decreased the bacteria load.

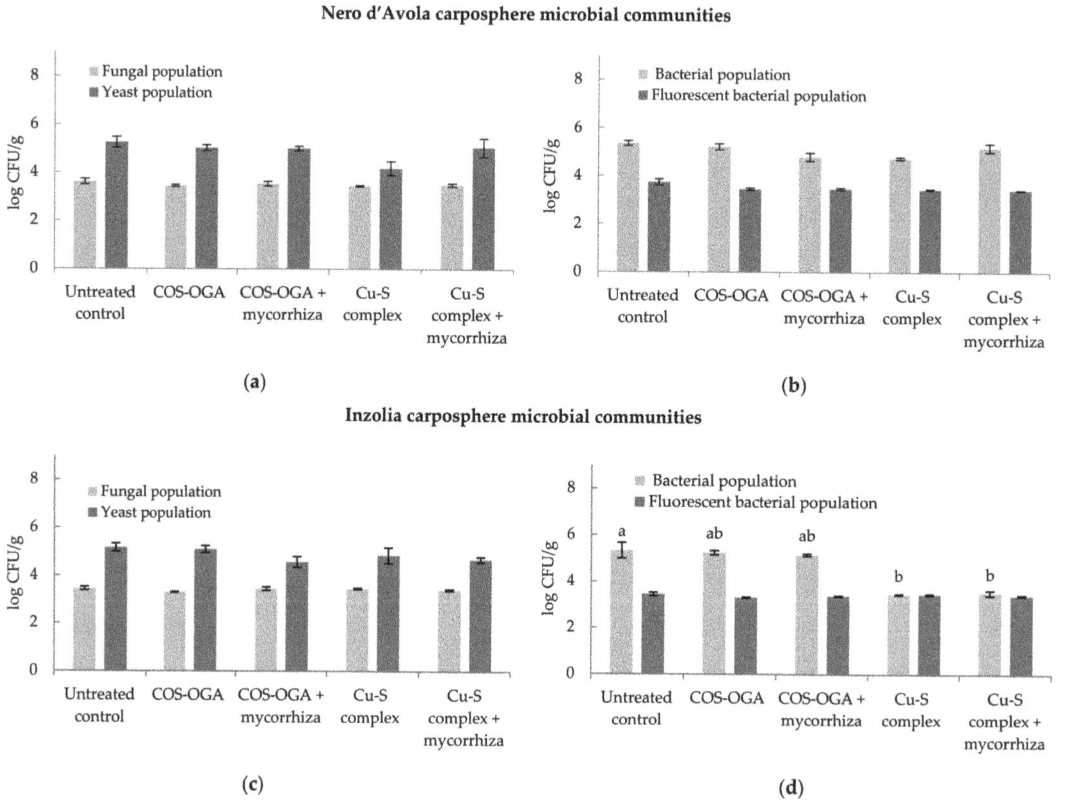

Figure 6. Fungal and yeast population on (**a**) Nero d'Avola and (**c**) Inzolia carposphere. Bacterial and fluorescent bacterial population on (**b**) Nero d'Avola and (**d**) Inzolia carposphere. Data presented as Log_{10} CFU/g of fresh weight. Bars show the standard error of the mean (±SEM). Bacterial population data followed by different letter(s) differs significantly according to Fisher's least significance differences test ($\alpha = 0.05$).

3. Discussion

Organic viticulture is currently dependent on the availability of Cu and S, which are crucial components of grapevine protection against the main diseases. The massive use over time of Cu and S has led to negative effects on environment health. Therefore, in the present paper, sustainable and ecofriendly alternative compounds were tested for the first time in Sicily against powdery mildew, gray mold and sour rot under field conditions on Nero d'Avola and Inzolia, two among the most worldwide appreciated Sicilian cultivars. The effectiveness of COS-OGA as RI, and its effects combined with arbuscular mycorrhizal fungi (AMF) were assessed in controlling grapevine diseases. The good exposure of the Sicilian organic vineyards to solar radiation and wind reduced humidity conditions, thus limiting *P. viticola* infections. Otherwise, the climatic conditions were very conducive for powdery mildew epidemic, confirming that Sicily is a high-risk area for *E. necator* infections.

Moreover, powdery mildew occurred with different disease pressures depending on the cultivar and, specifically, Nero d'Avola was most susceptible. On this cultivar whitish powdery efflorescence was associated with necrotic reticulation and suberization of the epidermal cells, often leading to berry cracking. Otherwise, a lower severity was recorded on the Inzolia cultivar. The symptoms consisted of necrotic reticulations, which generally do not evolve into berry cracks. Concerning gray mold and sour rot, the rainfalls that occurred from middle July up to early September (BBCH 81–89) resulted in low (Inzolia) and high (Nero d'Avola) disease pressures. As a consequence of berry cracking, Nero d'Avola was also severely affected by gray mold and sour rot. Diversely, lower levels of gray mold and sour rot decays have been recorded on Inzolia, reduced about to one third if compared to red wine grape cultivar. To the best of our knowledge, this study reports for the first time that Nero d'Avola is more susceptible to powdery mildew, gray mold and sour rot than Inzolia cultivar.

Although the Cu–S complex plus mycorrhiza and Cu–S complex always provided the best results against major fungal diseases of wine grapes, COS-OGA-based treatments gave also noteworthy performances. In detail, COS-OGA combined with mycorrhiza application followed by COS-OGA alone were effective in reducing severity of powdery mildew attacks on Inzolia, whereas only COS-OGA plus AMF combination were able to significantly reduce the powdery mildew amount on Nero d'Avola. Our data showed that the elicitor was more effective in reducing severity than incidence and AMF could have mitigated *E. necator* infections. Since the colonization ability of AMF was not evaluated in this paper, further studies should be performed to confirm the performances of this treatment combination. However, present results are in accordance with findings obtained by van Aubel et al. [26] in French and Spain vineyards. This is probably due to the elicitors action mode, that do not involve direct toxic effects against pathogens but triggers natural host defenses, leading to a reduction of disease amounts; this could be the reason why disease incidence is less well-controlled than severity under high disease pressure [26]. In addition, COS-OGA applied alone and combined with mycorrhizal fungi always proved to be effective against gray mold and sour rot both on Nero d'Avola and Inzolia. In particular, on Inzolia their performances were comparable with Cu–S complex, which is the standard product for many organic wine growers.

Little is known about the effects of the organic vineyard management on the wine-making process [1], but frequent Cu and S treatments could probably compromise the composition and the sensory properties of the wine [9,35]. Moreover, Cu and S repeated applications in vineyards could influence the species diversity of fermentation microbiota [7,36], including indigenous yeasts and other environment-related microorganisms, which can contribute to define wine regional typicality [1]. Definitely, the replacement or reduction in chemical inputs by using of sustainable and eco-friendly compounds might lead to benefits for the postharvest stages and microbial communities. To this regard, as the present study demonstrates, COS-OGA is able to reduce yield losses and simultaneously maintains the main chemical grape characteristics—defining technological maturity—in full respect of carposphere microorganisms. Our findings showed that oenological parameters (sugar content, pH and total acidity) were not affected by the tested treatments, including COS-OGA application in agreement with previous data reported by Rantsiou et al. [7]. These authors evaluated the effect of chemical products, bioproducts and RIs (COS-OGA mixed with Cu and metiram) on yield and parameters involved in the technological and phenolic maturity of the grapes at the harvest time. Unlike Rantsiou et al. [7], the grape production varied considerably among the tested products being COS-OGA-based treatments always able to enhance the average yields and to preserve microbial communities on Nero d'Avola and Inzolia wine grapes. Moreover, COS-OGA-based applications did not negatively influence aerobic bacteria communities on Inzolia carposphere, whereas Cu–S-based treatments failed. Based on these data, COS-OGA application can be encouraged on a large scale since this compound does not exert negative pressure on carposphere microorganisms.

Although RIs are rarely used in viticulture, due to variable efficacy depending on several biotic and abiotic factors [23], COS-OGA compounds present many additional advantages if compared to Cu and S-based products. Their application, for example, does not imply any reentry interval for growers in vineyard, pre-harvest interval time, residues harmful for final consumers or the arising risk of resistance phenomena [26]. Moreover, the protection provided by elicitors as COS-OGA, is not specific and can potentially manage a wide spectrum of targeted phytopathogens as reported in the literature [37]. Although costs relative to COS-OGA applications are almost comparable to those reported for Cu–S, slight differences in the cost–benefit evaluation are justified by the above reported positive aspects. These data should be confirmed under different operative conditions. However, COS-OGA alone or combined with other organic control measures could be proposed to enhance the sustainability of the management of main airborne diseases and grape production. These efforts will allow a significant decrease over time in the use of Cu in organic viticulture according to the global green policies.

4. Materials and Methods

4.1. Field Experiments

Two field experiments were performed in duplicate during 2020 in two 10-year-old organic vineyards of *Vitis vinifera* L., i.e., on red berried Nero d'Avola and on white berried Inzolia wine grape cultivars, both grafted onto 140 Ru (Ruggeri) rootstock. Two trials for each cultivar were conducted in different sites of vineyards located at Rodì Milici (Messina province, Italy, lat. 38°06′ N; long. 15°08′ E, altitude of about 100 m a.s.l). The plants were spaced by 1 m in the rows, with 2.5 m between the rows, and they were grown according to the Guyot trellis system, leaving 5–6 buds per grapevine, with grass cover between the rows. The height of the fruiting cane was about 60–65 cm from the soil surface. The vineyards were not irrigated and the natural organo-mineral fertilizer (Vigna Pro, NPK 3-6-12, TerComposti S.p.A.) was distributed at the rate of 500 kg ha^{-1} banded under the grapevine in the winter, according to the common practices for the cultivation area. Moreover, arbuscular mycorrhizal fungi (Table 8) were applied as soil treatment at vine budbreak of the previous crop seasons (in 2018, 2019 and 2020).

Table 8. Integrated strategies adopted in the application of Cu and S formulation and alternative products (COS-OGA and mycorrhizae) and dates of the grapevine foliar applications for Nero d'Avola and Inzolia cultivars in different sites.

Treatment/Active Compound	Dosage	Product and Company	N. and Timing of Applications *
Untreated control	-	-	2 on May 2nd and 25th; 2 on June 15th and 30th; 1 on July 22nd and 1 on August 13th
COS–OGA	3 L ha^{-1}	Ibisco®, Gowan Italia S.r.l.	
COS-OGA + mycorrhiza	3 L ha^{-1} + 5 kg ha^{-1}	Ibisco®; Micosat F® MO, CCS Aosta S.r.l.	
Cu–S complex 3%	5 L ha^{-1}	Heliocuivre®–Heliosoufre®, CBC Europe S.r.l.	
Cu–S complex 3% + mycorrhiza	5 L ha^{-1} + 5 kg ha^{-1}	Heliocuivre®–Heliosoufre®; Micosat F® MO	

* Application data referred only to COS-OGA and Cu–S complex, while mycorrhiza (AMF) were applied only once on 5 June 2020.

A randomized complete block design with 4 replicates each consisting of 12 plants for treatment was adopted for each of the 4 experimental trials. Moreover, buffers were always inserted among plots differently treated. Four treatments for each cultivar were included and compared with an untreated control. All treatments were applied from the start of May until the end of August, with a total of 6 applications for each site (Table 8). At the time of the first application, grapevines were at the phenological stage of inflorescences swelling (BBCH 55) and the shoots were about 20 cm long. The COS-OGA and Cu–S applications were done by spraying a volume equivalent to 500 L ha^{-1}, using a backpack sprayer (Volpi UNI mod. 78P) in treatments 1, 2, 3, 4 and by airblast sprayer in treatments

5 and 6, whereas AMF was sprayed only once onto the canopy (Table 8). Grapes were harvested at their optimum technological maturity (8 September). Thereafter, Nero d'Avola grapes were destemmed, left to macerate, and then crushed, whereas Inzolia ones were only destemmed and crushed.

4.2. Climate Data

The weather parameters, i.e., average temperature (°C), relative humidity (%) and rainfall (mm) were obtained from the data provided by the weather stations of Novara di Sicilia and Patti (Messina province) of Servizio Informativo Agrometeorologico Siciliano (SIAS), Sicily region. These data were implemented with phenological stages of vine budbreak, bloom and veraison.

4.3. Assessment of Disease Symptoms

Disease incidence and severity were determined by the assessment of symptoms in each plot of vineyards. Powdery mildew (*E. necator*), gray mold (*B. cinerea*) and sour rot (yeasts plus acetic acid-producing bacteria) [5,38] infections were directly evaluated on bunches. Powdery mildew infection was assessed at five different monitoring times (12April, 15 June, 6 July, 5 August and 8 September, 2020), whereas gray mold and sour rot infections were assessed in September 2020 at the grape harvesting time. Disease symptoms were evaluated according to general EPPO guidelines [39] on four bunches for each plant (five plants for each plot). A five-point scale was used for both diseases with class '0' being no symptoms and class '4' being the highest damage. Class values corresponded to percentage infections range on the bunches—where class 0 = 0%, class 1 = 1–25% of infected berries on single bunch, class 2 = 25.1–50% of infected berries on single bunch, class 3 = 50.1–75% of infected berries on single bunch, class 4 = 75.1–100% of infected berries on single bunch. Data processing involved the calculation of the percentage of symptomatic bunches on the total number of examined bunches (= disease incidence, DI) and the average class (weighted mean) value of examined bunches (= disease severity, DS) for each plot. Moreover, the infection index (or McKinney's index = I_{MK}), which combines both the incidence and severity of the disease, was calculated according to the following equation:

$$I_{MK} = \frac{\Sigma\,(d \times f)}{N \times D} \times 100.$$

where d = category of disease class scored for the grape bunches; f = disease frequency; N = total number of examined bunches; D = highest class of disease intensity that occurred on the empirical scale [40].

Following disease assessment, a representative number of bunch samples were recovered randomly within each plot and transferred to the laboratory to identify and confirm the causal agents using microscope (Olympus–Bx61, Tokyo, Japan) observation (hyphae and conidia for powdery mildew and greyish layers containing conidiophores and conidia for gray mold) and isolation (producing typical colonies of *B. cinerea*) onto potato dextrose agar (PDA, Oxoid, Basingstoke, UK). The streaking technique was used to recover yeasts and acid acetic-producing bacteria (responsible of sour rot) from macerated berries on yeast peptone dextrose agar (YPDA, 10 g L^{-1} of yeast extract, 10 g L^{-1} of peptone, 20 g L^{-1} of dextrose, 20 g L^{-1} of bacteriological agar) supplemented with 100 mg L^{-1} of chloramphenicol and nutrient agar (NA, Oxoid, Basingstoke, UK) supplemented with 100 mg L^{-1} cycloheximide (sour rot).

4.4. Chemical Analysis of Grape Must

The quality parameters of must, including the total soluble solids (°Brix), total acidity (g L^{-1} of tartaric acid) and pH were determined through laboratory analysis carried out by Istituto Regionale del Vino e dell'Olio (IRVO) (Milazzo, Italy). Berry sampling was performed in a random way for all replicates in order to increase their representativeness, accordingly to the exposure and position of the berries on the bunch. The analyses were performed with the OenoFoss^{TM} instrument (FOSS Italia S.r.l. PD, Italy). Moreover, the average grape yield (weight of harvested bunches expressed as kg per plant) per treatment was recorded at harvest time.

4.5. Microbiological Analysis of Grapes

Carposphere microorganisms were evaluated on berries collected at harvest time. From each sample, approximately 100 g of healthy berries for each treatment were randomly removed from the bunches, placed in sterilized flasks with 500 mL of buffered peptone water (BPW, pH = 7.0 ± 0.2 at 25°C, Biolife, Milan, Italy) and 0.02% Tween 80 (VWR Chemicals, Solon, Ohio, USA), and then subjected to orbital shaking at 150 rpm for 1 h. Culturable bacteria, fungi and yeasts were assessed by plating tenfold serial dilutions in triplicate on four culture media: PDA and YPDA supplemented with 100 mg L^{-1} of chloramphenicol (AppliChem GmbH, Darmstadt, Germany) to inhibit bacterial growth; NA and King'B agar [41] supplemented with 100 mg L^{-1} cycloheximide (AppliChem GmbH, Darmstadt, Germany) to inhibit yeast and mold growth. Four different microbial communities were studied: total aerobic bacteria, fluorescent bacteria, fungi and yeasts. After incubation at 25°C for 2–5 days, colony forming units (CFUs) per unit of berry weight (CFUs g^{-1}) were calculated.

4.6. Statistical Analyses

Data from field trials were subjected to analysis of variance by using the Statistical 10 package software (StatSoft Inc., Tulsa, OK, USA) to determine significant differences among the tested treatments in field performances against *E. necator*, *B. cinerea* and sour rot. Data obtained from two trials were compared and analyzed for each cultivar. Initial analyses of disease incidence (DI), severity (DS) and McKinney's index (I_{MK}) were conducted by calculating F and the correspondent P-value associated with the main source of variation (treatment, evaluation time, and cultivar) and with the interactions among them. Thus, arithmetic means of DI, DS and I_{MK} of the two trials for each cultivar were calculated, averaging the values determined for the single replicates of treatments. Percentage data (DI and I_{MK}) were previously transformed using the arcsine transformation ($\sin^{-1}$ square root x). Post-hoc comparisons among different treatments were achieved by means of Fisher's least significant difference test at $\alpha = 0.05$. Similarly, significant differences in yield, chemical characteristics and microbial communities of wine grape were also assessed. The effects of the treatments on all tested parameters were also referred to the untreated control by using the Abbott's formula [42].

$$I(\%) = \frac{C - T}{C} \times 100$$

where I = percentage reduction data, C = mean parameter value in the untreated control plots and T = mean parameter value in the treated plots.

5. Conclusions

Comprehensively, COS-OGA applications could be considered as ecofriendly alternatives to Cu and S treatments since it manages natural infections of botrytis bunch and sour rots and, to a lesser extent, those of powdery mildew occurring in organic vineyards. For the latter disease, their performances depend on disease pressure and cultivar susceptibility. Although on average, less effective than Cu-S applications, COS-OGA alone or combined with AMF is able to reduce yield losses and to maintain postharvest grapes quality. Nevertheless, a further validation of this treatment combination is necessary since herein the colonization by AMF was not evaluated. Under high-disease pressure, COS-OGA represents an option to be integrated with already existing organic measures since it has a wide pathogen spectrum; no negative effects on carposphere microbiota; no harmful implications for environment, farmers or consumers; no residue and no fungicide resistance risks in total respect of current GND policies.

Author Contributions: Conceptualization, A.V. and G.C.; methodology, F.C. and A.V.; software, F.C. and A.V.; validation, F.C., A.V., S.P., M.F.L. and G.C.; formal analysis, A.V.; investigation, F.C.; resources, G.C.; data curation, F.C. and A.V.; writing—original draft preparation, F.C. and A.V.; writing—review and editing, A.V. and G.C.; visualization, F.C., A.V., S.P., M.F.L. and G.C.; supervision, G.C. and A.V.; project administration, G.C.; funding acquisition, G.C. All authors have read and agreed to the published version of the manuscript.

Funding: This research was supported by the Organic PLUS project that has received funding from the European Union's Horizon 2020 research and innovation program under grant agreement no. 774340-Organic PLUS and by University of Catania, Italy (PIA.CE.RI. 2020–2022 Linea 2: Research Project MEDIT-ECO) and Fondi di Ateneo 2020–2022—University of Catania (Italy), linea Open Access.

Data Availability Statement: The data presented in this study are available on request from the corresponding author.

Acknowledgments: Authors expressed deeply gratitude to Tenuta Lacco winery for supporting the research.

Conflicts of Interest: The authors declare no conflict of interest.

References

1. Provost, C.; Pedneault, K. The organic vineyard as a balanced ecosystem: Improved organic grape management and impacts on wine quality. *Sci. Hortic.* **2016**, *208*, 57–77. [CrossRef]
2. Rockström, J.; Williams, J.; Daily, G.; Noble, A.; Matthews, N.; Gordon, L.; Wetterstrand, H.; DeClerck, F.; Shah, M.; Steduto, P.; et al. Sustainable intensification of agriculture for human prosperity and global sustainability. *Ambio* **2017**, *46*, 4–17. [CrossRef] [PubMed]
3. D'Amico, M.; Di Vita, G.; Monaco, L. Exploring environmental consciousness and consumer preferences for organic wines without sulfites. *J. Clean. Prod.* **2016**, *120*, 64–71. [CrossRef]
4. SINAB (Sistema d'Informazione Nazionale Sull'agricoltura Biologica) 2020. *Uffici SINAB c/o MiPAAF*; Italian Ministry of Agriculture: Rome, Italy, 2020; Available online: https://www.sinab.it (accessed on 9 November 2021).
5. Zoecklein, B.W.; Williams, J.M.; Duncan, S.E. Effect of sour rot on the composition of white Riesling (*Vitis vinifera* L.) grapes. *Small Fruits Rev.* **2000**, *1*, 63–77. [CrossRef]
6. Gessler, C.; Pertot, I.; Perazzolli, M. *Plasmopara viticola*: A review of knowledge on downy mildew of grapevine and effective disease management. *Phytopathol. Mediterr* **2011**, *50*, 3–44.
7. Rantsiou, K.; Giacosa, S.; Pugliese, M.; Englezos, V.; Ferrocino, I.; Río Segade, S.; Monchiero, M.; Gribaudo, I.; Gambino, G.; Gullino, M.L.; et al. Impact of chemical and alternative fungicides applied to grapevine cv Nebbiolo on microbial ecology and chemical-physical grape characteristics at harvest. *Front. Plant Sci.* **2020**, *11*, 700. [CrossRef]
8. Dagostin, S.; Schärer, H.J.; Pertot, I.; Tamm, L. Are there alternatives to copper for controlling grapevine downy mildew inorganic viticulture? *Crop Prot.* **2011**, *30*, 776–788. [CrossRef]
9. Griffith, C.M.; Woodrow, J.E.; Seiber, J.N. Environmental behavior and analysis of agricultural sulfur. *Pest Manag. Sci.* **2015**, *71*, 1486–1496. [CrossRef]
10. Cabús, A.; Pellini, M.; Zanzotti, R.; Devigili, L.; Maines, R.; Giovannini, O.; Mattedi, L.; Mescalchin, E. Efficacy of reduced copper dosages against *Plasmopara viticola* in organic agriculture. *Crop Prot.* **2017**, *96*, 103–108. [CrossRef]
11. La Torre, A.; Iovino, V.; Caradonia, F. Copper in plant protection: Current situation and prospects. *Phytopathol. Mediterr.* **2018**, *57*, 201–236.
12. Redl, M.; Sitavanc, L.; Hanousek, F.; Steinkellner, S. A single out-of-season fungicide application reduces the grape powdery mildew inoculum. *Crop Prot.* **2021**, *149*, 105760. [CrossRef]
13. Moth, S.; Walzer, A.; Redl, M.; Petrovi ́c, B.; Hoffmann, C.; Winter, S. Unexpected effects of local management and land-scape composition on predatory mites and their food resources in vineyards. *Insects* **2021**, *12*, 180. [CrossRef]
14. European Commission. Commission Regulation (EC) No. 473/2002. *Off. J. Eur. Comm.* **2002**, L75/21. Available online: https://eur-lex.europa.eu/legal-content/EN/TXT/PDF/?uri=CELEX:32002R0473&from=EN (accessed on 16 November 2021).
15. European Commission. Commission Implementing Regulation (EC) No. 1981/2018. *Off. J. Eur. Comm.* **2018**, L317/16. Available online: https://eur-lex.europa.eu/legal-content/EN/TXT/PDF/?uri=CELEX:32018R1981&rid=3 (accessed on 16 November 2021).
16. European Commission. Commission Implementing Regulation (EC) No. 84/2018. *Off. J. Eur. Comm.* **2018**, L16/8. Available online: https://eur-lex.europa.eu/legal-content/EN/TXT/PDF/?uri=CELEX:32018R0084&from=EN (accessed on 16 November 2021).
17. Rotolo, C.; De Miccolis Angelini, R.M.; Dongiovanni, C.; Pollastro, S.; Fumarola, G.; Di Carolo, M.; Perrelli, D.; Natale, P.; Faretra, F. Use of biocontrol agents and botanicals in integrated management of *Botrytis cinerea* in table grape vineyards. *Pest Manag. Sci.* **2018**, *74*, 715–725. [CrossRef]

18. Cataldo, E.; Fucile, M.; Mattii, G.B. Biostimulants in viticulture: A sustainable approach against biotic and abiotic stresses. *Plants* **2022**, *11*, 162. [CrossRef]
19. Cruz-Silva, A.; Figueiredo, A.; Sebastiana, M. First Insights into the Effect of Mycorrhizae on the Expression of pathogen effectors during the infection of grapevine with *Plasmopara viticola*. *Sustainability* **2021**, *13*, 1226. [CrossRef]
20. Bruisson, S.; Maillot, P.; Schellenbaum, P.; Walter, B.; Gindro, K.; Deglene-Benbrahim, L. Arbuscular mycorrhizal symbiosis stimulates key genes of the phenylpropanoid biosynthesis and stilbenoid production in grapevine leaves in response to downy mildew and grey mould infection. *Phytochemistry* **2016**, *131*, 92–99. [CrossRef]
21. Smith, S.E.; Read, D.J. *Mycorrhizal Symbiosis*, 3rd ed.; Academic Press: London, UK, 2008; pp. 1–787.
22. Cesaro, P.; Massa, N.; Bona, E.; Novello, G.; Todeschini, V.; Boatti, L.; Mignone, F.; Gamalero, E.; Berta, G.; Lingua, G. Native AMF communities in an Italian vineyard at two different phenological stages of *Vitis vinifera*. *Front. Microbiol.* **2021**, *12*, 676610. [CrossRef]
23. Delaunois, B.; Farace, G.; Jeandet, P.; Clément, C.; Baillieul, F.; Dorey, S.; Cordelier, S. Elicitors as alternative strategy to pesticides in grapevine? Current knowledge on their mode of action from controlled conditions to vineyard. Environ. *Sci. Pollut. Res. Int.* **2014**, *21*, 4837–4846. [CrossRef]
24. Aziz, A.; Poinssot, B.; Daire, X.; Adrian, M.; Bezier, A.; Lambert, B.; Joubert, J.M.; Pugin, A. Laminarin elicits defense responses in grapevine and induces protection against *Botrytis cinerea* and *Plasmopara viticola*. *Mol. Plant-Microbe Interact.* **2003**, *16*, 1118–1128. [CrossRef]
25. Aziz, A.; Trotel-Aziz, P.; Dhuicq, L.; Jeandet, P.; Couderchet, M.; Vernet, G. Chitosan oligomers and copper sulfate induce grapevine defense reactions and resistance to gray mold and downy mildew. *Phytopathology* **2006**, *96*, 1188–1194. [CrossRef]
26. Van Aubel, G.; Buonatesta, R.; Van Cutsem, P. COS-OGA: A novel oligosaccharidic elicitor that protects grapes and cucumbers against powdery mildew. *Crop Prot.* **2014**, *65*, 129–137. [CrossRef]
27. Romanazzi, G.; Mancini, V.; Feliziani, E.; Servili, A.; Endeshaw, S.; Neri, D. Impact of alternative fungicides on grape downy mildew control and vine growth and development. *Plant Dis.* **2016**, *100*, 739–748. [CrossRef]
28. Romanazzi, G.; Mancini, V.; Foglia, R.; Marcolini, D.; Kavari, M.; Piancatelli, S. Use of Chitosan and Other Natural compounds alone or in different strategies with copper hydroxide for control of grapevine downy mildew. *Plant Dis.* **2021**, *105*, 3261–3268. [CrossRef]
29. Singh, R.R.; Chinnasri, B.; De Smet, L.; Haeck, A.; Demeestere, K.; Van Cutsem, P.; Van Aubel, G.; Gheysen, G.; Kyndt, T. Systemic defense activation by COS-OGA in rice against root-knot nematodes depends on stimulation of the phenylpropanoid pathway. *Plant Physiol. Biochem.* **2019**, *142*, 202–210. [CrossRef]
30. EFSA (European Food Safety Authority). Conclusion on the Peer Review of the Pesticide Risk Assessment of the Active Substance COS-OGA. *EFSA J.* **2014**, *12*, 3868. Available online: https://www.efsa.europa.eu/it/efsajournal/pub/3868 (accessed on 19 November 2021). [CrossRef]
31. European Commission. Commission Regulation (EC) No. 543/2015. *Off. J. Eur. Comm.* **2015**, L90/1. Available online: https://eur-lex.europa.eu/legal-content/EN/TXT/PDF/?uri=CELEX:32015R0543&from=EN (accessed on 19 November 2021).
32. Cabrera, J.C.; Boland, A.; Cambier, P.; Frettinger, P.; Van Cutsem, P. Chitosan oligosaccharides modulate the supramolecular conformation and the biologica activity of oligogalacturonides in Arabidopsis. *Glycobiology* **2010**, *20*, 775–786. [CrossRef]
33. Van Aubel, G.; Cambier, P.; Dieu, M.; Van Cutsem, P. Plant immunity induced by COS-OGA elicitor is a cumulative process that involves salicylic acid. *Plant Sci.* **2016**, *247*, 60–70. [CrossRef] [PubMed]
34. Clinckemaillie, A.; Decroës, A.; van Aubel, G.; Carrola dos Santos, S.; Renard, M.E.; Van Cutsem, P.; Legrève, A. The novel elicitor COS-OGA enhances potato resistance to late blight. *Plant Pathol.* **2017**, *66*, 818–825. [CrossRef]
35. Clark, A.C.; Scollary, G.R. Determination of copper in white wine by stripping potentiometry utilising medium ex-change. *Anal. Chim. Acta* **2000**, *413*, 25–32. [CrossRef]
36. Grangeteau, C.; David, V.; Hervé, A.; Guilloux-Benatier, M.; Rousseaux, S. The sensitivity of yeasts and yeasts-like fungi to copper and sulfur could explain lower yeast biodiversity in organic vineyards. *FEMS Yeast Res.* **2017**, *17*, fox092. [CrossRef]
37. Sharathchandra, R.G.; Raj, S.N.; Shetty, N.P.; Amruthesh, K.N.; Shetty, H.S. A Chitosan formulation Elexa™ induces downy mildew disease resistance and growth promotion in pearl millet. *Crop Prot.* **2004**, *23*, 881–888. [CrossRef]
38. Hall, M.E.; Loeb, G.M.; Cadle-Davidson, L.; Evans, K.J.; Wilcox, W.F. Grape sour rot: A four-way interaction involving the host, yeast, acetic acid bacteria, and insects. *Phytopathology* **2018**, *108*, 1429–1442. [CrossRef]
39. EPPO (European and Mediterranean Plant Protection Organization). 2022. Available online: https://www.eppo.int/RESOURCES/eppo_standards/pp1_list (accessed on 24 November 2021).
40. McKinney, H.H. Influence of soil temperature and moisture on infection of wheat seedlings by *Helminthosporium sativum*. *J. Agric. Res.* **1923**, *26*, 195–218.
41. King, E.O.; Ward, M.K.; Raney, D.E. Two simple media for the demonstration of pyocyanin and fluorescin. *J. Lab. Clin. Med.* **1954**, *44*, 301–307.
42. Abbott, W.S. A method of computing the effectiveness of an insecticide. *J. Econ. Entomol.* **1925**, *18*, 265–267. [CrossRef]

MDPI
St. Alban-Anlage 66
4052 Basel
Switzerland
Tel. +41 61 683 77 34
Fax +41 61 302 89 18
www.mdpi.com

Plants Editorial Office
E-mail: plants@mdpi.com
www.mdpi.com/journal/plants

www.ingramcontent.com/pod-product-compliance
Lightning Source LLC
LaVergne TN
LVHW070729100526
838202LV00013B/1199
* 9 7 8 3 0 3 6 5 6 9 5 3 6 *